THE JEAN AND ALEXANDER
HEARD LIBRARY

JHA

VANDERBILT
UNIVERSITY
NASHVILLE, TENNESSEE

Stevenson Library
3200 Stevenson Center
419 21st Avenue South
Nashville, TN 37240

ASTRONOMY AND ASTROPHYSICS LIBRARY

Series Editors: M. Harwit, R. Kippenhahn, V. Trimble, J.-P. Zahn

Advisory Board:

J. N. Bahcall
P. L. Biermann
S. Chandrasekhar
S. M. Faber
L. V. Kuhi
P. G. Mezger
P. A. Strittmatter

ASTRONOMY AND ASTROPHYSICS LIBRARY

Series Editors: M. Harwit, R. Kippenhahn, V. Trimble, J.-P. Zahn

Tools of Radio Astronomy
By K. Rohlfs

Physics of the Galaxy and Interstellar Matter
By H. Scheffler and H. Elsässer

Galactic and Extragalactic Radio Astronomy 2nd Edition
Editors: G. L. Verschuur, K. I. Kellermann

Observational Astrophysics
By P. Léna

Astrophysical Concepts 2nd Edition
By M. Harwit

The Sun An Introduction
By M. Stix

Stellar Structure and Evolution
By R. Kippenhahn and A. Weigert

Relativity in Astrometry, Celestial Mechanics and Geodesy
By M. H. Soffel

The Solar System
By T. Encrenaz and J.-P. Bibring

Physics and Chemistry of Comets
Editor: W. F. Huebner

Supernovae
Editor: A. Petschek

T. Encrenaz J.-P. Bibring

The Solar System

With the Participation of M. Blanc

Translated by S. Dunlop

With 157 Figures

Springer-Verlag Berlin Heidelberg New York
London Paris Tokyo Hong Kong

Dr. Thérèse Encrenaz

Laboratoire de Recherche Spatiale
Observatoire de Paris, 5 place Jules Janssen
F-92195 Meudon, France

Jean-Pierre Bibring

Université Paris XI
Laboratoire René Bernas
F-91406 Orsay, France

Michel Blanc

Observatoire Midi-Pyrénées
F-31400 Toulouse, France

Translator: Storm Dunlop

140 Stocks Lane, East Wittering
West Sussex, P020 8NT, England

Series Editors

Martin Harwit

The National Air and Space
Museum, Smithsonian Institution
7th St. and Independence Ave. S.W.
Washington, DC 20560, USA

Virginia Trimble

Astronomy Program
University of Maryland
College Park, MD 20742, USA
and Department of Physics
University of California
Irvine, CA 92717, USA

Rudolf Kippenhahn

Max-Planck-Institut für
Physik und Astrophysik
Institut für Astrophysik
Karl-Schwarzschild-Straße 1
D-8046 Garching, Fed. Rep. of Germany

Jean-Paul Zahn

Université Paul Sabatier
Observatoires du Pic-du-Midi
et de Toulouse
14, Avenue Edouard-Belin
F-31400 Toulouse, France

Cover picture: The planet Mars and its satellite Phobos
(© McEwen, USGS/NASA – JPL Promospace. Montage: Olivier de Goursac, Promospace)

Title of the original French edition: *Astrophysique: Le Système Solaire*
© InterEditions et Editions du CNRS, Paris 1987

ISBN 3-540-18910-6 Springer-Verlag Berlin Heidelberg New York
ISBN 0-387-18910-6 Springer-Verlag New York Berlin Heidelberg

Library of Congress Cataloging-in-Publication Data. Encrenaz, T. (Thérèse), 1946–[Système solaire. English] The solar system / T. Encrenaz, J.-P. Bibring ; with the participation of M. Blanc ; translated by S. Dunlop. p. cm. –– (Astronomy and astrophysics library) Translation of: Le système solaire. Bibliography: p. Includs index. ISBN 0-387-18910-6 (U.S.) 1. Planetology. I. Bibring, J.-P. (Jean-Pierre), 1947–II. Title. III. Series. QB601.E5313 1989 523.2––dc20 89-11605

This work is subject to copyright. All rights are reserved, whether the whole or part of the material is concerned, specifically the rights of translation, reprinting, re-use of illustrations, recitation, broadcasting, reproduction on microfilms or in other ways, and storage in data banks. Duplication of this publication or parts thereof is only permitted under the provisions of the German Copyright Law of September 9, 1965, in its version of June 24, 1985, and a copyright fee must always be paid. Violations fall under the prosecution act of the German Copyright Law.

© Springer-Verlag Berlin Heidelberg 1990. Printed in Germany

The use of registered names, trademarks, etc. in this publication does not imply, even in the absence of a specific statement, that such names are exempt from the relevant protective laws and regulations and therefore free for general use.

Printing and binding: Konrad Triltsch, Graphischer Betrieb, Würzburg.
2156/3150-543210 – Printed on acid-free paper

Preface

More than ever, planetology is a science in full development. The year 1989 has seen two important achievements: the exploration of Mars by the Phobos spacecraft, and the encounter of Neptune and its satellite Triton by the Voyager 2 probe. The first event marked the beginning of a new Mars exploration program, which will hopefully lead to a Martian return sample mission within the next fifteen years. The exploration of Venus will continue with Magellan, just launched by NASA. The second event ends the first stage of the space exploration of the giant planets. It has demonstrated the amazing variety of the objects encountered – planets, rings and satellites – which reflect the variety and the complexity of their formation and evolution processes.

The next era in outer solar system exploration has just started with the launch of the Galileo mission, which will explore Jupiter and its Galilean satellites in 1995–1996. Hopefully, more and more ambitious missions will follow, in a fully international cooperative effort. The Cassini mission will be devoted to the Saturn system, with special emphasis on its satellite Titan. The space exploration of the small bodies will continue with rendezvous missions to cometary nuclei – in particular the CRAF mission (Comet Rendez-vous and Asteroidal Fly-by) – and possibly, asteroids, in preparation for the ultimate Comet Nucleus Sample Return mission, planned for the beginning of the next century. On the other hand, interplanetary medium studies will greatly benefit from the Ulysses and Soho-Cluster missions.

Finally, new opportunities will be provided by Earth-orbiting satellites – such as the Hubble Space Observatory and the Infrared Space Observatory – as well as large ground-based telescopes equipped with ever more sophisticated instrumentation. Exciting new discoveries are expected to come in the next decade.

We hope that the present book will be helpful as a reference for planetologists, students and research scientists.

Our grateful thanks are due to Mr. S. Dunlop for his willing cooperation and for providing such an excellent translation.

September 1989 *T. Encrenaz · J.-P. Bibring*

Preface to the French Edition

Planetology – or the study of the objects in the Solar System – enjoys a place apart in the science of astronomy. From being predominant in past centuries (the planets and nearby comets were the brightest objects in the sky), it marked time in the middle of the 20th century, whilst solar and stellar astronomy, and then galactic and extragalactic astronomy, continued to develop. From the sixties, planetology blossomed again, largely thanks to the development of space research.

The era of space exploration opened the way for the in-situ study of objects relatively close to the Earth, which led to a fantastic harvest of new results. Among the technical advances that have also contributed to the development of planetology mention should be made of infrared astronomy, which, whether from the ground, aircraft or balloons, is particularly well suited to the study of the cold bodies that form the objects in the Solar System.

What is equally notable about the development of planetology is its enlargement to embrace related disciplines. To begin with, the closer links between astronomers and physicists allowed the physical study of the objects, not just the descriptive observation that had prevailed previously. Investigation of data relating to planetary surfaces and atmospheres has likewise resulted in planetologists and geophysicists working together, whilst the development of planetary spectroscopy has only been possible thanks to increasing collaboration between planetologists and laboratory spectroscopists. Finally, investigation of complex prebiotic molecules present on the giant planets and Titan has reinforced the links that existed between planetologists, chemists and biochemists. At the end of the eighties, planetology is sustained by a very varied space programme and appears as an actively growing, multidisciplinary science.

Most of the results regarding the physical and chemical composition of objects in the Solar System have been acquired in the last twenty years. By way of example, at the beginning of the sixties, only three molecules were known on Jupiter. All the other species, about fifteen of them, have been discovered since 1972. Another remarkable result is the discovery of rings around the giant planets, and the multiplicity of rings and small satellites that orbit Saturn. Another revelation has been the discovery of the sheer range of surfaces encountered, from those of the inner planets to those of the Galilean and other satellites, particularly the volcanic activity on Io. In another area, the study of the isotopic composition of carbonaceous chondrites and the discovery of isotopic anomalies have brought fundamental knowledge of the composition of the presolar nebula. Many other examples could be cited.

In the light of all these results, our understanding of the origin and the history of the Solar System has progressed remarkably over the last twenty years. A model derived from Laplace's nebular model is universally accepted. Although numerical estimates may differ from author to author, the nature of the successive processes is no longer in doubt. Our idea of the past history of the Solar System is taking shape.

This progress in our understanding still leaves plenty of questions unanswered, and certainly raises new ones. Although the overall evolution is evident, there is still plenty to learn about the specific evolutionary processes that have taken place on each object. This is true both for the giant planets and their systems, and for the inner planets and the "small bodies", the comets and the minor planets. In particular, it appears more and more that the systematic study of the small bodies, which for a long time was deferred, it not being feasible, could provide us with important information about the history of the Solar System, and the ways in which it has evolved.

This book aims to discuss the current state of planetology. In the first four chapters the Solar System is treated as a whole. The various objects are studied, class by class, in the seven following chapters. The last chapter describes the way in which contemporary planetology is developing.

The preparation of this book has benefited from comments made by numerous collaborators. We should particularly like to thank E. Schatzman, J.-C. Pecker, F. Praderie, M. Festou, F. Combes, P. Encrenaz, E. Falgarone, N. Meyer and A. Roux.

A table of useful constants is given on the front end-paper of the book, a general bibliography by chapter can be found at the end.

Contents

1. General Features of the Solar System

The Solar System may be defined as consisting of those objects that are governed by the Sun's gravitational field. Other effects arising from the proximity of the Sun could equally well be used as criteria, such as radiation pressure or interaction with the solar wind. With any of these definitions the Solar System extends out to about two light-years; the closest star, Proxima Centauri, lies at a distance of slightly more than four light-years. Our knowledge of this region of space certainly does not reach as far as this, however, because the most distant Solar-System objects that we know about, the comets, seem to originate no more than 50 000 astronomical units[1] away, or less than a third of the total distance, whilst the other Solar-System bodies known to us lie within 50 AU. Our study is therefore confined to what is primarily the central region of the Solar System.

The first thing to note is that the total mass of the objects in the Solar System represents a negligible fraction of the mass of the Sun itself (less than 0.0015); and the second is that most of these objects orbit close to the plane of the solar equator. The Solar System thus forms an essentially empty disk within which the planets and their satellites, the minor planets and the comets orbit at considerable distances from one another.

The *planets* are the most massive bodies in the Solar System. In ancient times the motion of these objects relative to the sphere of fixed stars was noted by the Greeks, who gave them the name "wandering stars" (planets). Planets have essentially circular orbits around the Sun and lie at various heliocentric distances of between 0.4 AU and 40 AU; their diameters range from a few thousand km to more than 100 000 km. Nine are known at present: *Mercury, Venus, Earth, Mars, Jupiter, Saturn, Uranus, Neptune*, and *Pluto*.

Between the orbits of Mars and Jupiter, a few AU from the Sun, there is a family of smaller bodies, the diameters of which range from a few tens to a few hundreds of km, also having essentially circular orbits around the Sun: these are the *asteroids* (or *minor planets*).

Even smaller are the *comets* whose nuclei do not exceed a few km in radius, and which move in elliptical paths that are often highly inclined relative to the plane of the Earth's orbit (the plane of the ecliptic).

Another class of objects is subject primarily to the action of the gravitational fields of the planets themselves; these are the *satellites*. The largest of them have dimensions comparable to those of the smallest planets. Finally, some planets also

[1] One astronomical unit is the semi-major axis of the Earth's orbit around the Sun, i.e. 149.6 million km. One light-year is the distance travelled by light in one year, i.e. 9.5×10^{15} m.

have systems of *rings*, consisting of particles of very different sizes, which may range from a few microns to a few metres.

The interplanetary medium between all these objects is not completely empty: there are dust grains, the dimensions of which may be reckoned in microns; and there is also a plasma of electrons and ions, which arises mainly in the solar corona: this is the *solar wind*.

1.1 Solar-System Mechanics

1.1.1 Kepler's Laws

As the total mass of the all the planets and satellites is very low when compared with that of the Sun, the mutual interactions of the planets may be neglected in calculating, to a first approximation, the orbits of the latter: this is the Keplerian approximation. It is thus reduced to a two-body problem, with the specific feature that the mass of one is negligible relative to that of the other, the Sun.

In the case of a planet and satellite system, the problem can also be reduced to a two-body case, to a first approximation, but here the mass of the satellite may not be negligible in comparison with that of the planet. The path of a planet, of a comet, or of an asteroid is governed only by the Sun's gravitational force, which is inversely proportional to the square of the heliocentric distance, in accordance with Newton's law of universal attraction.

It can thus be shown that the path is a conic section with the Sun at one focus, and one may derive the three empirical laws formulated by Kepler at the beginning of the 17th century for the motion of the planets:

1. The orbit of a planet is an ellipse, with the centre of the Sun at one of the foci. The movement of the object may be described as follows:

$$r = a\frac{1 - e^2}{1 + e\cos\theta} \tag{1.1}$$

where r is the heliocentric distance, a is the semi-major axis of the ellipse, e is its eccentricity (that is to say the ratio of the distance between the two foci to that of the major axis $2a$), and θ is the angle, measured in radians, relative to the point that is closest to the Sun, and known as perihelion (where $\theta = 0$). The most distant point from the Sun is known as aphelion ($\theta = \pi$).

2. The radius vector joining the centre of the Sun to the planet sweeps out equal areas in equal periods of time. This is the law of areas, which is expressed as:

$$\frac{dA}{dt} = \frac{1}{2}\left(r^2\frac{d\theta}{dt}\right) = \frac{h}{2} \tag{1.2}$$

where A is the area swept out by the Sun-planet radius, and h is the area constant.

3. The ratio of the cube of the semi-major axis to the square of the period is the same for all the planets:

$$\frac{a^3}{P^2} = 1 \qquad (1.3)$$

where P is the orbital period in years, and a is the semi-major axis, in astronomical units.

In the general case for the gravitational attraction between two bodies of masses m and m', (1.1) and (1.3) may be written, respectively, as:

$$\frac{1}{r} = \frac{1}{p}(1 + e \cos \theta) \quad \text{and} \qquad (1.4)$$

$$\frac{a^3}{P^2} = \frac{G(m + m')}{4\pi^2} \qquad (1.5)$$

where p is defined by:

$$p = \frac{h^2}{G(m + m')} \quad . \qquad (1.6)$$

Using a system of units where the semi-major axis a, the period P (time for one orbit), and the masses m and m' are expressed in metres, seconds, and kilogrammes respectively, G has a value of 6.67×10^{-11}.

The planets and asteroids have elliptical orbits that are essentially circular, with a few exceptions. Periodic comets have elliptical orbits. Satellites equally have essentially circular orbits around their parent planets. In the case of non-periodic comets, which seem to come from a reservoir known as the "Oort cloud", situated at a distance of several tens of thousands of AU, the orbit around the Sun is to all intents and purposes a parabola.

The orbits of planets, minor planets and comets are referred to the ecliptic as a reference plane. In the case of satellites, the reference plane is the equatorial plane of the planet concerned. The path of an object cuts the ecliptic at two points, known as the *ascending* and *descending nodes*. The line of intersection of the plane of the orbit and the reference plane is known as the *line of nodes*.

An orbit is determined by five *elements*:

1. the semi-major axis, a;
2. the eccentricity of the ellipse, e;
3. the inclination i of the plane of the orbit with reference to the plane of the ecliptic;
4. the longitude Ω of the ascending node (i.e. the one that is crossed from south to north; see Fig. 1.1), measured from the vernal equinox. The latter is defined by the intersection of the planes of the ecliptic and of the terrestrial equator for 1950. The vernal equinox corresponds to the position of the Earth at the autumnal equinox;
5. the argument of perihelion ω, (the angle between the perihelion position and the line of nodes; see Fig. 1.1).

The position of a point on the orbit is additionally defined by the time t of perihelion passage.

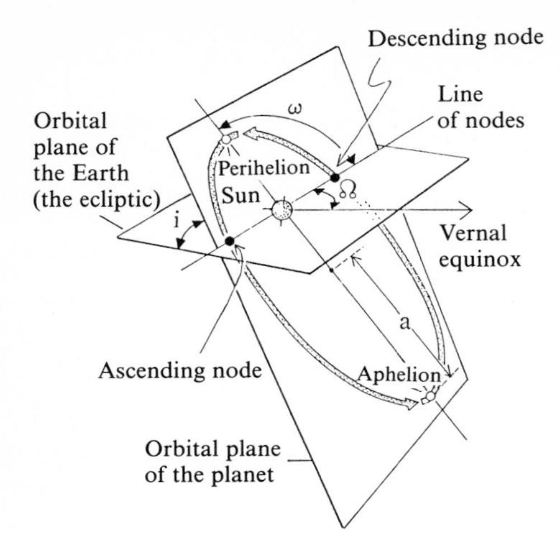

Fig. 1.1. The definition of orbital elements

In the case of comets, the following parameters may be used:

1. the perihelion distance, $q = a(1 - e)$;
2. the aphelion distance, $Q = a(1 + e)$;

which express the heliocentric distances of the comet at perihelion and at aphelion respectively.

Because of mutual interaction between objects, these parameters are not perfectly constant over time, as would be the case if the movement were truly Keplerian. As far as the planets are concerned, these variations are very slow: the line of nodes sweeps round the plane of the orbit over a period of several tens of thousands of years. In the case of comets, however, the perturbations produced in an orbit by close approaches to planets may be very considerable.

1.1.2 The Planets

The Titius-Bode "Law"

As we have seen, the distances of the nine planets now known range from 0.4 AU in the case of Mercury, to about 40 AU for Pluto. In the 18th century, the German astronomers Titius and Bode, and later Wolf, showed that the average heliocentric distances of the planets approximately followed an empirical law. Originally applied to the planets from Mercury to Saturn, Uranus being added later, the law is given by:

$$D = 0.4 + 0.3 \times 2^n \tag{1.7}$$

where D is the heliocentric distance expressed in AU, n takes the value of $-\infty$ for Mercury, 0 for Venus, and is increased by 1 for each successive planet. The law is valid, with a moderate degree of error (of the order of 5 %), as far as Uranus; but for Neptune the error is 22 %, and for Pluto 49 %. Note that the value $n = 3$ does not correspond to a planet as such, but to the asteroidal belt. It is possible

Table 1.1. Orbital characteristics of the planets

Name	Semi-major axis (AU)	Eccentricity	Inclination to the ecliptic	Sidereal period	Mean orbital velocity	Synodic period[1]
Mercury	0.387 1	0.206	7°00′	87.969 d	47.89	115.9 d
Venus	0.723 3	0.007	3°24′	224.701 d	35.04	1 yr 218.7 d
Earth	1	0.017	0°	365.256 d	29.80	–
Mars	1.523 7	0.093	1°51′	1 yr 321.73 d	24.14	2 yrs 49.5 d
Jupiter	5.202 6	0.048	1°19′	11 yrs 314.84 d	13.06	1 yr 33.6 d
Saturn	9.554 7	0.056	2°30′	29 yrs 167.0 d	9.64	1 yr 12.8 d
Uranus	19.218 1	0.046	0°46′	84 yrs 7.4 d	6.80	1 yr 4.4 d
Neptune	30.109 6	0.009	1°47′	164 yrs 280.3 d	5.43	1 yr 2.2 d
Pluto	39.438 7	0.246	17°10′	247 yrs 249.0 d	4.74	1 yr 1.5 d

1 year (yr) = 365.256 days (d)
[1] The synodic orbital period is the time separating two successive, identical configurations of the planet-Sun-Earth system, for example between two oppositions.

that the asteroids consist of partially accreted material that was not able to form a single body. This hypothesis would explain the existence of the asteroidal belt at the heliocentric distance predicted by the Titius-Bode law.

The orbital parameters and the physical properties of the planets are given in Tables 1.1 and 1.3.

Visibility of Planets from the Earth

Like the other bodies in the Solar System, the planets are relatively cold objects: their temperatures do not exceed a few hundred K. Their own individual visible radiation is therefore very feeble compared with the solar radiation that they reflect towards us. As a result, their visibility from the Earth depends upon their position relative to the Sun. Two principal types of configuration can be distinguished.

1. In the case of the inferior planets, where the heliocentric distance is less than 1 AU, the objects – Mercury and Venus – are always seen close to the Sun, and therefore at nightfall or daybreak. The illuminated fraction of their surface varies according to their position as a function of phase angle[2]: they therefore show a *phase effect*, similar to that of the Moon. When an inferior planet passes between the Earth and the Sun, it is said to be at *inferior conjunction* and a transit of the planet across the Sun's disk may be observed when the three bodies are exactly aligned; when it passes behind the Sun, it is said to be at *superior conjunction*.
2. In the case of superior planets, where the heliocentric distance is greater than 1 AU, observation is possible in the middle of the night. When these planets are on the opposite side of the Earth to the Sun, they are said to be at *opposition* and their distance is at a minimum (for that particular apparition). The distance is at a maximum when the planet is behind the Sun, at *conjunction*. The phase effect is less the farther the planet is away; in the case of Jupiter the maximum phase angle is 11°.

[2] The phase angle is the angle Sun-Object-Earth.

1.1.3 The Satellites

With the exception of Mercury and Venus, the planets possess one or more satellites, their number being particularly high in the case of the most massive planets. The Earth has one large satellite, the Moon, and Mars has two small ones, Phobos and Deimos. Jupiter, Saturn and Uranus, which have been explored by the Voyager spaceprobes, each have more than ten known satellites in a considerable range of sizes, and it is probable that even smaller ones remain to be discovered. Neptune has eight known satellites.

As for the outermost planet, Pluto, it in effect consists of a double system, because its satellite Charon has a diameter that is at least equal to a third of that of Pluto, and orbits at a distance of about 6 diameters only from the planet.

The overall characteristics of the satellites are given in Table 1.2.

The Three-body Problem: the Lagrangian Points

The determination of the path of a satellite around a planet, in the presence of other satellites, and taking the effect of the Sun's gravitational force into account, requires the resolution of the N-body problem, which, in general, does not possess a mathematical solution. In the case of the Moon, one resorts to the perturbation method, similar to that used in the calculation of planetary orbits to take the effects of other planets into account.

In a number of cases one can come down to a three-body problem in describing the movement of a small body in the gravitational fields of the Sun and a planet, or in those of a planet and a satellite. There then exist a certain number of equilibrium positions, which are known as *Lagrangian points*. Three of these positions lie on the planet-satellite (or Sun-planet) line, and the other two lie at the corners of equilateral triangles that have the planet-satellite side as a base (Fig. 1.2). The two latter solutions may be stable. This is the case with the arrangement found for the Trojan minor planets, which move in Jupiter's orbit at the two Lagrangian points, 60° in front of, and behind, the planet. This also seems to be the case with Saturn's satellite 1980 S6, recently discovered in the orbit of Dione.

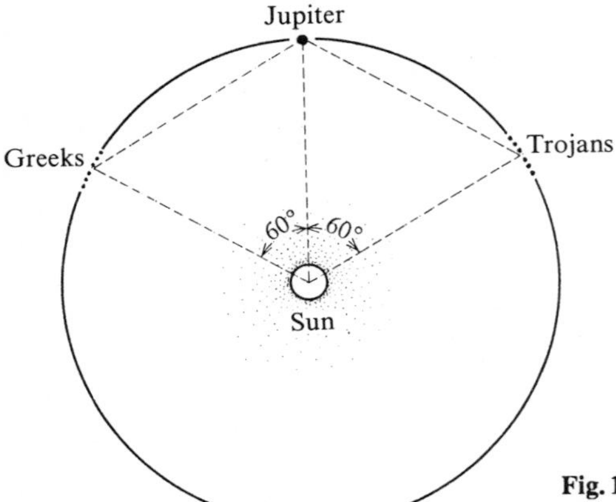

Fig. 1.2. The location of the Trojan asteroids at Jupiter's Lagrangian points

Specific Configurations in the Earth-Moon System

If the plane of the Moon's orbit were identical with the ecliptic, there would be an eclipse of the Sun or of the Moon each time that the Moon is in conjunction (New Moon), or in opposition (Full Moon). As the Moon's orbit is inclined about 5° from the ecliptic, the phenomena only arise if the Moon is also at one of the nodes of its orbit. The frequency with which a specific identical configuration recurs has been calculated as being approximately 18 years 10 days. This is known as the *Saros*.

An *eclipse of the Moon* must occur at opposition if the centre of the Moon is less than 9° from one of the nodes of its orbit; it may occur if the centre is less than 12.5° from a node. According to the geometrical configuration, there is an eclipse either by the umbra or by the penumbra (Fig. 1.3); and an umbral eclipse may be total or partial. An eclipse of the Moon may last as long as 1 hr 45 min.

An *eclipse of the Sun* must occur at conjunction if the centre of the Sun is less than 13.5° from one of the nodes of the lunar orbit; it may occur if the centre is less than 18.5° from a node. By a remarkable coincidence the angular diameters of the Sun and the Moon are very similar (about 30 minutes of arc). One can be slightly larger than the other, depending on the Earth-Sun and Earth-Moon distances. According to the specific circumstances (see Fig. 1.4) one may see either a total eclipse (when the Earth is relatively farther from the Sun or closer to the Moon), or an annular eclipse (when the Earth is relatively closer to the Sun or farther from the Moon); in both cases a partial eclipse can be seen. The maximum duration of a total eclipse of the Sun is of the order of 7 minutes.

It may be calculated that there are between four and seven eclipses every year, with at least two eclipses of the Sun and two eclipses of the Moon (including penumbral eclipses).

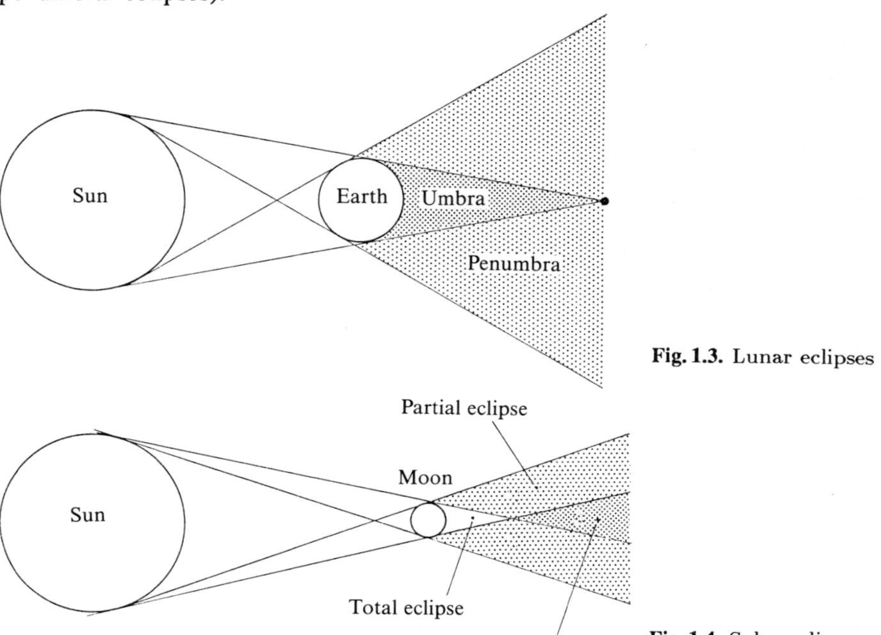

Fig. 1.3. Lunar eclipses

Fig. 1.4. Solar eclipses

Table 1.2. Planetary satellites

Name	Year of discovery	Discoverer	Sidereal orbital period[1] (days)	Semi-major axis of the orbit (10^3 km)	(planetary radii)	Orbital eccentricity	Orbital inclination[2] (°)	Diameter[3] (km)	Density (g/cm^3)
EARTH									
Moon			27.322	384.40	60.268	0.055	5.1	3476	3.33
MARS									
Phobos	1877	A. Hall (USA)	0.319	9.38	2.76	0.017	[1.1] (2)	27×21×19 (4)	~2
Deimos	1877	A. Hall (USA)	1.262	23.48	6.91	0.003	[0.9]–[2.7]	15×12×11 (4)	~2
JUPITER									
Metis	1979	S. P. Synnott (USA)	0.294	128.2	1.80	(0)	(0)	(40)	?
Adrastrea	1979	D. Jewitt, E. Danielson (USA)	0.297	128.4	1.80	(0)	(0)	(30)	?
Amalthea	1892	E. Barnard (USA)	0.489	181.0	2.52	0.003	[0.4]	270×170×150	?
Thebe	1979	S. P. Synnott (USA)	0.678	222.4	3.11	(0)	(0)	(80)	?
Io	1610	Galileo (Italy)	1.769	421.8	5.78	0.004	0.027	3632	3.42
Europa	1610	Galileo (Italy)	3.551	670.9	9.40	0.000	0.468	3138	3.03
Ganymede	1610	Galileo (Italy)	7.155	1070.0	14.99	0.001	0.183	5262	1.93
Callisto	1610	Galileo (Italy)	16.689	1883.0	26.33	0.007	0.253	4800	1.83
Leda	1974	C. Kowall (USA)	240.0	11094	156.0	0.146	26.1	2×14	?
Himalia	1904	C. Perrine (USA)	250.6	11480	159.8	0.158	27.6	180	?
Lysithea	1938	S. Nicholson (USA)	260.0	11720	163.2	0.107	28.8	6×32	?
Elara	1905	C. Perrine (USA)	260.1	11737	163.4	0.207	24.8	80	?
Ananke	1951	S. Nicholson (USA)	631 (r)	21200	290.0	0.169	146.7	6×28	?
Carme	1938	S. Nicholson (USA)	692 (r)	22600	313.0	0.207	163.4	8×40	?
Pasiphae	1908	P. Melotte (UK)	735 (r)	23500	326.3	0.380	145.0	8×46	?
Sinope	1914	S. Nicholson (USA)	758 (r)	23700	332.0	0.280	153.0	6×36	?
SATURN									
Atlas	1980	Voyager 1	0.602	137.6	2.28	0.002	[0.3]	20×40	
Prometheus	1980	–	0.613	139.4	2.31	0.004	[0]	80×140×100	
Pandora	1980	–	0.628	141.7	2.35	0.004	[0.05]	70×110×90	
Epimetheus	1980	–	0.694	151.4	2.51	0.009	[0.3]	100×120×140	?
Janus	1966	A. Dollfus (Fr.)	0.695	151.5	2.51	0.007	[0.1]	160×200×220	?
Mimas	1789	W. Herschel (UK)	0.942	185.5	3.07	0.020	[1.5]	390	1.4

Enceladus	1789	W. Herschel (UK)	1.370	238.0	3.95	0.004	[0]	500	1.2
Telesto	1980	B. Smith (USA)	1.888	294.7	4.88	0	–	26×28×34	
Calypso	1980	B. Smith (USA)	1.888	294.7	6.27	0	–	22×22×34	
Tethys	1684	J. D. Cassini (Fr.)	1.888	294.7	4.88	0	[1.9]	1050	1.2
Helene	1980	J. Lecacheux (Fr.)	2.737	377.4	6.27	–	–	30×32×36	?
Dione	1684	J. D. Cassini (Fr.)	2.737	377.4	6.27	0.002	0.02	1120	1.4
Rhea	1672	J. D. Cassini (Fr.)	4.518	527	8.74	0.001	[0.4]	1530	1.2
Titan	1655	C. Huygens (Holl.)	15.945	1221.8	20.25	0.029	[0.3]	5140	1.9
Hyperion	1848	W. Bond (USA)	21.280	1481	24.55	0.104	[0.4]	220×260×410	?
Iapetus	1671	J. D. Cassini (Fr.)	79.33	3560	59	0.028	14.7	1440	1.2
Phoebe	1898	W. Pickering (USA)	550.45 (r)	12954	216	0.163	177	220	?
URANUS									
Cordelia	1986	Voyager 2	0.33 (r)	49.7	1.96	?	?	(40)	?
Ophelia	1986	Voyager 2	0.37 (r)	53.8	2.12	?	?	(50)	?
Bianca	1986	Voyager 2	0.43 (r)	59.2	2.33	?	?	(50)	?
Cressida	1986	Voyager 2	0.46 (r)	61.8	2.43	?	?	(60)	?
Desdemona	1986	Voyager 2	0.47 (r)	62.7	2.47	?	?	(60)	?
Juliet	1986	Voyager 2	0.49 (r)	64.6	2.54	?	?	(80)	?
Portia	1986	Voyager 2	0.51 (r)	66.1	2.60	?	?	(80)	?
Rosalinda	1986	Voyager 2	0.56 (r)	69.9	2.75	?	?	(60)	?
Belinda	1986	Voyager 2	0.62 (r)	75.3	2.96	?	?	(60)	?
Puck	1985	Voyager 2	0.76 (r)	86.0	3.39	?	?	170	?
Miranda	1948	G. Kuiper (USA)	1.41 (r)	129.9	5.11	?	?	484	1.3
Ariel	1851	W. Lassel (UK)	2.52 (r)	191.0	7.52	?	?	1160	1.6
Umbriel	1851	W. Lassel (UK)	4.14 (r)	266.0	10.47	?	?	1190	1.4
Titania	1787	W. Herschel (UK)	8.70 (r)	436.3	17.18	?	?	1610	1.6
Oberon	1787	W. Herschel (UK)	13.46 (r)	583.4	22.97	?	?	1550	1.5
NEPTUNE[5]									
1989N1	1989	Voyager 2	1.1223	~ 117.5	~ 4.8	?	?	?	?
Triton	1846	W. Lassel (UK)	5.877 (r)	355.3	14.6	0	160	(4000)	
Nereid	1949	G. Kuiper (USA)	360.2	5510	229	0.749	27.5	(300)	
PLUTO									
Charon	1978	J. W. Christy (USA)	6.3867	20	13	0	(100)	≥1200	

[1] (r) indicates that orbital motion is retrograde
[2] Brackets indicate the inclination to the planet's equator
[3] Parentheses indicate that the value is an estimate only
[4] Triaxial ellipsoid of revolution
[5] Voyager 2 has discovered 5 further satellites

1.2 Physics of the Solar System

The essential difference between stars and planets lies in their mass. A contracting object cannot be transformed into a star unless its central temperature is sufficiently high for the first thermonuclear reactions to take place (in the proton-proton, or proton-deuterium cycles). In order for the centre of the object to be heated to the several million degrees required, the objects must have a mass m of at least one-twentieth of that of the Sun, i.e. $m = 10^{32}$g. The largest planet in the Solar System, Jupiter, however, only amounts to one-thousandth of a solar mass. In the absence of thermonuclear reactions, Solar-System objects only have small amounts of internal energy, which may, moreover, derive from more than one source (radioactivity in the case of the terrestrial planets, contraction or internal differentiation in the case of the giant planets). As a result, the surface temperature of these objects is primarily a function of the solar flux that they receive; this is why their temperature falls with increasing distance from the Sun, from about 500 K at the surface of Mercury, to about 40 K at the surface of Pluto. The physics of the Solar System therefore shares a number of features with other cold objects in the universe, such as the interstellar medium and molecular clouds.

1.2.1 Thermal Radiation and Reflected Solar Radiation

Solar-System objects emit thermal radiation that is a function of their temperature and its variation with wavelength is given by Planck's Law (the radiation of a black body at a temperature T):

$$B(\nu) = (2h\nu^3/c^2)(\exp(\hbar\nu/kT) - 1)^{-1} \tag{1.8}$$

$$B(\lambda) = (2hc^2/\lambda^5)(\exp(\hbar c/\lambda kT) - 1)^{-1} \quad . \tag{1.9}$$

There is also the relationship

$$\int_0^\infty B(\nu)d\nu = \int_0^\infty B(\lambda)d\lambda = \sigma T^4 \tag{1.10}$$

σ being Stefan's constant.

The units generally used in astrophysics correspond to the CGS system: $B(\nu)$ is the flux emitted per unit frequency per unit solid angle (in erg s^{-1} cm^{-2} Hz^{-1} sr^{-1}). $\hbar$ is Planck's constant, k Boltzmann's constant, c the velocity of light (in cm s^{-1}), ν the frequency (in Hz), and T the temperature (in K). $B(\lambda)$ is the flux emitted per unit wavelength per unit solid angle (in ergs^{-1}, cm^{-2} cm^{-1} sr^{-1}); is the wavelength expressed in cm. Two other units are currently used in astrophysics:

1. cm^{-1} (wave-number), a frequency unit. The reciprocal of the wavelength, expressed in cm, i.e. the number of waves per unit distance.
2. the micron (μm), a unit of wavelength, 10^{-6} m, 10^{-4} cm. A wavelength of 100 μm ($= 10^{-2}$ cm) corresponds to a wave-number of 100; a wavelength of 1 μm ($= 10^{-4}$ cm) to a wave-number of 10000.

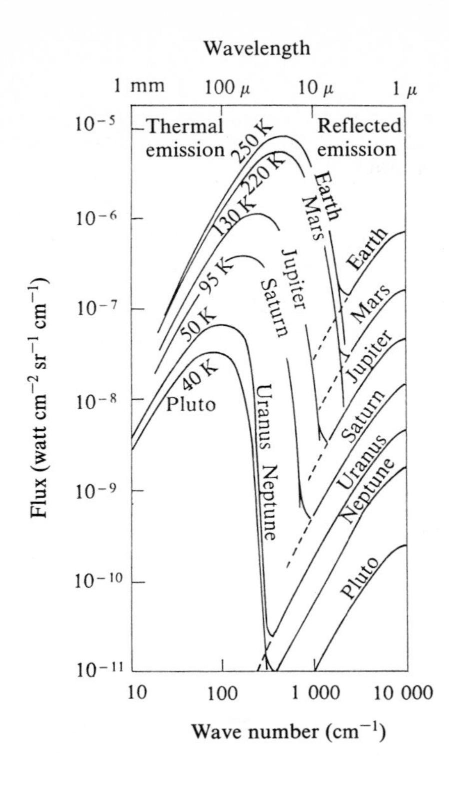

Fig. 1.5. Planetary radiation flux (reflected and thermal components). [After Th. Encrenaz: Space Science Reviews **38**, 35 (D. Reidel Publishing Company 1984)]

It can be shown that there is a simple relationship between the temperature of a black body and the frequency ν_0 of maximum emission for the $B(\nu)$ function:

$$T\nu_0 = 0.5099\,\mathrm{cm\,K} \quad . \tag{1.11}$$

Similarly, if λ_m is the wavelength at which $B(\lambda)$ is a maximum, then

$$\lambda_m T = 2880\,\mu\mathrm{m\ K} \quad . \tag{1.12}$$

As a result, the colder a body is, the longer the wavelength of its maximum emission: for planets this maximum varies from around 9 μm for Mercury to 110 μm for Pluto (see Fig. 1.5). In particular, this shows why the planets' own emission in the visible region between 0.4–0.8 μm is negligible.

Solar-System objects are nevertheless visible in the optical region only because they reflect light from the Sun: this is the second component of their emission (Fig. 1.5). In fact a solar photon intercepted by a Solar-System body may be either absorbed or reflected back into space. In the first case it is converted into thermal energy and contributes to the infrared radiation that we have just described. In the second case it may be either reflected directly or diffused (by a planetary atmosphere, for example) before being emitted into space. In what follows, we shall consider the reflected component to consist of the sum of all radiation that is not absorbed.

These two spectral components, thermal and reflected, are observed for all the planets and also for comets; for the latter in particular the maximum of the thermal

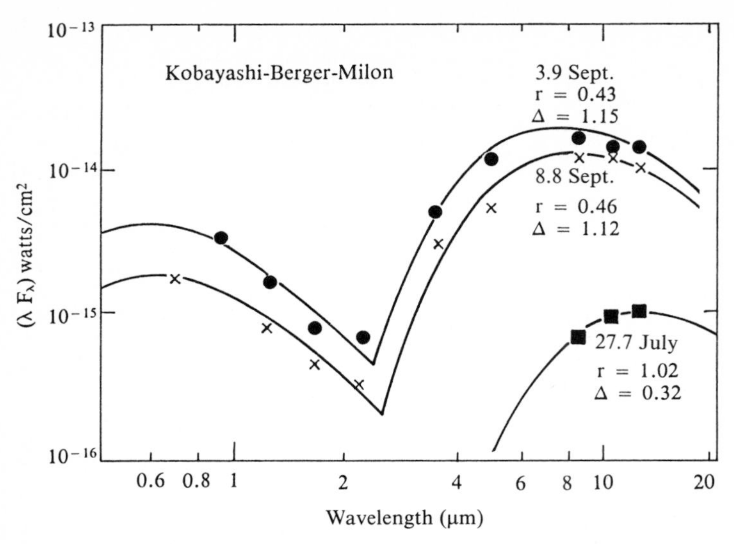

Fig. 1.6. The flux from a comet as a function of heliocentric distance r (r = heliocentric distance, Δ = geocentric distance. [After E. Ney: *Comets*, ed. by L. Wilkening (University of Arizona Press, Tucson 1982)]

radiation is observed to shift in wavelength with a change in heliocentric distance (Fig. 1.6).

The intensity of the reflected component of solar radiation does not depend on the object's temperature, but on a quantity known as its *albedo*. According to Bond's definition (1861), the albedo is the ratio of the flux reflected by a sphere in all directions to the flux of parallel light illuminating it. The albedo is therefore related to the coefficient of reflection of the surface or of the cloud layer that returns the solar radiation. In the case of an icy surface (that of the satellites of Saturn, for example), the albedo may be higher than 0.5, whereas for certain minor planets that are covered in carbonaceous material, the albedo may be only a few percent.

In the case of the terrestrial planets and minor bodies, the internal sources of energy (the radioactive disintegration of uranium and thorium) do not play an important part in determining the surface energy-balance. As a result, the thermal energy emitted by these objects arises from the conversion of that part of the solar radiation that is not reflected, but is absorbed by the object. There is therefore a simple relation between the albedo (that is to say the percentage of solar energy reflected) and the temperature of the object. The temperature of a black body that would emit the same overall thermal energy as a particular object is known as the *effective temperature*. In the absence of internal energy, the effective temperature is defined by the following relationship:

$$\frac{\theta_{\odot}}{D} \cdot \pi R^2 \cdot (1 - A) = \sigma T_e^4 \cdot 4\pi R^2 \tag{1.13}$$

if the object is in rapid rotation (which is the case with the planets Jupiter and Saturn, for example), and

$$\frac{\theta_{\odot}}{D} \cdot \pi R^2 \cdot (1 - A) = \sigma T_e^4 \cdot 2\pi R^2 \qquad (1.14)$$

if the object is rotating slowly (as for example in the case of Venus).

In these equations $\theta_{\odot}$ is the solar flux received by unit solid angle at a heliocentric distance of 1 AU; D is the heliocentric distance of the object (in AU), R its radius, A its albedo, and T_e its effective temperature. The factor of 4 that appears in the first equation arises from the fact that for a rapidly rotating body, the solar flux intercepted and absorbed by the area πR^2 is later re-radiated by the whole surface, i.e. by an area $4\pi R^2$. With slow rotation, the solar flux absorbed is re-emitted over half of the total surface area.

One can see from (1.13) and (1.14) that T_e depends on the albedo and the heliocentric distance, but not on the size of the object: for a given albedo and heliocentric distance one obtains the same effective temperature T_e for the smallest minor planets and for the largest planets. (This is not valid, however, for grains of dust, because the size of these is of the same order as the wavelength, and the calculation has to be made taking Mie's theory into account.)

Effective temperature is expressed as follows:

$$T_c = (1 - A)^{1/4} \frac{273}{D^{1/2}} \qquad (1.15)$$

for an object in rapid rotation; and

$$T_c = (1 - A)^{1/4} \frac{324}{D^{1/2}} \qquad (1.16)$$

for an object in slow rotation.

In the case of the most massive planets, however, in compiling the energy balance, account must be taken of the contribution from other, internal sources, which may, as for Jupiter and Saturn, equal the component arising from absorbed solar energy.

1.2.2 The Planets

Terrestrial Planets and Giant Planets

From the physical point of view, two principal families of planets can be distinguished:

1. The *terrestrial planets* (Mercury, Venus, Earth, and Mars), closest to the Sun, are small in size: they are all smaller than the Earth, but their mean density is, on the other hand, relatively high (from 3 to 6 g/cm^3). They have few or no satellites, and they do not have rings. The terrestrial planets all have solid surfaces and overlying atmospheres, very dense in the case of Venus, and very tenuous in the case of Mercury.
2. The *giant planets* (Jupiter, Saturn, Uranus and Neptune) more closely resemble one another. Lying farther from the Sun, all four are considerably larger than the terrestrial planets. Their mean density is low and they mainly consist of hydrogen and helium. Towards their interiors, their atmospheres reach considerable pressures – of the order of a million atmospheres. They probably

possess denser central cores. In addition they have several satellites – more than 15 in the cases of Jupiter and Saturn.

Jupiter and Saturn, which are very bright, have been observed since antiquity; Uranus and Neptune, farther away, were only detected comparatively recently: Uranus by Herschel in 1781; and Neptune by Le Verrier in 1846.

Finally, *Pluto*, beyond Neptune, remains in a class of its own and cannot be assigned to either family: it is at the same time very small and low in density, and mostly resembles the satellites of the giant planets.

The gross characteristics of this classification can be explained in terms of a very general scheme governing the formation of the planets. It is generally accepted nowadays that the planets formed by the accretion of material into larger and larger clumps. At considerable distances from the Sun (several AU), the temperature at the time of the origin of the Solar System was low enough for ices not to sublime. Large and relatively low-density cores were therefore able to form. At positions close to the centre of the nebula (of the order of one AU), on the other hand, as soon as the Sun was formed, the temperature was much higher. The only elements that could exist in solid form were therefore dense, refractory phases, so the bodies formed were smaller and denser. In qualitative terms, therefore, we can understand the division of planets into two major classes, the terrestrial and the giant planets.

Primitive and Secondary Atmospheres

In order to understand the origin of the chemical composition of planetary atmospheres, it is necessary to introduce the concept of *escape velocity*. For any body of mass m, subject to the gravitational field of a planet of mass M, and at a distance R from the centre, the escape velocity v_{esc} is given by the equation

$$\frac{1}{2}mv_{\mathrm{esc}}^2 = \frac{mMG}{R} \tag{1.17}$$

G being the universal gravitational constant. From this we have that

$$v_{\mathrm{esc}} = \sqrt{\frac{2GM}{R}} \quad . \tag{1.18}$$

Consider a planetary atmosphere at a temperature T, subjected to the planet's gravitational field: in accordance with the Maxwellian distribution of velocities, the most likely velocity v_{th} for a molecule of mass m is

$$v_{\mathrm{th}} = \sqrt{\frac{2kT}{m}} \quad . \tag{1.19}$$

The probability that a molecule will escape from a given atmosphere thus depends on the relationship between v_{th} and v_{esc}. The escape velocity, which does not depend on the mass of the particle, is larger the more massive the planet. The thermal velocity is larger the higher the value of T and the lower the value of m. From this we see that molecules escape more easily if the planet is small and its temperature is high. This explains why the giant planets have been able to retain all their elements, even the lightest. Their atmospheres, whose composition reflects that of the original gaseous nebula, are, at least in part, "primitive", consisting of approximately 90 % hydrogen and about 10 % helium; other elements exist in reduced form (CH_4, NH_4, etc.). The terrestrial planets, on the other hand, have not been able to retain the lightest elements. Their atmospheres, based on C, N and O, are secondary atmospheres, probably arising from degassing, or by chemical or biological evolution. These processes, specific to each planet, explain the great differences that exist between the chemical components in the atmospheres of the terrestrial planets. A list of various physical and atmospheric properties for the planets is given in Table 1.3.

1.2.3 The Satellites

Apart from the three exceptions formed by the Moon and the two satellites of Mars, Phobos and Deimos, all the satellites belong to giant planets, or in the case of Charon, to Pluto.

Although the *Moon* has perhaps been, since ancient times, the most-studied astronomical object, its origin is far from being understood. Its size, when compared with that of the Earth, is remarkably great (see Table 1.2). Several theories have been advanced to explain its origin: separation of a part of the Earth, capture by the Earth, or initial formation as a member of a double system. Because of its low mass, the Moon cannot retain a dense atmosphere; its surface, which is covered by dark and light patches – incorrectly called "seas" (maria) and "continents" (terrae) – is covered with craters formed by meteoric impact.

Phobos and *Deimos*, which orbit Mars, are two rocky bodies that are small in size and irregular in shape; their respective sizes are of the order of 25 km and 15 km. Their surfaces, very ancient and covered with impact scars, bear witness to the very high number of meteoritic impacts that have occurred during the course of the Solar System's history.

The satellites of the giant planets may be classified in different groups. Jupiter has four large satellites, discovered by Galileo at the beginning of the 17th century: Io, Europa, Ganymede, and Callisto, known as the *Galilean satellites*. The surface and interior of Io are subject to violent movements caused by the tides raised owing to the proximity of Jupiter. This is the source of the volcanism discovered on Io. The other three Galilean satellites, unlike the satellites of the Earth and Mars, consist of a mixture of ice and rock; none of them possess atmospheres. The other satellites of Jupiter rarely exceed one hundred kilometres in diameter. Saturn has, apart from several smaller satellites, a dozen satellites with diameters between 300 and 1500 km, without atmospheres, and with surfaces scarred by meteoritic impacts. Essentially, they consist of ice. Several small satellites have been discovered close to the rings. Finally, one satellite of Saturn is markedly different to the others: this is Titan, the largest satellite in the Solar System (after Ganymede). It is surrounded by a thick atmosphere in which complex organic molecules have been discovered. It is possible that Titan could have a surface covered with a liquid ocean. At present these seem to be the only Solar-System objects that resemble the Earth in this respect.

Table 1.3. Physical characteristics of the planets

Name	Symbol	Equatorial diameter relative to the Earth	Equatorial diameter (km)	Flattening	Mass relative to the Earth[1]	Mean density	Surface gravity (in ms^{-2})	Escape velocity (in $km\ s^{-1}$)	Sideral rotation (in days, or in hours, minutes and seconds)	Inclination of the equator to the orbital plane	Principal atmospheric components
Mercury	☿	0.382	4878	0	0.055	5.44	3.78	4.25	58,646 d	0°	H, He, Ne (solar wind)
Venus	♀	0.949	12104	0	0.815	5.25	8.60	10.36	243 d (r)[2]	2°07′	CO_2 (97%)
Earth	♁	1	12756	0.003 353	1	5.52	9.78	11.18	23 h 56 min 04 s	23°26′	N_2 (78%) O_2 (21%)
Mars	♂	0.533	6794	0.005	0.107	3.94	3.72	5.02	24 h 37 min 23 s	23°59′	CO_2 (95%)
Jupiter	♃	11.19	142800	0.062	317.80	1.24	24.8	59.64	9 h 50 min to 9 h 56 min	3°04′	H, He, CH_4, NH_3
Saturn	♄	9.41	120000	0.0912	95.1	0.63	10.5	35.41	10 h 14 min to 10 h 39 min	26°44′	H, He, CH_4, NH_3
Uranus	♅	3.98	50800	0.06	14.6	1.21	8.5	21.41	17 h 06 min[3]	98°	H, He, CH_4, NH_3
Neptune	♆	3.81	48600	0.02	17.2	1.67	10.8	23.52	15 h 48 min	29°	H, He, CH_4, NH_3
Pluto	♇	~0.2	~3000	?	0.002	1 ?		?	6 d 9 h 18 min	?	–

[1] $m_E = 5.976 \times 10^{24}$ kg
[2] (r) indicates that the rotation is retrograde
[3] rotation period of the magnetic field.

1.2.4 The Rings

It should perhaps first be noted that the distinction between rings and satellites is primarily historic. We now know that Saturn's rings consist of particles of various sizes, the largest of which may be several kilometres across. Such dimensions are not far from those of Deimos, the smaller of the two Martian satellites. It is possible that there is no real physical difference between these two classes of object.

It was Galileo who, at the beginning of the 17th century, first noted changes in the appearance and the amount of light emitted from the vicinity of Saturn. A little later, Huygens solved this puzzle: there was a disk situated in the equatorial plane of the planet, and the amount of light that it reflected towards the Earth varied according to the position of the latter with respect to the plane of the rings. At the end of the 17th century, Cassini discovered the division that carries his name, suggesting for the first time that the rings were not uniform, but consisted of a swarm of small satellites. This idea was confirmed, on the basis of calculation of celestial mechanics, by Laplace at the end of the 18th century, and then by Maxwell.

Thanks to exploration by the Voyager spacecraft, we now know that the rings consist of innumerable small bodies, probably composed of ice and refractory grains, which revolve independently in concentric orbits. The origin of this system of rings is still poorly understood: it may have resulted from the fragmentation of a satellite, or it may, on the other hand, represent the residue left over from the formation of Saturn as a planet.

With the discovery of Uranus' system in 1977, and then that of Jupiter in 1979, the existence of rings has ceased to be exceptional in the Solar System. The difference between the three systems is nevertheless striking: Uranus' system consists of extremely narrow rings, which have an albedo amounting to a few percent at most. It is thus hardly likely that they consist of H_2O-ice, as is the case with Saturn's system. The ring discovered around Jupiter is also thin and dark, and lies close to the planet. The presence of a ring system around Neptune was confirmed in 1989. These successive discoveries lead one to think that the formation of a system of rings could form one stage in the process of planetary accretion; other systems of rings perhaps remain to be discovered, notably around certain minor planets.

1.2.5 The Asteroids

The discovery of the first asteroids dates from the beginning of the 19th century, after astronomers realised that there was a planet missing at the distance predicted by Bode's law for $n = 3$ [see (1.7)]. The largest asteroids thus discovered were (in order): *Ceres, Pallas, Juno,* and *Vesta*. By now several thousand minor planets have been catalogued; it is estimated that the total number larger than 1 km in diameter in the Solar System is several hundred thousand. Table 1.4 summarizes the characteristics of the largest asteroids (also called minor planets).

Most of the minor planets orbit at distances of between 2.2 and 3.4 AU: this is the *asteroidal belt*. A small number lie in the orbits of Mars and Jupiter: this is the case with the *Trojans*, at Jupiter's Lagrangian points (see Sect. 1.3.1). Others have strongly elliptical orbits and may approach the Earth: these are the *Apollo-Amor* type.

Table 1.4. Characteristics of some of the brightest minor planets

Classification number	Name	Year of discovery	Semi-major axis (AU)	Orbital period (years)	Orbital eccentricity	Orbital inclination	Diameter in (km)	Rotation period (hours)
1	Ceres	1801	2.77	4.60	0.078	10.6	1032	9.1
2	Pallas	1802	2.77	4.61	0.234	34.8	588	10.1
3	Juno	1804	2.67	4.36	0.258	13.0	248	7.2
4	Vesta	1807	2.36	3.63	0.089	7.1	576	10.6
5	Astra	1845	2.58	4.13	0.190	5.4	120	16.8
6	Hebe	1847	2.42	3.78	0.202	14.8	204	7.3
7	Iris	1847	2.39	3.69	0.230	5.5	208	7.1
8	Flora	1847	2.20	3.27	0.156	5.9	162	13.6
9	Metis	1848	2.39	3.69	0.122	5.6	158	5.1
10	Hygiea	1849	3.14	5.55	0.120	3.8	430	18.0
15	Eunomia	1851	2.64	4.30	0.185	11.8	260	6.1
16	Psyche	1852	2.92	5.00	0.134	3.1	248	4.3

From a physical point of view, asteroids are classified by their surface composition and, in particular, their albedo. The most important groups are type C (very dark objects, probably with carbonaceous surfaces, and with albedos below 0.06); type S (silicate surface, albedo around 0.2, and with spectra suggesting the presence of silicates); and type M (objects with albedos of about 0.1, thought to be rich in metals).

The origin of the asteroids is still poorly known. As has been mentioned earlier (Sect. 1.2.1), one theory largely accepted nowadays is that the asteroids are small condensations from the primitive solar nebula that were unable to form into a single body, doubtless because of the gravitational instability caused by the presence of Jupiter.

1.2.6 Comets

Like the asteroids, comets are "primitive" Solar-System objects. But whilst the asteroids, because of their masses, underwent various processes of differentiation, it would seem that cometary nuclei, which are very small and which evolved in a very cold region, represent unaltered, primitive Solar-System material. Apart from the absence of differentiation, because of the low temperature prevailing in their environment, comets have preserved intact not only the matter that condensed, but also volatile elements. This explains why comets are richer in ice than the minor planets.

Comets have been known since prehistory. Their sudden, unpredictable appearance and their rapid movement across the sky caused them to be regarded for many centuries as supernatural, and harbingers of bad fortune. The appearance of a comet can indeed be spectacular. At large heliocentric distances it only consists of an inert nucleus, invisible to the naked eye and probably consisting of dust and ice, with a diameter of no more than a few kilometres. As the comet's orbit brings the object closer to the Sun, however, the nucleus is heated by the Sun's radiation and begins to sublime, to degas, and to eject dust particles. It is the solar radiation reflected and

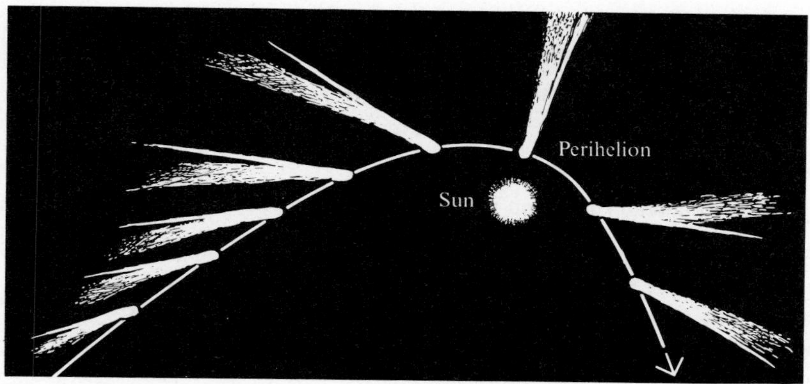

Fig. 1.7. Degassing of a comet is a function of its heliocentric distance

diffused by this dust and gas that produces the spectacular phenomenon that we see, with maximum brightness occurring at perihelion passage or slightly after. As the comet continues to move along its orbit it recedes from the Sun, degassing declines and then ceases, and the comet returns to invisibility (Fig. 1.7). At each passage a comet loses its external layers and so it has a limited lifetime. For historical reasons, comets are divided into two principal categories:

1. *short-period comets* have elliptical orbits with periods of less than 200 years; in some cases the period may be only a few years;
2. *long-period comets* have periods greater than 200 years; their orbits may be elliptical, parabolic or hyperbolic, according to the planetary perturbations that they have undergone; their discovery cannot be predicted.

About 900 comets have been observed up to the present. If one takes into account the perturbations that the approximately 80 comets with hyperbolic orbits have undergone, primitive cometary orbits, prior to perturbation, are found to be elliptical, with semi-major axes of the order of 40 000 AU. Comets therefore come from an immense and very distant reservoir, known as the *Oort cloud*. This fundamental result implies that all the comets observed now originated in the Solar System and not in interstellar space. The Oort cloud must in fact contain millions of these cometary bodies, of which we only see an infinitesimal part - those bodies that, as a result of perturbations, perhaps by nearby stars or molecular clouds, are deflected in towards the Sun. These objects, the comets that we see, thus have a very short lifetime when compared with that of the Solar System.

Table 10.1 summarizes the orbital characteristics of some of the best-known comets.

1.2.7 The Interplanetary Medium

The Solar System is not just confined to the relatively massive bodies that we have described so far: these bodies are not moving in an absolute vacuum. The medium in which they orbit does not have a density of zero. It is, nevertheless, well below the best vacuum obtainable in the laboratory and is, in particular, too low to perturb the motion of the solid bodies in any noticeable way. The material in the interstellar medium consists of two components:

1. an ionized-particle flux, originating in the Sun: the *solar wind* (and to a much lesser extent, solar cosmic rays, to which must be added galactic cosmic rays);
2. a solid component, consisting of dust.

In addition, the interplanetary medium is bathed in radiation, mainly of solar origin.

The Solar Wind

The discovery of the solar wind is relatively recent. It was, in fact, in the fifties that L. Biermann concluded from the systematic effects on the ionized tails of comets that there was a continuous flow of ionized particles emanating from the Sun. In 1958, E. Parker put forward a theory to describe the flow of this hot plasma, directly linking it with the solar corona and the latter's high thermal conductivity. The temperature of the corona (of the order of a million degrees) is such that a stationary, stable state requires continuous expansion of the corona, in a "wind" at a uniform rate. Beyond a few solar radii, the velocity of the plasma becomes supersonic. It is approximately 400 km/s at 1 AU and has a flux of approximately 2×10^8 particles $cm^{-2}s^{-1}$, which corresponds to an average density of 5 ions per cm^3. At times of strong solar activity the density may rise by a factor of 10. Away from the Sun, the density of ions decreases as the square of the heliocentric distance. The ionic composition is approximately 95 % protons and 5 % helium nuclei.

The solar wind is a very high-temperature plasma, and is therefore very conductive. It carries with it an interplanetary magnetic field, also originating in the Sun. At 1 AU the field entrained is about 1 to 10 gammas[3]. This field may grow by a factor of 10 when there are energetic solar eruptions, which are the source of the solar cosmic-ray emission.

The interaction of the solar wind with the bodies within the Solar System gives rise to different configurations according to whether a particular body has a gaseous, but nonmagnetic, envelope (this is the case with Venus, Mars, Titan, and the comets), or has neither atmosphere nor magnetic field (like Io, the Moon, and most other satellites) or, finally, has its own magnetic field (as with Mercury, the Earth, Jupiter and Saturn).

Interplanetary Dust

The existence of interplanetary dust is shown by *meteors* and *meteorites* that continuously encounter the Earth's atmosphere. One speaks of a meteorite when a solid object reaches the Earth's surface, and of a meteor when it disintegrates completely in the terrestrial atmosphere. The latter case gives rise to the "shooting stars" that are so well-known. Some bombardments take place periodically and in this instance the meteors arise from the cloud of dust that is spread along the orbit of a comet, or from the remnants of a comet. The dust arises from the gradual erosion or the disin-

[3] 1 gamma $= 1\,\gamma = 10^{-5}$ gauss.

tegration of cometary nuclei. This is the case with the Leonids, which are associated with Comet Temple; the Taurids, associated with Comet Encke; the Orionids, associated with Comet Halley, and so on. It is thought that meteors, which are particles ranging from 200 μm to a few cm in diameter, are of cometary origin.

Bodies that reach the surface of the Earth without being completely destroyed are the meteorites, with initial masses above one kilogramme. They lose a portion of their mass during their path through the atmosphere, and reach the ground with a frequency of the order of 2 to 10 per day. A meteorite may be metallic, may consist of a metal-silicate mixture, or be silicate. The asteroids are probably the meteorites' parent bodies.

Another class of particles may penetrate the Earth's atmosphere without being destroyed. These are the micrometeorites, with diameters below ten microns, and with masses below 10^{-7} g. As these particles are not changed by their passage through the atmosphere, they have the same physical characteristics as interplanetary dust, most of which probably has a cometary origin. Micrometeorites comprise the bulk of extraterrestrial material that passes through the terrestrial atmosphere.

Apart from the extraterrestrial material that encounters the Earth, the existence of interplanetary dust is also shown by observation of the *zodiacal light*. This is formed by sunlight reflected by dust particles present in the Solar System. It can be seen along the ecliptic, hence its name. In the antisolar direction the phenomenon is known as the *Gegenschein*. In 1983, observations by the infrared satellite IRAS showed that the dust is not confined to the ecliptic, but that clouds of dust occur at certain well-defined ecliptic latitudes.

1.3 The Variability of Objects in the Solar System

One peculiarity of the study of objects in the Solar System, in comparison with that of other astronomical observations, is that it involves objects that are primarily variable in their luminosity over the course of time. These variations may be recurrent or sporadic. The causes of variability are many: they may be linked with variations on the Sun, to that of the objects themselves or, finally and most frequently, to the relative motion of the Sun, the Earth, and the object being studied. The variations in Solar-System bodies can often be used profitably to study the physical properties of the objects themselves.

1.3.1 Variations Caused by the Sun

Any variation in the solar radiation flux received by an object in the Solar System has immediate repercussions at several levels: (1) the intensity of the UV and visible spectrum reflected by the object; (2) in the case of planetary and cometary ionospheres, the intensity of the fluorescence lines excited by solar UV and visible radiation; (3) the photochemistry, the photodissociation and the ionization of planetary and cometary atmospheres; (4) the thermal, infrared energy re-emitted that is derived from the absorbed solar radiation.

The first two effects involve solar fluctuations at every rate, including the most rapid; on the other hand, variations in the level of internal energy can only be observed by means of long-period variations in the solar flux.

The solar wind may also vary significantly apart from the periodic variation caused by the Sun's 26-day rotation. These variations have repercussion in the interplanetary medium, and thus on every ionized atmosphere.

1.3.2 Variations Having an Internal Origin Within an Object

Objects in the Solar System may show their own variations caused by the physics of their interiors. This is the case, for example, with the dynamical phenomena that have an internal source in the giant planets. It is also the case with the volcanism of the Earth and Io; and with the non-gravitational forces produced by cometary nuclei.

1.3.3 Motion of a Body with Respect to the Sun

The most obvious change in a Solar-System object is that caused by its position relative to the Sun. It is by studying this that the orbital parameters and the mass of the object are derived. Many physical phenomena are linked with variations in the heliocentric distance of the body: the best example of this is the degassing of comets when close to perihelion.

The movement of the body itself around the Sun also has physical implications. Atmospheric circulation on the planets is a consequence of their rotation, and in the same way, the oscillation of the polar axis of Uranus, and its inclination relative to the Sun, have implications for the planet's internal and atmospheric physics.

Variation in the heliocentric distance of a body can have less immediate effects: thus the variation in the Doppler shift of the Fraunhofer lines in the solar spectrum induces a variation in the fluorescence spectrum of comets, with repercussions into the radio region; this is the Swings effect.

1.3.4 Motion of a Body with Respect to the Earth

Variation in the geocentric distance of an object obviously alters its visibility. Minor planets and comets cannot generally be observed except at specific, well-determined times.

The rotation of the bodies themselves, already mentioned, is similarly seen from the Earth and permits the observation of the whole of the surface or of the atmosphere of the object.

Observation of the movements of satellites around planets is used to define their orbital parameters and to determine the mass of the satellites (a recent example is the detection of the Pluto-Charon system).

1.3.5 Specific Configurations of the Sun-Earth-Object System

We have already mentioned, in discussing the planets and the Moon, specific configurations known as conjunction and opposition (see Sect. 1.1.2). Apart from these extreme cases, observation of an object as its phase-angle varies gives us information about the physical properties of the particles that are on its surface or in its atmosphere. In the case of the Moon, solar eclipses are observed in order to study the solar corona and chromosphere.

1.3.6 Specific Configurations of an Earth-Planet-Satellite or Sun-Planet-Satellite System

The passage of the Earth through the equatorial plane of a giant planet produces some notable phenomena. In the case of Jupiter, the observation of mutual occultations by the Galilean satellites allows their astrometric parameters to be refined; while with Saturn, observation of the system of rings edge-on allows their thickness to be estimated and small satellites, normally invisible, to be detected. The phenomena occur every six years in the case of Jupiter, and every sixteen for Saturn.

When the Sun passes through the equatorial plane of a giant planet, eclipses of the satellites can be observed. Measurements at these times allow the solar and thermal components to be separated out from averaged infrared observations. In the case of Titan, the effects of the absence of solar radiation on the aeronomy can be studied.

The study of all these classes of variation entails, for telescopic observation, a certain number of observational constraints that are specific to Solar-System objects. They have to be observed in a repetitive fashion, at well-defined times. Objects close to the Sun can only be observed at the beginning or the end of the night at low elevations and, finally, they are subject to very considerable proper motion.

2. Methods of Studying the Solar System

Because of the brilliance of certain of its objects, the Solar System has been studied since antiquity. For centuries these observations were restricted to what was visible with the naked eye. Successive advances have been made, since the beginning of the 17th century, by the use of larger and larger telescopes, then by photographic observations taking over from visual ones. During the 20th century, astronomical observational techniques have undergone a veritable revolution. First of all, observations from space allowed access to ultraviolet and infrared regions of the spectrum, as well as to X- and γ-rays. In addition, the century has seen the beginning of radio astronomy and, above all, beginning in the sixties, the start of "in situ" observations of the Solar System, with the launch of spaceprobes to the Moon and planets. This chapter aims to review the observational techniques that have so far been used to study the Solar System. A detailed discussion of astrophysical instrumentation can be found in the book by P. Léna: *Méthodes physiques de l'observation* [InterEditions/Editions du CNRS (Paris 1986); English edition: *Observational Astrophysics* (Springer-Verlag, Berlin, Heidelberg 1988)].

2.1 Determination of Geometrical and Physical Properties

2.1.1 Distance Determination

With objects that are relatively close to the Earth, their geocentric distances can be measured by the geometrical parallax method, the terrestrial radius – known from the work of geodetic experts – being chosen as the reference length. One may proceed to measure the Earth-Sun distance by determining the geocentric distance of one of the inner planets at the time of opposition (Fig. 2.1), and then applying Kepler's third law (see Sect. 1.1.1). This method was first employed by J.D. Cassini in 1672, when there was a favourable opposition of the planet Mars. More recently, a more precise measurement was obtained by means of the minor planet Eros, which passes very close to the Earth.

Since the beginning of the space age, more precise determinations of distances within the Solar System have been obtained by radar echoes from the Moon, the nearest planets and spaceprobes. In the case of the Moon, radar has even been replaced by laser ranging, which increases the accuracy of measurement even further, to just a few centimetres.

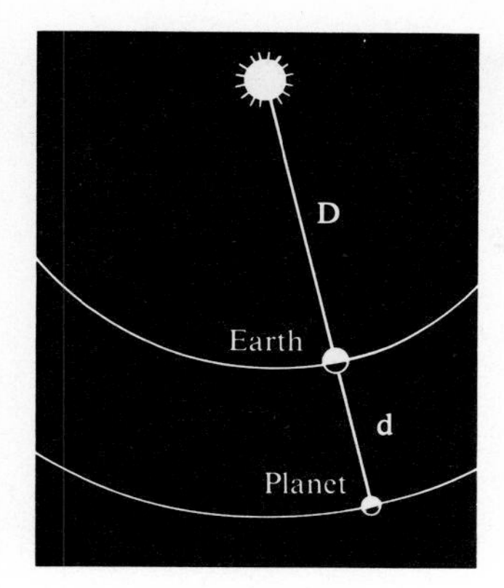

Fig. 2.1. The measurement of the Earth-Sun distance at the time of opposition of a terrestrial planet

2.1.2 Determination of Diameters

The first diameter measurements were obtained from the apparent diameter (i.e. the angle that an object subtends at the Earth). But this method is not very accurate, because the angular diameter of the object is always small; moreover, the result is only true, strictly speaking, for bodies without atmospheres. With planets that have a dense atmosphere the diameter measured applies to a cloud layer of unknown depth. Three other types of measurement may be undertaken from Earth: (1) for nearby objects (inner planets, certain asteroids and some very close comets), radar measurements give the diameter of the solid body; (2) with small bodies whose albedo may be independently estimated, measurement of the amount of flux reflected enables the mean diameter to be determined; (3) for distant objects (asteroids or giant planets) the method of measurement that uses the observation of stellar occultations gives a remarkably precise value for the length of the occulting chord (see Fig. 2.2). If the astrometry of the event is sufficiently accurate, it is possible to deduce the diameter of the object. We should note, however, that in the case of the giant planets, this diameter refers to a layer at a specific atmospheric pressure, beyond which the stellar flux is absorbed. This level therefore depends both on the atmospheric composition and on the wavelength employed for the observation.

Since the introduction of spaceprobes, very precise measurements of diameters have been obtained by radar. In the case of the inner planets and the principal satellites, altimetry of the surface has been possible by using radar equipment on the probes themselves.

2.1.3 Determination of Masses

The determination of the mass of a planet is relatively easy when it has one or more satellites. Assuming that the satellite undergoes Newtonian motion around the planet, it is possible to deduce the mass of the planet as a function of the velocity of the satellite and of its distance from the planet (see Sect. 1.1.1). For Mercury, Venus, and planetary satellites, measurement of the mass from observations made from Earth is much more difficult; this mass can be estimated from the perturbations exerted on neighbouring objects, but the results are not very precise. In the cases

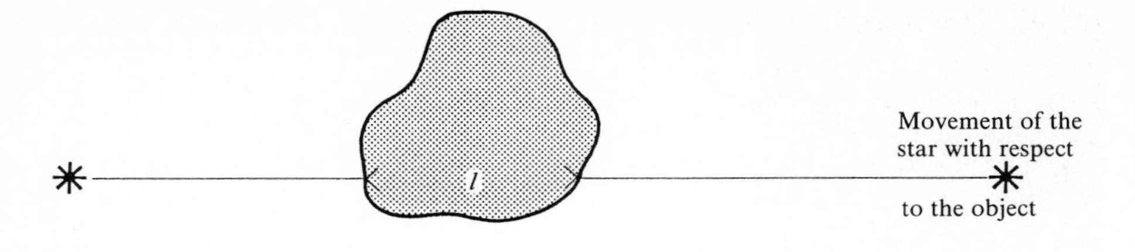

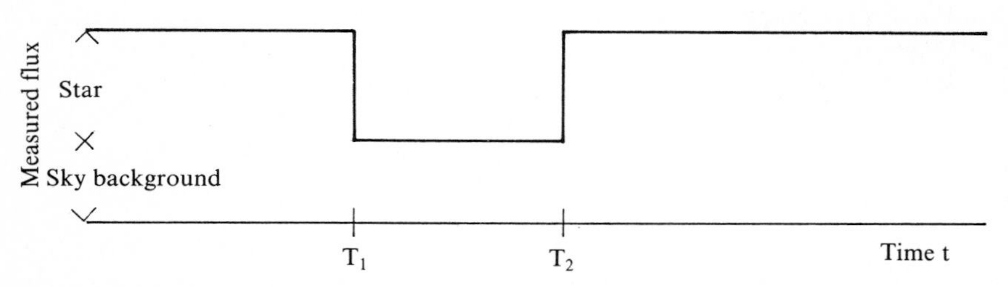

If l is the chord to be measured and V the speed of the object, $l = V(T_2 - T_1)$

Fig. 2.2. Measurement of the size of an object by means of the stellar-occultation method

of Mercury and Venus, and also of the Galilean satellites, precise determination of their mass has been possible thanks to the observation of the path of spaceprobes that have passed close to these objects.

2.1.4 Determination of Rotational Periods

If a Solar-System object has an angular diameter of a few seconds of arc or more, its rotational period can be determined by visual or photographic observation of the surface. It should be noted that this period may correspond to that of a cloud layer, seen at visible wavelengths, rather than that of the surface itself. If different atmospheric layers are examined by using different wavelengths, the periods determined may also differ. This applies, in particular, to the planet Venus.

Where objects relatively close to the Earth are concerned, radar measurements may equally be used. For objects with small angular diameters, one may determine the light-curves as a function of time; the period is given by the periodic variation in the photometric curve. This method is applied to minor planets in particular. Again it should be noted that interpretation of the light-curve is not always easy; minor planets are, in general, not spherical but irregular, so the light-curve is not simply sinusoidal in shape.

2.2 Physical and Chemical Analysis of Gas and Particles

2.2.1 Analysis of Radiation by Remote Sensing Spectroscopy

Until the advent of in-situ experiments, it was mainly just remote sensing spectroscopy – that is, the analysis of radiation as a function of its wavelength – that gave us information about the chemical composition and physical conditions found on Solar-System bodies. Gaseous components provide spectral signatures at characteristic wavelengths, which enable them to be identified in planetary and cometary spectra. Relative intensities and widths of the different lines in a specific molecular band give information about the temperature and pressure of the medium involved.

With a solid body, the spectrum of the reflected sunlight may show very wide absorption bands, characteristic of the minerals present on the surface of the body. Finally, the study of dust particles, their sizes and their composition is also possible by spectroscopy and by polarimetry.

Radiation Transfer in Planetary Atmospheres

We have already mentioned (Sect. 1.2.1) that radiation from Solar-System objects consists of two components: the reflected solar flux and the thermal flux. In the case of planetary atmospheres, the reflected solar component is dominant from the UV to about 3 μm, whilst the thermal component becomes important beyond 3 μm (Fig. 1.5). In order to analyze this radiation it is necessary to understand the thermal structure of the atmosphere under consideration.

The Thermal Structure of a Planetary Atmosphere

First let us summarize the nomenclature used. The lower part of an atmosphere, within which all the constituents that do not condense are evenly mixed, is called the *homosphere*. Above this region and separated from it by the *mesopause* is the *heterosphere* in which each gas diffuses separately within the gravitational field. The homosphere itself is divided into several layers according to the means of energy transfer that predominates within them. In the lower atmosphere we find the *troposphere*, characterized by convective transfer, then, above the *tropopause*, the *stratosphere*, where energy transfer is radiative. Figure 5.16 shows the various layers found in the Earth's atmosphere.

We may use parallel planes to model the atmosphere by assuming that locally it may be represented by a succession of superimposed, homogeneous, plane-parallel layers, and also by assuming axial symmetry around the atmosphere's vertical axis. The radiation transfer equation may now be written as:

$$\mu \frac{dI_\nu}{d\tau_\nu} = I_\nu - J_\nu \tag{2.1}$$

where $\mu = \cos\theta$, θ being the angle between the line of sight at the vertical; I_ν is the specific intensity (a function of frequency ν); τ_ν is the optical thickness (also a function of ν) above a level at altitude z; J_ν is the source function. If K_ν is the coefficient of absorption and ϱ the density, we have:

$$\tau_\nu = \int_z^\infty K_\nu \varrho \, dz \quad . \tag{2.2}$$

For planetary atmospheres, which are relatively cold and dense environments, it is found that at pressures greater than one mb, there is *local thermodynamic equilibrium* (LTE): the source function is the Planck function. Each atmospheric layer radiates like a black body at the appropriate temperature; this is Kirchhoff's law:

$$J_\nu = B_\nu(T) \quad . \tag{2.3}$$

In order to determine the thermal structure of the atmosphere (that is to say, the value of T at each atmospheric level) it is necessary to solve an energy-conservation equation reduced to radiative equilibrium. In other words, for each atmospheric level the flux divergence must be zero. The integral over frequency of the flux absorbed that arrives from higher and lower levels, together with the incident solar flux and that from any internal source, is equal to the integral of the flux emitted by the layer. The latter is linked with the temperature of the layer by Stefan's law. The temperature at a level z is given by

$$\sum_\nu \sum_{z'} \sum_\mu B(T') \mathrm{e}^{-\tau'/\mu} = \sigma T^4 \tag{2.4}$$

where τ' is the optical thickness between layers at altitudes z and z', and T' is the temperature of the layer z':

$$\tau' = \int_z^{z'} K_\nu \varrho \, dz \tag{2.5}$$

σ being Stefan's constant.

Apart from the radiative equilibrium equation, atmospheric properties are linked by other constraints:

1. The law of hydrostatic equilibrium:

$$dP = -\varrho g \, dz \tag{2.6}$$

where P is the pressure and G the gravitational constant. As the atmosphere has a very small thickness when compared with the planetary radius, the factor g does not depend on z.

2. The perfect gas law:

$$P = k \varrho T / \mu \tag{2.7}$$

where k is Boltzmann's constant and μ the mean molecular mass.

From these last two equations it follows that for an isothermal atmosphere:

$$P(z) = P(z_0) \exp[-(z - z_0)/H] \quad . \tag{2.8}$$

H is the scale height, defined by:

$$H = kT/\mu g \quad . \tag{2.9}$$

z_0 may be either the surface or, in the case of the giant planets, a specific reference level.

If the mean molecular mass μ (and thus the atmospheric composition of the planet) is known from other methods, it is possible, by numerical integration and successive iterations, to determine the thermal structure as a function of altitude.

The Reflected Solar Component

In the case of a planet having a true surface and a tenuous atmosphere, like Mars, the calculation is quite simple. The radiation from the Sun is returned towards the Earth after scattering by the surface. It is generally accepted that this scattering conforms to Lambert's law:

$$F(\theta) = F \cos \theta_0 \tag{2.10}$$

where F is the flux arriving at the surface, θ_0 the angle of incidence, measured relative to the normal to the surface, and θ the angle between the normal and the direction of scattering. This law implies isotropic scattering of light, irrespective of the direction in which one observes the radiation. If one observes at a wavelength corresponding to a transition of one of the atmospheric components, an absorption line is observed in the planetary spectrum. In the visible and the near infrared, the lines correspond to vibrational-rotational transitions. Knowing the angle made by the incident radiation with the normal to the surface, it is easy to deduce the number of molecules along the line of sight. Comparison of the intensities of several absorption lines from the same component provides an estimate of the average temperature of the medium. If the spectral resolution of the equipment is sufficiently great, a measure of the width of the line can provide an estimate of the mean pressure in the medium, if this is sufficiently dense for the line to be pressure-broadened. When the measurement is made with low spectral resolution, the average temperature may be deduced from the general shape of the absorption band. With a very dense atmosphere, the situation is more complicated. Radiation no longer propagates in a straight line through the atmosphere, but is scattered either by the molecules (Rayleigh diffusion) or by the solid particles suspended in the atmosphere or present in clouds. The construction of an atmospheric model therefore requires one to know, for each type of scattering particle, its density, its distribution as a function of altitude, its coefficients of absorption and extinction, and its scattering index (that is, the distribution of emergent flux as a function of the scattering angle). In the absence of all these values, it has long been the practice to use the "reflecting layer model" (RLM), in which a dense cloudy layer is considered equivalent to a surface covered by a clear atmosphere without scattering. This simple method has, in a certain number of cases, given good qualitative results, in particular in measuring the abundance of methane in the atmospheres of the giant planets.

The Thermal Component

This component is more difficult to analyze, because it is found in the infrared, which is less easily observed than the visible region. The terrestrial atmosphere is in fact opaque to infrared radiation, apart from certain localized "windows" (4.7–5.2 μm; 7–13 μm; 20 μm). On the other hand it does have the advantage of being easier to study from a theoretical point of view, because infrared radiation is less affected by scattering than UV and visible radiation. Indeed, according to Mie's theory, which is applicable as soon as the wavelength becomes comparable to the size of the particles, the scattering extinction coefficient is inversely proportional to the wavelength. Therefore, to a close approximation, one can ignore scattering phenomena and calculate the emergent flux at a frequency ν under the LTE theory by the transfer equation:

$$\phi_\nu = \int_\mu \int_{z_0}^{\infty} B(z,\nu)\exp[-\tau_\nu(z)/\mu]\, d\tau(\mu)/\mu \quad . \tag{2.11}$$

The altitude z_0 may refer to the surface level, where the atmosphere is tenuous, or to a level that is sufficiently deep for $\exp[-\tau_\nu(z)/\mu]$ to be negligible: this applies in dense atmospheres, in particular those of the giant planets.

A useful quantity for measuring ϕ_ν is the *brightness temperature* $T_B(\nu)$: this is the temperature of a black body that would emit the same flux at a given wavelength. The theory of radiative transfer shows that there is a simple relationship between the brightness temperature and the optical thickness $\tau_\nu(z)$; this is the Barbier-Eddington approximation:

1. for the specific intensity emitted along the vertical, the brightness temperature is, to a first approximation, the temperature of the atmospheric layer for which the optical thickness is equal to 1;
2. for the flux emitted by the whole of a disk (i.e. integrated over the angle θ), the brightness temperature is, again to a first approximation, the temperature of the atmospheric layer for which the optical thickness is equal to 0.66.

In other words, the radiation received at a given frequency primarily originates from a well-determined atmospheric layer, defined by its optical thickness. This layer is also that for which the weight function defined by

$$FP(\nu,z) = \exp[-\tau_\nu(z)/\mu]\frac{1}{\mu}\frac{d\tau_\nu(z)}{dz} \tag{2.12}$$

is a maximum. The flux can, in fact, be expressed as a function of $FP(\nu,z)$:

$$\phi_\nu = \int_\mu \int_{z_0}^{\infty} B(\nu,z)FP(\nu,z)dz \quad . \tag{2.13}$$

In the case of an atmosphere where the temperature gradient changes sign with height (which applies to all the planets), the infrared spectrum may show molecular bands, either in emission (if the gradient is positive), or in absorption (if the gradient is negative), according to the intensity of the line considered and the abundance and distribution of the absorbing gas (see Fig. 2.3).

In practice, one assumes the thermal profile of the planet, and determines the abundances and vertical distribution of minor components detected spectroscopically in planetary infrared spectra, by comparison between observations and the predictions of theoretical models.

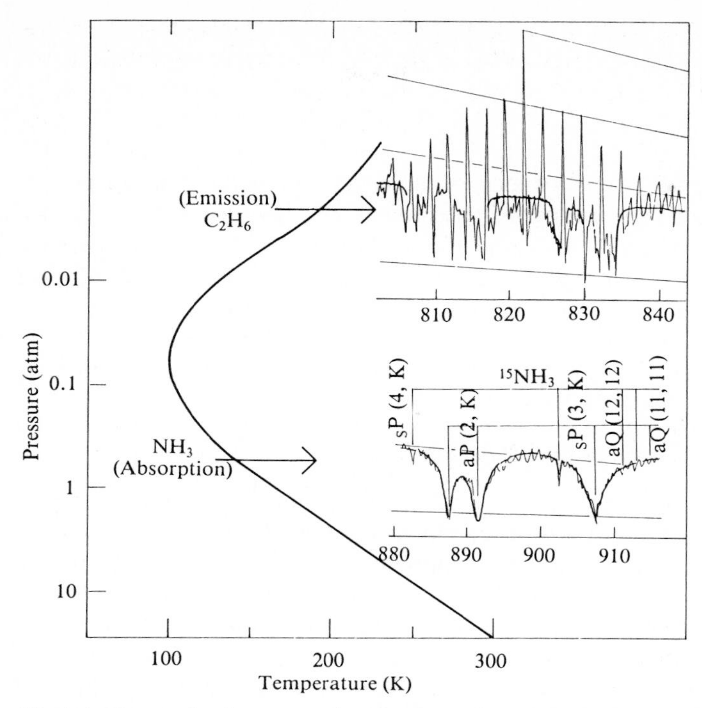

Fig. 2.3. The mechanism governing the formation of spectral lines in a planetary atmosphere. [After Th. Encrenaz: Space Science Reviews **38**, 35 (D. Reidel Publishing Company 1984)]. The spectrum on the right is that of Jupiter, between 810 and 910 cm^{-1} (11–12.3 μm) [A.T. Tokunaga, R.F. Knacke, S.T. Ridgway, L. Wallace: Astrophys. J. **232**, 603, Univ. of Chicago Press (© The American Astronomical Society 1979)]

Fluorescent Emission in Atmospheres and Ionospheres

When an atmospheric component (ion, atom, radical or molecule) moves from its ground state to an excited state, either by absorption of a solar photon (UV, visible or IR), or from the effects of particle showers, it may return to the ground state either directly by emitting a photon of the same wavelength, or by cascading through a series of intermediate energy levels. In the first case there is resonant fluorescence: an emission line is observed in the spectrum of the object at a wavelength corresponding to the transition under consideration. In the second case a spectrum is observed that has fluorescent lines at wavelengths greater than that of the source of the excitation. These wavelengths are such that the sum of the corresponding transition energies is equal to the energy of the transition that produced the excitation.

For objects in the Solar System, the source of pumping is mainly solar radiation. With fluorescence, the rate of pumping per second is given as follows:

$$g_{\mathrm{hl}} = \frac{\Omega_\odot \omega_{\mathrm{h}}}{4\pi\omega_{\mathrm{l}}} A_{\mathrm{hl}} \left(\exp\left(\frac{\hbar\nu}{kT_\odot} \right) - 1 \right)^{-1} . \tag{2.14}$$

$\Omega_\odot$ is the solid angle subtended by the Sun, and $T_\odot$ is its brightness temperature; ω_{h} and ω_{l} are the respective statistical weights of levels higher and lower than the transition, A_{hl} in s^{-1} is the Einstein coefficient for spontaneous emission at the transition, and ν is the transition frequency, in Hz. The intensity of the emission line is then given by:

$$I = g_{hl} \hbar \nu N \tag{2.15}$$

where N is the number of atoms (or ions, or molecules) along the line of sight.

Fluorescence spectra are primarily observed in the UV and visible regions. This is because of the shape of the solar spectrum, which is the principal source of excitation. It applies to ionospheres and to the outer atmospheres of the giant planets, and also to comets. A particularly notable emission is that of Lyman α, at 1 216 Å. With comets, the theory predicts that molecular fluorescence lines may be observable in the infrared, if the spectral resolution is sufficient. Such emissions have been observed in Comet Halley.

Physical Chemistry of Surfaces and Dust Particles

For objects far from the Earth, whose surfaces have not yet been explored by spaceprobes, spectroscopy from a distance remains a powerful method of studying the mineralogy of the surface layer. This method uses the broad spectral signatures present in the spectra of sunlight reflected back from the surface of the body concerned. These spectral signatures are specific to certain mineral phases. As they have no significant features requiring high spectral resolutions, they are easy to observe even for faint objects such as minor planets or distant satellites. First, the observed spectra are compared with spectra from laboratory samples, in particular from various meteorites. Then, in a second stage, the spectral signatures are identified, and used to diagnose the presence of specific mineral phases.

The spectroscopic study of dust is more complex. The intensity of the flux observed at a given wavelength (whether it be reflected solar radiation or the thermal flux) does not, in fact, just depend on the composition of the grains, but also on their size. When the latter is comparable – to an order of magnitude – with the wavelength itself (that is to say from a few tenths to a few tens of microns), diffusion becomes very important. The problem may be tackled by Mie theory, if the diffusion and refraction indices are known. For example, the silicate emission band at 10 μm may, or may not, be visible in the spectrum of cometary dust, according to the size of the particles involved.

2.2.2 Analysis of the Radiation as a Function of Phase Angle

As the radiation received from dust is dominated by the process of scattering, an important method of studying scattering materials consists of measuring the flux received at a function of phase angle. This method has been used for the study of the dust responsible for the zodiacal light, of cometary dust, and also of the clouds and aerosols present in planetary atmospheres. In this way one obtains an estimate of the scattering indices for the particles, and information about their density, their size, and their composition. In a similar way, the variation with phase angle of the polarization of the scattered light is linked to the mean albedo of the surface of the body being studied.

2.2.3 Analysis of the Thermal Structure of an Atmosphere by the Stellar Occultation Method

When a planet passes in front of a sufficiently bright star (having a visual magnitude of 10 or more), it is possible to study the physical chemistry of its atmosphere by analyzing the light-curves of the star at the times of immersion and emersion, when the star's radiation passes through the planet's atmosphere. The starlight is affected by atmospheric refraction (Fig. 2.4). By working backwards from the light-curve it is possible to determine the refractive index of the atmosphere. If the mean molecular mass is known from other methods it is possible to obtain the temperature profile as a function of altitude. In the case of the giant planets only the upper atmosphere can be probed from Earth. Several observations of Jupiter, Uranus and Neptune have been made in the visible and near-infrared regions. Radio-occultation experiments have also been carried out on Jupiter, Saturn and Uranus from the Pioneer and Voyager probes.

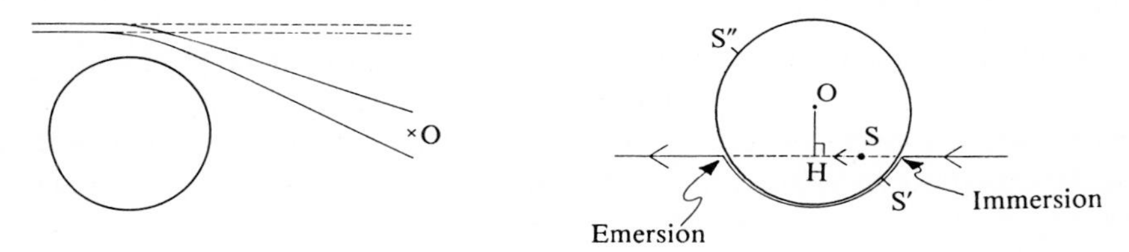

Fig. 2.4. Investigation of an atmosphere by the method of stellar occultations and refraction by a planetary atmosphere. O is the centre of the planet, S is the star, the light from which is observed at S′ and S″. OH is the impact parameter

2.2.4 In-situ Analysis

The direct analysis of mineral samples is not new: it began with the study of meteorites. The arrival of the space age has greatly enlarged this sphere of research, at first by the return of lunar samples, then by the analyses of the soils of Mars and Venus. Spaceprobes have also enabled direct studies of planetary atmospheres and interiors to be carried out.

The Atmospheres of the Inner Planets

Several spaceprobe programmes have been devoted to in-situ research of the atmospheres of the inner planets: Venera and Pioneer for Venus, and Viking for Mars. A typical spacecraft consists of a mother craft, the orbiter, and of a smaller probe, which is released by the orbiter, and drops into the planetary atmosphere; in the case of Venus, balloons have also been dropped into the atmosphere. Measurements are returned by telemetry during the descent. The instrumentation on this sort of probe may include a radar for altimetry, temperature and pressure sensors, one or more mass spectrometers, a chemical chromatograph for determining atmospheric gases, and a nephelometer for studying the location and composition of clouds.

For Mars and Venus it has also been possible to measure, to a high degree of precision, the thermal profile at a few points on each planet, and to determine the atmospheric pressure at different altitudes, and the mixing ratios of the atmospheric components.

Methods for the Physical Analysis of Planetary Surfaces by Spaceprobes

Determination of the chemical composition of the soil on a planetary body is made by the analysis of the energy carried by radiation or particles emitted by the surface. Depending upon the particular case, this emission is caused either by radiation or particles originating in the Sun, by cosmic radiation, or by radiation from the planet itself. In some cases, it is necessary to stimulate this emission by in-situ irradiation, using ion guns, lasers, or radioactive sources. We shall describe briefly some of the techniques that have been used so far, or which have been proposed for future experiments.

X-ray Spectroscopic Analysis

When a sample of material is bombarded by energetic particles or radiation, it emits X-rays, the energy of which is representative of the electronic transitions that have been induced. Analysis of the energy of this radiation enables one to deduce the elemental composition of the soil being examined. Bearing in mind the mean path of X-rays in the material, this analysis applies to the outermost few microns of the soil. When planetary bodies are devoid of atmospheres, solar X-rays are sufficient to induce such transitions. A detector in orbit can therefore pick up the secondary X-rays. This property has been used, in particular, on the Apollo 15 and 16 flights, when the module in orbit analyzed the lunar surface by means of three proportional counters filled with a mixture of argon, carbon dioxide, and helium, and with windows of Be 25 μm thick. The energy resolution and the sensitivity of these detectors were such that the Al, Mg and Si content of the surface near the lunar equator could be determined, with a spatial resolution of some twenty kilometres. These analyses were able, in particular, to delineate the lunar maria and highlands, the latter being characterized by a high Al/Si ratio, because of the presence of feldspar, contrasting with the lunar maria, rich in basalts and with a low Al/Si ratio, linked with a high Mg content.

Because of its high degree of reliability, X-ray spectroscopy was also used for determining the composition of the Martian soil, as well as that of Venus, from the landers Viking 1 and 2 (for Mars), and Venera 13 and 14, and Vega (for Venus). The atmospheres of these two planets are opaque to solar X-rays, which means that the emission has to be induced by radioactive sources. The American probes each carried ^{55}Fe and ^{109}Cd sources, while the Soviet probes used ^{55}Fe and ^{238}Pu sources. In both cases the instruments were able to measure the mineralogical composition of selected samples, down to concentrations of the order of one percent, with a precision of 10 to 30%.

γ-ray Spectroscopic Analysis

In general, a planetary surface emits γ-radiation that has two principal sources. The first is the natural disintegration of radioactive elements with long half-lives (U, Th, K). For example, the transformation of ^{40}K into ^{40}A leads to the emission of radiation with an energy of 1.46 MeV, just as the disintegration of ^{208}Th, which itself derives from that of ^{238}Th, produces emission at 2.62 MeV. The second possible origin for the emission comes from the irradiation of the surface by high-energy cosmic rays, mainly by galactic cosmic rays. The general process is as follows: protons of 10^8 to 10^9 eV produce neutrons by interactions with the component elements of the soil, such as Mg, Al, Si.... These neutrons react in their turn with atoms within the particles, either by inelastic diffusion, or by capture. In the first case, they excite those atoms, which emit characteristic γ-rays in the de-excitation process, such as 0.84 MeV for Fe. In the second case, the binding energy is released, particularly in the form of γ-radiation (for Fe, this radiation is at 7.64 MeV. Spectrometry therefore permits the composition of the soil to be determined, on the one hand with respect to its radioactive elements, and on the other with respect to its major elements. The depth of soil analyzed, corresponding to the characteristic diffusion depth for γ-rays, is a few cm. This is therefore a technique that is potentially more interesting than X-ray fluorescence, although of much lesser sensitivity. It has been used for the analysis of lunar soil from Apollo modules in orbit, as well as for that of the surface of Venus by the Venera probes.

Analysis by Infrared Spectrometry

Ultraviolet and visible spectroscopy of a sample of dust only reveal a few spectral signatures characteristic of the composition. The infrared region on the other hand, from the near infrared (0.8 μm) to the intermediate infrared, covers a spectral range where there are numerous signatures from mineral components. In general, we are dealing with molecular transitions associated with the vibrational and rotational modes of the most common bonds. A silicate, for example, shows a characteristic structure with a principal band, centred close to 10 μm, accompanied by a second band near to 18 μm. These bands are wide (several tens of microns) in contrast to the narrow bands of molecular transitions that are found in gaseous compounds. The position of the centre of the band, measured with an accuracy of about 0.1 μm, allows the determination of the stoichometry of the mineral being analyzed. For example, a variation in the SiO/SiO_2 ratio results in a change in this position from 9.7 μm (pure SiO) to 10.2 μm (pure SiO_2). In more general terms, the characteristics of this band at 10 μm allow the main families of silicates to be distinguished. This identification is easier, however, in the near infrared. Figure 6.3 shows that in the 0.8–2 μm region, the principal silicates give rise to an absorption structure, the position and the shape of which can be identified immediately. Whilst pyroxene has a narrow band at 1 μm and a wide band at 2 μm, olivine produces an asymmetrical band between 1 μm and 1.5 μm, and feldspar a band at 1.25 μm. The band at 2 μm that appears in the spectrum of pyroxene is apparently caused by the presence of iron in the form Fe^{++}, that is iron oxide, FeO. It has been shown in particular that the ratio of intensities of the bands at 2 μm and 1 μm is directly correlated with pyroxene's content of this compound. Finally it should be noted that the presence of water in a sample being analyzed is shown by the existence of a band centred close to 3 μm, as can be seen in curve (d) in Fig. 6.3. Overall, a spectral resolving power close to 30 suffices to carry out spectroscopy of a planetary soil at a distance, and leads to the determination of its mineralogical composition.

Meteorites and Interplanetary Dust

Even before the advent of the space age, the analysis of extraterrestrial material collected on Earth had been carried out for about a century: about 500 tonnes of meteorites have been identified on Earth. Nowadays, most of the collection of meteorites is undertaken on the surface of the polar ice caps, where meteorites have been found preserved in the ice and thus protected from atmospheric pollution. The chemical analysis of samples has become very sophisticated thanks to progress in making isotopic measurements in the laboratory. The isotopic ratios in meteorites, measured with mass spectrometers, gives us information about the age of meteorites as well as about the intensity of cosmic rays in different regions and at different times in the history of the Solar System.

Interplanetary grains larger than about 100 microns are vaporized in their passage through the Earth's atmosphere (as meteors). Smaller grains are decelerated in the region of the bottom of the stratosphere, where they can be collected by aircraft flying at altitudes of 15 to 20 km. Collection may be made elsewhere, such as from spacecraft (Salyut or Mir), or at ground level, from the polar ice caps, where the grains are protected from atmospheric pollution.

2.3 Analysis of Plasmas and Magnetic Fields

In the cases of both the interplanetary medium and the ionized envelopes surrounding the various bodies in the Solar System, methods of observation are divided into two groups: those undertaken from the ground, and those made from space.

2.3.1 Observations from the Ground

Monitoring the solar wind involves monitoring solar eruptions. This is primarily carried out in the X-ray region, or, more easily from the surface of the Earth, in the radio region. At centimetric and decimetric wavelengths the flux arises from synchrotron radiation from electrons spiralling at high speed around lines of force in the magnetic field. The solar magnetic field may also be measured by the Hanlé or Zeeman effects using polarimetric measurements.

The individual magnetic fields of magnetic objects in the Solar System may be observed in the same way; in the case of Jupiter in particular, synchrotron radiation is observed at centimetric and decimetric wavelengths, and in addition decametric radiation is observed, emitted sporadically in the form of bursts. Jupiter's magnetic field – like that of Saturn – may also be indicated by auroral phenomena, characterized by intense, non-thermal emission, localized around the magnetic poles. This emission is produced at the wavelength of a transition of one of the components in the upper atmosphere. It may occur in the ultraviolet, or the visible and infrared as well. Finally it should be noted that one striking example of a magnetosphere is that of the Earth, which is particularly well-known from observations made in situ.

The ionized envelopes of Solar-System objects are analyzed by classical spectroscopy at a distance, mainly in the visible and ultraviolet regions.

2.3.2 Observations from Space

Many missions in space have measured the characteristics of the solar wind (the density of protons and electrons, temperatures, magnetic field components) since the first Luna 1 and 2 missions demonstrated its existence. The solar wind has been analyzed at heliocentric distances ranging from 0.5 to about 20 AU. So far, however, all the measurements have been made close to the plane of the ecliptic.

The magnetospheres of Jupiter and Saturn have been studied by the Pioneer and Voyager probes, which explored the planets' environments and were able to establish precisely the directions of the lines of force. The probes measured the magnetic-field components and particle densities. In particular, the Voyager missions showed the existence of a plasma torus around the orbit of Jupiter's satellite Io. For Saturn, they found that the flux of highly energetic particles trapped in the magnetosphere is strongly influenced by the presence of the rings and the satellites. The magnetometer on the Voyager 2 probe determined the distinctive configuration of the magnetic field of Uranus.

2.4 The Future of Solar-System Exploration

It is undeniable that our knowledge of the Solar System has progressed in a spectacular manner during the last fifteen years. Although this success is linked in an obvious way to the development of interplanetary missions, they are not solely responsible. The progress in astronomy carried out from Earth orbit must also be mentioned; in particular, studies in the ultraviolet and infrared regions. Note should also be paid to the increasing interaction between planetology and general astrophysics, and to the development of theoretical models. Planetology has ceased to be descriptive; it now aims at integrating all the observational data into a coherent whole that tries to explain the origin and the evolution of the Solar System.

2.4.1 Observations from the Ground and from Earth Orbit

Observations from the ground in the visible region, although they may have begun long ago, are by no means outdated. The continual progress made in two-dimensional detector techniques has meant that the minimum detectable flux has been repeatedly lowered. For the Solar System, these new techniques have allowed observation and spectroscopy of more and more faint objects (asteroids, comets, and distant satellites).

Still with ground-based observations, millimetric astronomy, which is in the midst of its development, is a wonderful tool for studying the stratospheric lines in planetary atmospheres, as well as for research into molecular bands in comets. This method takes advantage of the spectral resolution provided by heterodyne detection, which allows extremely narrow spectral lines to be resolved.

As far as astronomy from Earth orbit is concerned, two major goals are in view for the nineties. First, the arrival of the Space Telescope (the Hubble Observatory), will increase the angular resolving power obtainable from Earth by a factor of at least 10 in the ultraviolet and visible regions. Priority programmes concern imagery of faint objects (minor planets and comets), and high-resolution spectroscopy of bright objects (planets). The second region of interest is in the infrared, inaccessible from the ground, i.e. mainly beyond 13 μm. Very important results have already been obtained in the last ten years, in particular from the Kuiper Airborne Observatory. This programme can only develop in the next few years, especially with the advent of infrared heterodyne receptors. After the success of the IRAS infrared satellite, two projects for infrared satellites have now been accepted or are under study by space agencies, ISO by ESA and SIRTF by NASA.

2.4.2 Space Missions

The launch into orbit of Sputnik 1, the first artificial satellite, on 4th October 1957, began the era of astrophysics from space: at a single step, it became possible to overcome the constraints imposed by the atmosphere, and to detect – at least in principle – the whole range of electromagnetic radiation emitted by stars, and not just the small fraction that passes through the atmosphere and reaches telescopes on the ground (see above). At the same time as these satellite observatories were put into orbit, at altitudes of a few hundred kilometres, we saw the start of "interplanetary voyages", using unmanned spacecraft, and, in the case of the Moon, with manned flights (the Apollo programme).

In 30 years, 8 of the 9 planets have been observed from nearby probes when Voyager 2 arrived at Neptune in August 1989. Only Pluto will remain inaccessible to planetary probes for at least another few decades. The first phase, that of exploration and discovery of planetary bodies, which until now were only known from telescopic observations, is therefore coming to a close.

We are now beginning the second phase, that of the systematic study of both the inner and giant planets. Among the first of these, Mars is being given priority in the programmes of the various space agencies. We see the strategy for some very ambitious missions being defined, quite apart from the Soviet PHOBOS mission (launched 7th and 12th July 1988) and the American MARS OBSERVER mission (scheduled for 1992). The return of martian samples collected "intelligently" from different regions of the surface by mobile robots is being considered. An extension to this that is being more and more seriously considered is a manned flight to Mars, to take place at some time in the next century. As far as the giant planets are concerned, NASA's GALILEO project is to carry out a comprehensive study of the jovian system, as will the CASSINI mission for Saturn. GALILEO was launched in 1989. CASSINI is in the planning stage, and will be a cooperative venture between NASA and the European Space Agency. The latter may take responsibility for providing an autonomous probe to study Titan: released from the main vehicle orbiting Saturn, it would enter Titan's atmosphere and analyze the constituents down to the surface.

One family of objects that has been largely neglected during the phase in which the various bodies within the Solar System was explored, is that of the "small bodies", the minor planets and the comets, the diversity of which reflects the range of components from which they were formed and their different dynamical origin. One of these objects, the nucleus of Comet Halley, was the subject of intense study by spacecraft (mainly VEGA and GIOTTO). Phobos, one of the satellites of Mars, is probably a captured asteroid. Its observation by the Soviet PHOBOS mission, would have been the first in-depth study of one of these objects. Beyond this, it is necessary to gain some knowledge of the specific properties of various classes of object in order to be able to determine their degree of evolution, and to place them within the context of the formation and early evolution of the Solar System.

NASA is developing a project known as CRAF (Comet Rendez-vous and Asteroidal Fly-by), which envisages a probe being put into orbit around a cometary nucleus at a considerable distance from the Sun, before perihelion and when the comet is still more or less inactive. The probe would follow the comet as it made its trip in towards the Sun. It would therefore be able to observe the progressive increase in the nucleus' activity up to perihelion. Finally, the European Space Agency has an even more ambitious project: the sending of a probe to land on a cometary nucleus, the collection of samples, and their return to Earth. This project, known as CNSR (Comet Nucleus Sample Return), might see the light of day during the first decade of the next century.

3. The Formation of the Solar System

Although the observation of objects in the Solar System has been practiced, to a high degree of precision, since ancient times, the problem of its origin was not really considered until after the Copernican revolution. This, which merely repeated the theory first advanced by Aristarchos of Samos, located the Sun at the centre of the System. The first models for the system's formation tried initially to explain, more or less in qualitative terms, the observed movements. Various factors had to be taken into account: (1) the orbits of the planets are close to the plane of the Earth's orbit; (2) the orbits are essentially circular (with the exception of that of Pluto, which was discovered in 1930); (3) the planets all rotate in the same sense, which is the same as that of the Sun; (4) the heliocentric distances of the planets obey the Titius-Bode law (see Sect. 1.1.2). These particular constraints applied to all theories developed up to the 19th century. At the end of the 19th century, and the beginning of the 20th, theoreticians started to pay attention to the problem of angular momentum. The angular momentum of the Sun, which contains 99.8 % of the mass of the Solar System, only has 2 % of the angular momentum possessed by all of the planets. Finally, the second half of the 20th century benefited from the contribution of new theories concerning stellar formation, as well as new data, on the one hand about the age of the various bodies, and on the other about their chemical, isotopic and crystallographic composition. All these new elements have allowed a choice to be made amongst all the different models proposed, and for the most likely one to be selected. The problem is far from being solved, however, and even if a coherent description of the system's formation is now becoming clear, all the various physical and chemical mechanisms are still not understood.

3.1 The History of Models of the Solar-System's Formation

3.1.1 The Advent of the Copernican Model

Among the Greek schools, there is just a single precursor: Aristarchos of Samos, who, around 280 B.C., proposed a heliocentric system in which the planets followed circular orbits around the Sun. His model, which contradicted the idea of a fixed Earth, as advanced by Aristotle's school, fell into oblivion. There it remained until the work of Nicholas of Cusa (1401–1464), who repeated Aristarchos' theory, but had no more success in convincing his contemporaries. We have Nicholas Copernicus (1473–1543) to thank for advancing the modern concept of the universe, where (1) the planets orbit the Sun, and also rotate about their own axes, and where (2) the dimensions of the Solar System are minute when compared with the distances of the stars.

The Copernican model, advanced in 1543 in a volume entitled *De revolutionibus orbium celestium libri VI*, would doubtless not have prevailed if it had not been for the scientific work of Kepler, Galileo and Newton. Johannes Kepler (1571–1630) – the pupil of Tycho Brahe (1546–1601), himself an exceptional observer – empirically discovered the laws that carry his name (see Sect. 1.1.1). Galileo Galilei (1564–1642), the first person to use an astronomical telescope, obtained a large number of planetary and stellar observations, which confirmed the theories of Copernicus. He also discovered the fundamentals of dynamics, notably the principle of inertia. Finally Isaac Newton (1642–1727) provided the mathematical proof of the Copernican model by publishing, in 1687, in his *Philosophiae Naturalis Principia Mathematica*, the law of universal gravitation. This law, which allowed Kepler's laws and Galileo's results to be explained, forms the true theoretical basis for the Copernican model.

3.1.2 Different Types of Models for the Formation of the Solar System

Models Based on Turbulence

René Descartes (1596–1650) was the first to try to find a scientific explanation for the existence of the Solar System, and the first to introduce the idea of evolution. In his *Théorie des vortex*, published in 1644, Descartes advances the theory that the universe, which is filled with aether and matter, is full of vortices of a whole range of sizes. This model was only qualitative, and one of the major objections to it was that it did not favour the plane of the ecliptic. It was abandoned after the discovery of Newton's laws, and now has historical interest only. The concepts of friction and turbulence introduced by Descartes, however, have been reconsidered by several authors in the 20th century (Von Weizsäcker, ter Haar, Kuiper, Whipple, McCrea).

The Nebular Theory

The concept of a primitive nebula, from which both the Sun and its system of planets were born, was first proposed by Kant (1724–1804) and Laplace (1749–1827). According to Laplace, the nebula contracts under the influence of gravitation, and its rotational velocity increases until it collapses into a disk. Subsequently, rings of gas are shed, which condense into planets and satellites.

This model had the merit of being able to explain all the observational phenomena known in the 18th century concerning the movements of the planets; which is why it escaped criticism for some time. Two grave objections appeared at the end of the 19th century. First, as Maxwell (1831–1879) showed, in this model it is difficult to explain the accretion of a planet from a ring of planetoids. The second objection is the problem, mentioned earlier, of the angular momentum. Most of the Solar System's angular momentum is, in fact, contained in the planets (see Table

1.1), particularly in Jupiter (60 %) and Saturn (25 %). Yet in Laplace's model, where all the mass of the nebula is found in the Sun and planets, the Sun, which has 99.8 % of the mass, ought to have retained most of the angular momentum. It would therefore be rotating at a very high rate, which is not the case. Its period of rotation is about 26 days at the equator, but if it were to have all the Solar System's angular momentum, it would rotate in half a day.

During the 20th century, numerous authors have modified Laplace's theory and expanded it in various ways to answer the objections that have been raised. Berlage considered the role of viscosity; Hoyle that of the magnetic field; and Schatzman that of the solar wind. A rotating star loses mass through the flow of particles along the lines of force associated with a centre of magnetic activity. This may transport the particles to distances a far greater than the stellar radius R. A small loss of mass may therefore lead to a large loss of angular momentum, and this is proportional to $(a/R)^2$. So if a/R is of the order of 10, the loss of a mass of only 0.003 $M_\odot$ is sufficient to brake the Sun down to its actual period of rotation. In addition to this mechanism, which is very efficient, there is also transport of matter by the stellar wind. Many observations of stars (such as the measurement of equatorial winds, and the intensity of spectral lines associated with magnetic activity, etc.) have shown that young stars appear to have intense magnetic activity and high mass-losses. As they age, the stars' rotation slows, and their magnetic activity declines. The Sun is therefore merely one example of this evolution. This explanation of magnetohydrodynamic braking of the Sun removes the objection about the angular momentum.

The overall outline of the formation of the Solar System that we accept nowadays has developed from these ideas, which have been the subject of a large number of papers. Models for the evolution of nebular disks may currently be divided into two main categories:

1. *The massive nebula model* (Cameron), which considers a viscous disk of about 1 $M_\odot$. A large fraction of this mass (85 %) is swept away by the solar wind in a very short time (10^5) years, whilst most of that remaining is accreted by the Sun. In this model, the planets are able to form directly from the gaseous nebula by gravitational instabilities.
2. *The low-mass-nebula model* (Safronov, Hayashi), where the mass of the disc after collapse is of the order of 10^{-2} $M_\odot$. The disk later cools, the dust accumulates in the central plane and forms low-mass ($\approx 10^{18}$ g) *planetesimals*, which then combine to form more massive bodies.

Elmegreen has established equations that describe the evolution of these two types of model. One example of a low-mass-nebula model is described later (Sect. 3.2.2).

The Tidal Theories

The first catastrophic theory of the formation of the Solar System was proposed by the naturalist Buffon (1707–1788), who suggested that the Solar System arose from an ejection of solar material caused by the collision of the Sun with a "comet" 70 000 years ago. (It should be remembered that the nature and the mass of comets were completely unknown at the time.) This theory, with no scientific basis, did not come to be accepted, but later, in view of the objections raised to the nebular model, catastrophic theories were re-examined. Bickerton in 1880, Chamberlain in 1901, and Moulton in 1905 replaced Buffon's comet with a star, and explained the formation of the planets by the condensation of material lost from the Sun. The gravitational force at the moment of closest encounter drew out a filament of solar material and this then went into orbit around the Sun with considerable angular momentum. The tidal theory was later developed by Jeans and Jeffreys, who numerically analyzed the effect of interaction between two stars. Other authors, however, objected that the condensation of a substantial filament into large planetary masses was difficult to explain (Nökle; Russell; Spitzer). As a result of this controversy, new models were proposed. It was suggested that the Sun was originally part of a binary system and that the planets were formed from the second star (Russell; Lyttleton; Hoyle). More recently Wolfson suggested that an encounter occurred between the Sun and a protostar, and that a filament from the latter condensed into planets. This last model had the advantage of taking into account new determinations of the chemical and isotopic composition of the planets, which show that the planetary material came from a cold medium, and not from a very hot stellar filament.

Accretion Theories

The final class of models considers the possibility of the Sun accreting interstellar material. In order to avoid this material from collapsing into the Sun, it is necessary to postulate another nearby star. According to Schmidt, this arrangement allow the interstellar material to condense into planets. Another theory is that of the models by Alfvén and Arrhenius, where the Sun is supposed to have encountered two nebulae, one consisting of non-volatile grains, from which the inner planets were formed, and the other consisting of hydrogen, which gave rise to the giant planets. According to Alfvén, collisions between particles would lead to the formation of jets of material, from which planets could have been formed. This idea has been rejected by other authors (Lynden-Bell and Pringle; Brahic; Goldreich and Tremaine), who have shown that collisions would lead to dispersion of the particles, except in the case where a ring was confined by one or more satellites.

3.1.3 Theoretical and Experimental Constraints upon the Various Types of Model

Amongst this wide range of models for the formation of the Solar System, proposed over two centuries, one could introduce a secondary classification based on the answers these models give to two fundamental questions. The first question concerns the age of the various objects: were the Sun and the Solar System formed together? To put it somewhat differently, did the Sun and its system of planets form at the same time? The second question concerns the nature of the objects: did the Solar System form from cold material, or from cooled solar material, that is material transformed by thermonuclear reactions? It is sensible to examine the problem in these terms, because exact observational evidence allows us to reply, without ambiguity, to these two questions. It has been possible to date planetary material to a high degree of accuracy, thanks to measurements made on the abundance of radioactive elements having very long half-lives in terrestrial rocks, lunar samples, and meteorites.

The principle of dating by measuring radioactive elements is described in Sect. 11.4.5. A parent radioactive element p disintegrates into a daughter element d:

$$\frac{dp}{dt} = \frac{-dd}{dt} = -\lambda p \tag{3.1}$$

where λ is the radioactive decay constant, the inverse of the interval. For a time t, we have:

$$d = d_0 + p\,(e^{\lambda t} - 1) \tag{3.2}$$

where d_0 is the initial abundance of the element d. If d' is a stable isotope of d, then at time t we have:

$$\frac{d}{d'} = \left(\frac{d}{d'}\right)_0 + \left(\frac{p}{d'}\right)(e^{\lambda t} - 1) \quad . \tag{3.3}$$

In order to measure the age of the Solar System one uses long-term clocks, mainly therefore the pairs (^{40}K, ^{40}A), (^{87}Rb, ^{87}Sr) and (^{238}U, ^{238}Pb), which have decay constants of 5.8×10^{-11}, 1.4×10^{-11} and 1.5×10^{-11} per year, respectively. Measurements made on these elements in meteorite samples have shown that the age of the Solar System is approximately 4.55×10^9 years, with an uncertainty of 10^8 years. Moreover, measurements of the abundances of plutonium-244 and of iodine-129 (which are radioactive elements with shorter half-lives) at the time of the solidification of the planetary material, show that no more than 10^8 years occurred between the isolation of the protosolar material from the interstellar medium and the formation of the planets. This time might correspond to the time for the protosolar cloud to pass between two of the Galaxy's spiral arms. This result implies that the Sun and the Solar System did form together, and might indicate that the Solar System formed in passing through a spiral arm.

The second question is resolved by measuring the deuterium/hydrogen ratio in the giant planets. Deuterium is present in the interstellar medium, but is destroyed in stars. In fact, in the proton-proton cycle that transforms hydrogen into helium, deuterium is transformed into helium-3 by the reaction

$$^2\mathrm{D} + {}^1\mathrm{H} \rightarrow {}^3\mathrm{He} + \gamma \tag{3.4}$$

on an average time-scale of a minute for a temperature of 10 million degrees. The other reactions in the proton-proton cycle themselves have average reaction time-scales greater than 10^6 years. Deuterium was therefore completely destroyed within the Sun as soon as nuclear reactions started. The value of D/H for the giant planets has been found to be higher than that prevailing in the interstellar medium. In the absence of any process capable of creating deuterium at great distances from the Sun, this result can be interpreted as indicating the abundance of interstellar deuterium 4.6 thousand million years ago. If the planetary material arose from solar material, its deuterium abundance would be zero, like that of the Sun. In the same way, the lithium abundance in meteorites is comparable with that of the interstellar medium, whilst that of the Sun is negligible. These very significant results therefore show that

Table 3.1. Abundance of elements in the Sun

Element	Relative abundance (by number)	Fraction of total mass
H	3.18×10^{10}	0.980 0
He	2.21×10^{9}	
C	1.18×10^{7}	
N	3.64×10^{6}	0.013 3
O	2.21×10^{7}	
Ne	3.44×10^{6}	0.001 7
Na	6×10^{4}	
Mg	1.06×10^{6}	
Al	8.5×10^{5}	
Si	10^{6} (standard)	0.003 65
S	5×10^{5}	
Ca	7.2×10^{4}	
Fe	8.3×10^{5}	
Ni	4.8×10^{4}	

(After *Formation des systèmes planétaires,* A. Brahic, ed., CNES/Cepadues, 1982.)

Table 3.2. Classification of models for the formation of the Solar System

	Planets formed of unaltered interstellar material	Planets formed of stellar material (raised to high temperatures)
The Sun and the planets formed at the same time	Kant, 1755 Laplace, 1796 Hoyle, 1960 Edgeworth, 1949 McCrea, 1960 Whipple, 1948 Urey, 1946 Von Weizsäcker, 1944 Kuiper, 1951 Ter Haar, 1950 Gurevich and Lebedinsky, 1950 Schmidt, 1959 Levin, 1972 Safronov, 1972 Cameron, 1962 Schatzman, 1963	Gunn, 1932 Lyttleton, 1940 Hoyle, 1944 Egyed, 1960
The Sun and the planets did not form at the same time	Berkeland, 1912 Berlage, 1927 Lyttleton, 1961 Hoyle, 1956 Sekiguchi, 1961 Schmidt, 1944 Alfvén, 1942 Woolfson, 1964	Buffon, 1745 Bickerton, 1818 Arrhenius, 1913 See, 1910 Jeans, 1916 Charmberlain, 1901 Moulton, 1905 Jeffreys, 1929 Lyttleton, 1937 Russell, 1935 Banerji and Srivastras, 1963

(After *Formation des systèmes planétaires,* A. Brahic, ed., CNES/Cepadues, 1982.)

the planets were not formed from material transformed by thermonuclear reactions at the centre of the Sun.

Table 3.2 shows how the different classes of models are divided on the basis of these two criteria. In general terms, the tidal models are to be discarded, because they imply that the planetary material is of solar origin. In the same way, the accretion models that imply that the Sun and the planets did not originate together, cannot be retained. It is only the improved versions of the nebular theory that can form a basis for our current view of the way in which the Solar System was formed.

3.2 Steps in Developing a Plausible Model for the Formation of the Solar System

3.2.1 The Observational Data

The Contribution from Meteoritic Data

We have already mentioned (Sect. 3.1.3) how the analysis of element and isotopic ratios in meteorites serves as a method of dating. This type of measurement has also produced information about "isotopic anomalies", which have fundamental significance for cosmology (see Chap. 11). For a certain number of elements, particularly oxygen, magnesium and neon, "anomalies" have been found that can, apparently, only be explained if one assumes the presence, in the primitive solar nebula, of "presolar" grains that had never been in volatile form, and which arose from material ejected by supernovae, novae or Wolf-Rayet stars. In particular, the excess of ^{26}Mg found in certain meteorite inclusions can only be accounted for by the presence of ^{26}Al in presolar grains directly ejected from a stellar explosion: ^{26}Al is only formed in this type of environment, and it decays radioactively into ^{26}Mg in a very short time ($1/\lambda = 7 \times 10^5$ years).

Observation of Stellar-formation Regions

Apart from observation and analysis of the Solar System itself, an important source of information is the observation of regions of our Galaxy where stellar formation is taking place. Such observations have undergone rapid expansion with the development of infrared astronomy: the analysis of the Orion Nebula is just one example. On the theoretical side, numerous models of the fragmentation and evolution of viscous disks have been developed (Elmegreen, Tscharnutter).

The contribution made by the examination of star-formation regions may be roughly summarized as follows. For a long time it has been thought that the principal sites of star formation correspond to the regions known as *OB associations*. Such a region has several O- and B-class stars, which are very massive and have very short lives (just a few million years). The presence of these stars establishes the age of the association. In external galaxies OB associations are found along spiral arms. In our Galaxy, they are found with the dark molecular clouds that lie in the Galaxy's central plane. When a cloud, orbiting in the Galaxy, encounters a spiral arm, it slows down, is compressed and becomes opaque, forming a protocluster cloud. This cloud itself consists of denser sub-clouds, containing several protostellar nebulae. The explosion of very short-lived stars as supernovae has important consequences for these sub-clouds. On the one hand, a shock-wave is formed, which can accelerate the process of contraction. On the other hand, heavy elements that have been synthesized in the supernovae are ejected, and are mixed in with the protostellar clouds. These heavy elements, which soon condense, introduce interstellar dust grains, with diameters of the order of $0.1\,\mu$m, into the nebulae. As a protostellar cloud contracts, the grains allow very efficient cooling of the gas, because they radiate in the far infrared. This process accelerates the contraction of the cloud, until a protostellar core is formed. This model has been developed by Cameron and Truran, in particular.

It does, however, now appear more and more likely that another type of site for stellar formation exists. This result is suggested by recent infrared observations by the IRAS satellite (1983), coupled with the study of the interstellar medium by molecular lines in the millimetre region. Maps obtained in this way (mainly in the ^{13}CO line, which has the advantage of not being saturated), show that dense regions in the interstellar medium (like the Orion region just described) are only a very small fraction of the total mass. Most of the material in the interstellar medium appears to be in the form of small fragments, of about one hundred solar masses, that have a low opacity to UV. According to the IRAS measurements, we seem to be observing gravitationally unstable molecular cores within these fragments, which will form stars of a few solar masses (Myers and Benson 1983).

On this model, the origin of short-lived radioactive elements like aluminium-26 in the protosolar nebula remains to be explained. According to Cameron (1984) is appears that these elements could come from the atmospheres of red giants of around one solar mass: the presence of one or more supernovae may not be necessary.

We obtain equally important information from the study of the early stages of stellar evolution. We mentioned earlier the importance of the magnetic activity of young stars, their stellar winds, and their mass-loss. We can assume that the Sun showed similar activity in its early stages (with higher values for luminosity, UV flux, temperature, and mass-loss). In addition, infrared observations more and more frequently reveal the existence of stars surrounded by accretion disks (see Sect. 3.3.3).

3.2.2 Formation of the Solar System: A Possible Sequence of Events

This section aims to describe the different stages in the formation of the Solar System, based on a model with a low-mass nebula. The detailed calculations can be found in the works of Schatzman, Brahic, Lattimer, Safronov, and Greenberg.

The Primitive Nebula

We start with a sphere of gas that is rotating and contracting, that has a radius of $10^4 R_\odot$ (which corresponds to the orbit of Pluto), a mean density of $\varrho \approx 10^{-12} \varrho_\odot$ (analogous to chromospheric densities). It must mainly consist of molecular hydrogen and dust. For a body in equilibrium:

$$\omega^2 R = g_{\text{eq}} \tag{3.5}$$

where ω is the velocity of rotation and g_{eq} is the gravitational force at the equator.

As the object contracts, the temperature rises, which leads to the disassociation of H_2. The coefficient of adiabatic compression

$$\gamma = \frac{d \ln p}{d \ln \varrho} \tag{3.6}$$

decreases. It can be shown (Schatzman, 1970) that if γ is less than 4/3 there is dynamical instability. A collapse ensues, which stops only when γ becomes greater than 4/3: this occurs for $R = 100\, R_\odot$ after a period of about 100 years.

It is possible to calculate the amount of mass left in the equatorial plane during the contraction phase. The fundamental dynamical relationship is

$$\frac{GM}{R^2} = \omega^2 R \quad . \tag{3.7}$$

The equation for the conservation of angular momentum is:

$$\frac{d}{dt}(KMR^2\omega) = R^2\omega\frac{dM}{dt} \tag{3.8}$$

which expresses the fact that the angular momentum is dissipated only in the equatorial plane; KMR^2 is the moment of inertia of the contracting star. Dividing the two terms in (3.8) by $KMR^2\omega$ we obtain

$$\left(1 - \frac{1}{K}\right)\frac{1}{M}\frac{dM}{dt} + \frac{2}{R}\frac{dR}{dt} + \frac{1}{\omega}\frac{d\omega}{dt} = 0 \tag{3.9}$$

which can be integrated as:

$$M^{1-(1/K)} R^2 \omega = \text{ const} \quad . \tag{3.10}$$

Using (3.7), we derive

$$\frac{M}{M_{\text{fin}}} = \left(\frac{R}{R_{\text{fin}}}\right)^p \quad \text{with} \tag{3.11}$$

$$\frac{1}{p} = \frac{2}{K} - 3 \quad \text{and} \tag{3.12}$$

$$\frac{M}{M_{\text{fin}}} = \left(\frac{\omega}{\omega_{\text{fin}}}\right)^q \quad \text{with} \tag{3.13}$$

$$\frac{1}{q} = \frac{3}{K} + 5 \quad . \tag{3.14}$$

In standard stellar models, K is taken to be equal to 0.04. The quantity of material that is not involved in the contraction of the Sun is of the order of 10%.

Let us now calculate the mass that is necessary to form the inner and giant planets. The total mass of the inner planets is twice the mass of the Earth, i.e. $6 \times 10^{-6}\, M_\odot$. These planets formed from solid material, so we should add to that figure the gases that escaped (hydrogen, helium, and components degassed from the planetary bodies), and which could have amounted to about 99 % of the total mass. That gives us a mass of $6 \times 10^{-4}\, M_\odot$. The contribution for the giant planets must also be added and this amounts to about $2 \times 10^{-3}\, M_\odot$. The total mass is therefore of the order of $3 \times 10^{-3}\, M_\odot$. It is obvious that the process just described provides the primitive nebula with enough mass for the formation of the planets.

Structure of the Protoplanetary Disk

We start with a disk that has an initial mass of $10^{-2}\, M_\odot$. We will assume that we are situated at a heliocentric distance of about 7 AU (i.e. between Jupiter and Saturn). The density at the surface is

$$\sigma = \frac{M_\odot}{\pi r_\odot^2} \sim 10^3 \mathrm{g\,cm^{-2}} \quad . \tag{3.15}$$

In order to calculate the thickness h of the disk at a distance $r_\odot$, we use the hydrostatic law at altitude z:

$$\frac{dp}{dz} = -\varrho_\mathrm{g} g_z \quad . \tag{3.16}$$

If we make the assumption (which can be verified *a posteriori*) that the vertical component g_z of g is principally caused by the solar gravitational field, we have:

$$\frac{dp}{dz} = -\varrho_\mathrm{g} \frac{GM_\odot}{r_\odot^2} \cdot \frac{z}{r_\odot} = -\varrho_\mathrm{g} \omega^2 z \tag{3.17}$$

where G is the gravitational constant and ω the angular velocity. On the other hand

$$p = C^2 \varrho_\mathrm{g} \quad . \tag{3.18}$$

where C is the speed of sound in the gas, h may be estimated as follows:

$$\frac{dp}{dz} \sim \frac{p}{h} \sim \frac{C^2 \varrho_\mathrm{g}}{h} \sim \varrho_\mathrm{g} \omega^2 h \quad \text{whence} \tag{3.19}$$

$$h \sim \frac{C}{\omega} \tag{3.20}$$

where $C \approx 10^5$ cm/s and $\omega \approx 10^{-8}\,\mathrm{s}^{-1}$.

We find that $h \approx 10^{13}$, and a tenth of the distance $r_\odot$. The initial disk is flattened, therefore, but its thickness in not negligible. We can verify that the gravitational acceleration $(g_z)_\mathrm{D}$ of the disk is small when compared with the solar component $(g_z)_\odot$

$$\frac{(g_z)_\mathrm{D}}{(g_z)_\odot} \sim \frac{G\sigma}{\omega^2 h} \sim \frac{\varrho_\mathrm{g}}{[M_\odot / r_\odot^3]} \sim \frac{h}{r_\odot} \sim \frac{1}{10} \quad . \tag{3.21}$$

The Roche Criterion

In 1847, Roche showed that a satellite in circular orbit around a central body would be destroyed by tidal forces if it came closer to the central body than a certain limit, the so-called Roche Limit (see Sect. 8.4). For a rigid, spherical object orbiting the Sun, this limit is:

$$\frac{a_R}{R_\odot} = 2.5\left(\frac{\varrho_\odot}{\varrho_g}\right)^{1/3} \quad . \tag{3.22}$$

From this relation it is possible to deduce that there is a critical density above which a gravitational instability will occur at a given heliocentric distance. It is found (Brahic 1982) that this density is of the order of 10 times the surface density ϱ_g. This result shows that, in the model with a low-mass nebula ($10^{-2}\,M_\odot$), the planets cannot form directly by gravitational instability. Such a sequence of events is possible only in the model with a high-mass nebula ($\sim 10^2\,M_\odot$); but then the problem is to find a process capable of eliminating all the mass that is not transformed into planets.

Formation of Grains: the Sequence of Condensation

Starting from a certain temperature, the nebula cools slowly, leading to the progressive condensation of different components, beginning with the most refractory. Starting with the abundances of elements measured in the Sun, and by employing the thermal and chemical data for all the components that can condense, and the reactions between gaseous and solid phases, it is possible to calculate the expected sequence of condensation, assuming chemical equilibrium. This was first done by Grossman (1972), followed by Larimer (1975).

For chemical equilibrium between different gaseous phases, we find the following relationship for the partial pressure P_i of the molecular gas i:

$$P_i = K_i \prod_j P_j^{\nu_{ij}} \quad . \tag{3.23}$$

K_i (a function of temperature) is the equilibrium constant for gas i (in accordance with the mass action law), the values for P_j are the partial pressures for monatomic gases j and ν_{ij} is the stoichometric coefficient, that is to say, the number of atoms of element j in phase i. The condensation of a solid k occurs when

$$A_k < \left[K_k \prod_j P_j^{\nu_{jk}}\right] / P_k \quad . \tag{3.24}$$

A_k is the activity of solid k; for an ideal case $A_k = 1$.

Finally, we have the mass conservation for each element j:

$$N_j = \frac{V}{kT}\left[P_j + \sum_i \nu_{ij} P_i\right] + \sum_k \nu_{kj} S_k \quad . \tag{3.25}$$

V is the volume of the system, and S_k the number of molecules of the solid.

For the solar nebula V is taken to be constant. The set of equations given above lead to the condensation sequence shown in Table 3.3 and in Fig. 3.1. The important points to note are:

1. the order of condensation of the elements is: Al, Ti, Ca, Mg, Si, Fe, Na, S. If one carries calculation to even lower temperatures, it is found that H, O, C and N only condense completely at temperatures lower than 200–300 K;
2. the phases that are stable around 1400 K, as well as the abundances of the principal elements are in very good agreement with the abundances measured from refractory inclusions in C3 chondrites (see Chap. 11), such as the Allende meteorite.

Table 3.3. Sequence of condensation

Case 1 C/O = 0.55 p = 10^{-3} atm			Case 2 C/O = 1.2 p = 10^{-3} atm		
Mineral	Condensation temperature	Solidification temperature	Mineral	Condensation temperature	Solidification temperature
Al_2O_3	1 743		TiC	1 893	1 025
$CaTiO_3$	1 677	1 409	SiC	1 742	1 154
melilite	1 625	1 438	C	1 732	871
$MgAl_2O_4$	1 533	1 391	Fe_3C	1 463	1 326
Fe−Si	1 458		AlN	1 390	1 234
$CaMgSi_2O_6$	1 438		CaS	1 385	1 040
Mg_2SiO_4	1 433		Fe−Si	1 326	
Ti_3O_5	1 409	1 274	Al_2O_3	1 235	1 229
$CaAl_2Si_2O_8$	1 392	1 068	$MgAl_2O_4$	1 229	1 077
$MgSiO_3$	1 351		Mg_2SiO_4	1 154	
Ti_4O_7	1 274	1 120	MgS	1 131	1 100
TiO_2	1 120	774	$CaMgSi_2O_6$	1 069	
Al_2SiO_5	1 068		$CaAl_2Si_2O_8$	1 057	1 044
$NaAlSi_3O_8$	1 028	780	$MgSiO_3$	1 054	
$NaAlSi_2O_6$	780		Al_2SiO_5	1 045	
$CaTiSiO_5$	774		TiN	1 025	910
			$NaAlSi_3O_8$	961	780
			Ti_4O_7	910	862
			TiO_2	862	770
			$NaAlSi_2O_6$	780	
			$CaTiSiO_5$	770	

(After J. M. Lattimer, *Formation des systèmes planétaires,* A. Brahic, ed., CNES/Cepadues, 1982.)

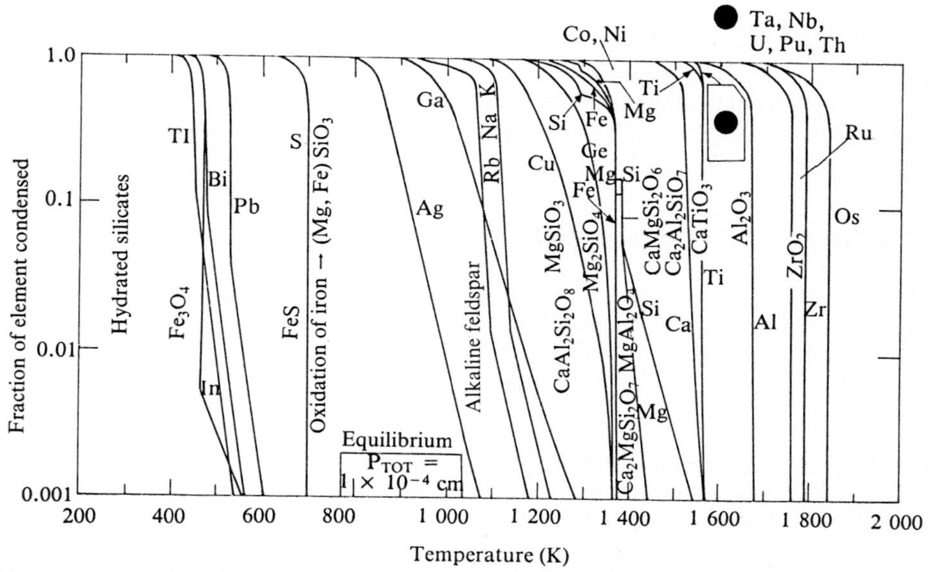

Fig. 3.1. The condensation sequence for a gas with solar composition. [After L. Grossman, J.W. Larimer: Review of Geophysics and Space Physics **12**, 71 (1974)]

It should be noted that the condensation sequence depends strongly on the state of oxidation of the initial gas. This can be seen by varying the C/O ratio. For the solar nebula it is taken to be 0.55 (see Table 3.3). If the C/O ratio is taken to be equal to 1.2, as found in the atmospheric envelopes of carbon stars, a completely different sequence appears (Table 3.3). The most refractory phases now consist of Ti, C and Si.

The Growth of Grains by Condensation

In order to gain an idea of the speed at which grains grow by condensation, we will assume that at every collision between a molecule and a grain, the molecule sticks to the grain. Let α (of the order of 10^{-2}) be the ratio between the condensed material and the gas, and A the molecular mass (of the order of 20). We then have

$$\frac{dm}{dt} \sim r^2 \cdot \alpha \varrho_g \cdot \frac{C}{A^{1/2}} \quad . \tag{3.26}$$

dm/dt is the rate of growth of a grain of mass m and radius r; $C/A^{1/2}$ is the thermal velocity of the molecules. The mass of the grain is

$$m \sim \varrho_p r^3 \tag{3.27}$$

ϱ_p being of the order of 1 g/cm^3. From this one can deduce that

$$\frac{dr}{dt} \sim \frac{\varrho_g}{\varrho_p} \frac{\alpha C}{A^{1/2}} \quad . \tag{3.28}$$

With the numerical values previously used, we obtain

$$\varrho_g = \sigma/h \sim 10^{-10} \mathrm{g/cm^3} \quad \text{and} \tag{3.29}$$

$$\frac{dr}{dt} \sim 10^{-8} \quad . \tag{3.30}$$

The rate of growth is of the order of one centimetre per year. This mechanism allows grains of the order of one micron to be formed very quickly (in a time that is of the order of one hour).

The Collapse of Grains Towards the Equatorial Plane

The solid particles, more massive than the gas, have a tendency to migrate towards the equatorial plane. Their equation of motion is given by:

$$\frac{d^2 z}{dt^2} = -g_z + \frac{F_D}{m} \tag{3.31}$$

F_D being the frictional force caused by the gas.

$$F_D \sim \varrho_g C \frac{dz}{dt} r^2 \quad . \tag{3.32}$$

As a first approximation, we may assume that there is equilibrium between the gravitational and frictional forces:

$$\omega^2 z = \frac{1}{m} \varrho_g C \frac{dz}{dt} r^2 = \frac{\varrho_g}{\varrho_p} \frac{C}{r} \frac{dz}{dt} \quad . \tag{3.33}$$

We thus obtain a characteristic time t for the particle to reach the equatorial plane:

$$t \sim \frac{\varrho_g C}{\varrho_p r \omega^2} \quad . \tag{3.34}$$

With the numerical values given above we find that $t \approx 100$ years. The grains therefore reach the equatorial plane in a very short time.

Once solid bodies have started to collect in the equatorial plane, inelastic collisions become more and more important. They have the effect of decreasing the thickness of the disk of dust until the thickness is only a few times the dimension of the bodies (Brahic).

The Formation of the Planets by Accretion

Inelestic collisions in the equatorial plane also have the effect of accelerating the growth of planetoids. As soon as the latter have a diameter of the order of a centimetre, they acquire mass by accreting particles that they encounter in their orbit. A recent summary of the problem has been given by Greenberg. Let m be the mass of the body, and r its radius. Its rate of growth can be expressed as

$$\frac{dm}{dt} = \pi l^2 \varrho_0 V \tag{3.35}$$

where ϱ_0 is the density of particles at the orbit of the body in the equatorial plane, and πl^2 is its effective cross-section. V is the relative velocity between the body and the particle. Because of the effect of gravity, its effective cross-section πl^2 is larger than its geometric cross-section πr^2. For an interaction with a particle of mass and radius m' and r' respectively (which are small relative to m and r), we may write (after Safronov):

$$\pi l^2 \sim \pi r^2 (1 + 2\,\theta) \quad \text{with} \tag{3.36}$$

$$\theta = \frac{Gm}{rV^2} \quad . \tag{3.37}$$

θ is a factor, the typical value of which is of the order of 3 to 7, that expresses the fact that the relative velocity is, on average, of the same order as the escape velocity of the larger body. The more massive the body, or the smaller the relative velocity, the larger the effective cross-section.

The product $\varrho_0 V$ that appears in (3.35) may be written as:

$$\varrho_0 V = \frac{4\sigma}{P} \tag{3.38}$$

where σ is the surface density and P the cloud's orbital period around the Sun.

As the planet accretes material lying in its orbit, the density of the latter decreases:

$$\sigma = \sigma_0 \left(1 - \frac{m}{Q} \right) \tag{3.39}$$

δ being the density of the body (assumed to remain constant). At 1 AU, the rate of growth

$$\frac{dm}{dt} = \frac{4\pi(1+2\theta)}{P}\sigma_0\left(1 - \frac{m}{Q}\right)r^2 \quad . \tag{3.40}$$

In the initial accretion stages, m is very small relative to Q and therefore the rate of growth is constant and the radius grows linearly as a function of time:

$$r = \frac{\sigma_0(1+2\theta)}{P\delta}t \tag{3.41}$$

δ being the density of the body (again assumed to be constant). At 1 AU, the rate of growth obtained is of the order of twenty centimetres per year (Safronov). It is therefore possible to accrete bodies several hundreds of km in size in 10^8 years.

To sum up, we have analyzed two processes that are capable of explaining the growth of planetoids: first the process of condensation, which enables grains up to a size of one micron to be formed, and second the gravitational attraction process, which enables actual planets to be formed from bodies of the order of one centimetre in size. In order to explain the growth of planetoids between one micron and one centimetre, Safronov suggests that, thanks to the relatively low velocities (less than 100 m/s), the collision process must have led to the clumping together of grains, and not their destruction. This process seems to be observed for clumps of lunar dust that have been previously irradiated, and whose size is of the order of one micron (Bibring and Maurette).

The Effect of the Solar Wind on Small-sized Particles

By analogy with what we observe currently in young stars, we may assume that the Sun, in the first stages of its existence, must have experienced intense magnetic activity. Equally, by comparison with the stellar winds observed in the *T-Tauri phases*, it is possible to estimate the intensity of the primordial solar wind as being about 10^8 times its current value. The effect of the current solar wind is about 10^3 times weaker than the radiation pressure exerted on a particle. We may therefore estimate that the pressure exerted on particles by the primordial solar wind was 10^5 times the current radiation pressure.

A particle subject to the pressure of solar radiation is also subject to the Sun's gravitational field, the two forces acting in opposite senses. The effective pressure on the particle, from radiation, is proportional to its cross-section, and thus to r^2, r being the radius. The gravitational force is proportional to the mass of the particle, and thus to r^3. There is therefore a value r_0 for r where the two forces are in equilibrium: for the situation now r_0 is of the order of one micron.

It will be seen that at the origin of the Solar System, when the pressure of the solar wind was 10^5 times stronger than the radiation pressure now, the radius r_0 was of the order of 10 cm. The early solar wind was therefore sufficiently strong to sweep away all particles smaller than a few centimetres in size.

The Titius-Bode Relation: Law or Coincidence?

As described earlier (Sect. 1.1.2) the Titius-Bode "law" is the empirical relationship formulated in the 18th century to account for the distances of the planets from the Sun

$$D = 0.4 + 0.3 \times 2^n \tag{3.42}$$

n taking the value of $-\infty$ for Mercury, and successive integers, beginning with 0, for Venus and the other planets. The agreement is very good as far as Uranus, but not beyond.

Many authors have tried to explain this relationship by model cosmogonies, hoping to be able to use the Titius-Bode relationship as a limiting test for the models. Attempts have also been made to extend this sort of relationship to the satellites of the giant planets. The most recent results have been negative, however. The discovery of numerous new satellites of Jupiter and Saturn since 1978 shows that a relationship of the Titius-Bode type cannot be applied to the satellites of the giant planets. In addition, Hénon (1969) and Lecar (1973) have shown that a random-number distribution could satisfy a relationship of the Titius-Bode type with appropriate values, with the sole constraint that these numbers should not be too close to one another. This constraint simply reflects the fact that two planets cannot form in orbits too close to one another, because there will not be sufficient material for them to accrete.

3.2.3 The Current State of Our Knowledge

Although no scenario is universally accepted, there is a considerable consensus of agreement on a number of points, which we shall list.

1. The models that seem most credible at present are those that are derived from Laplace's nebular model.
2. The formation of the Solar System could have taken place when the protosolar cloud passed through one of the Galaxy's spiral arms. The last enrichment of the cloud with material produced by nuclear synthesis could have occurred at the cloud's previous passage through a spiral arm, some 10^8 years earlier. This is suggested by dating methods that use iodine-129.
3. The Solar System may have formed either inside an OB association after the formation of very massive stars, or in a smaller molecular cloud, where lower-mass stars are formed. Whichever was the case, there was an injection of material into the primordial nebula from the explosion of supernovae or from other stars. This result is obtained from measurement of the isotopic anomalies in meteorites (of ^{16}O, ^{26}Al and ^{20}N in particular).
4. The slow rotation of the Sun now can be explained by magnetohydrodynamic braking that is also observed in other stars of classes F, G and K.
5. Among the different models proposed to explain the formation of planetary bodies, that with a low-mass primordial nebula ($10^{-2}\ M_\odot$) is widely accepted. In this model, after the collapse of the nebula into a disk, the particles fall into the equatorial plane in a very short period (of the order of one hundred

years), and accrete by collision with a growth rate that is of the order of one centimetre per year.

6. During the early stages of the Sun's existence, the solar wind must have been very intense. One effect of this solar wind would be to expel particles that had not accreted into bodies with a diameter of the order of at least one centimetre.

3.3 Observational Tests of Models for the Formation of the Solar System

Very diverse models for the formation of the Solar System may have seen the light of day in the past, but this was because at the time when they were proposed it was not really possible to challenge them with well-established observational facts. We have seen (Sect. 3.1.3) how dates obtained from meteorites and measurements of the deuterium abundance in the giant planets have allowed us to exclude several classes of models. It is useful for us to try now to refine the models that appear the most plausible, by confronting them with appropriate observations. These tests are of two types. On the one hand they concern the non-evolved material that is still found in the Solar System. On the other, the study of differentiated material gives us information about the different processes that occurred at the time of the Solar System's formation, and which produced that differentiation (see Sect. 12.2).

3.3.1 Primordial Material in the Solar System

Where in the Solar System should we look for primordial material that undergone no, or very little, alteration? In fact it occurs in two main forms.

In the gaseous phase, it is found in the atmospheres of the giant planets; it is also found in the inner regions of comets, where parent molecules are ejected from the nucleus by degassing when a comet approaches the Sun. In the case of the giant planets, the tests are the determination of the abundance ratios of elements and isotopes. Apart from deuterium, which we have already discussed, the importance of helium and the elements C, N and O, and their isotopes must be mentioned. At present these measurements have been carried out primarily for Jupiter and Saturn; by extending them to Uranus and Neptune we should be able to study the behaviour of the various properties as a function of heliocentric distance. In the case of comets, the composition of the parent molecules is a direct indication of the composition of the nucleus. This nucleus, because of its small size and the low temperature of its environment, most probably consists of undifferentiated primordial material.

It is believed that solid bodies in the Solar System have less chance of being differentiated the smaller they are. The reason is that the principal mechanism for heating the interiors of small bodies is that of radioactivity. The radioactive element that provides the most energy may be aluminium-26, the existence of which in the bodies at the time of their formation appears to be demonstrated by isotopic measurements on carbonaceous chondrites (see Sect. 11.3.6).

The temperature $T(0)$ at the centre of a body as a function of the energy released by radioactivity and as a function of the surface temperature $T(R)$ may be estimated by:

$$T(R) = T(0) - \left(\frac{\alpha}{6\gamma}\right) R^2 \quad , \tag{3.43}$$

where α is the output per unit volume and γ the thermal conductivity.

In order that no differentiation should take place, the difference $T(0) - T(R)$ should remain less than about 1800 K for a silicate body, and less than about 150 K for an icy one. In both cases one finds that the limit is of the order of a few kilometres. This indicates that remnants of the primordial solar nebula may be found among the *small bodies* in the Solar System (the minor planets and comets).

Statistical examination of the minor planets provides us with valuable information about their dynamics. Unfortunately the small bodies are still insufficiently known, because of their small size and low magnitudes. We can envisage that in the years to come more powerful instrumental methods will enable us to undertake systematic study of their physical and mechanical properties (masses, diameters, orbital parameters, rotation periods, number of craters, etc.), which will permit us, in particular, to determine the mass and diameter distribution functions. This information will have to be taken into account in the models that describe the collision and accretion of the smaller bodies.

3.3.2 Evolved Material in the Solar System

Several types of problem can be envisaged beginning with the observation of the Solar System today.

The analysis of secondary atmospheres that have been degassed from the bodies after the escape of the primary atmospheres, is full of information. For the inner planets, it gives us information about the physical conditions that prevailed on the body at the time of degassing; for Titan it gives information about the probable nature of the body itself, and thus about the condensation mechanism at great distances from the Sun. Finally, for comets, the observation of the secondary products formed by the radicals, photodisassociated by solar UV radiation when the comets approach the Sun, is an additional tool in trying to work back to the composition of the parent molecules and thus of the nucleus itself.

3.3.3 Observations Outside the Solar System

We have seen the importance of theories of stellar formation for the understanding of the formation of the Solar System. These theories have been developed as a result of intensive observation of regions of stellar formation, in particular that in Orion, conducted in the infrared and millimetre regions of the spectrum. We may expect important progress to be made in this field in the near future, thanks to instrumental developments.

Another field of research that is expanding rapidly is the search for extrasolar planets. Such a discovery, either around one or several nearby stars would allow us to answer a fundamental question: is the formation of planets a normal stage in the process of stellar formation, or is the Solar System an exception? Research carried out so far, mainly by astrometric methods, has not given a definitive answer. On the other hand, the envelopes representing protoplanetary disks have been detected, first in the infrared thanks to the IRAS satellite, and then in the visible by CCD cameras. These disks could be the preliminary stages in the formation of planets. Important progress in this field can be expected when the Space Telescope is launched, as this should permit us to obtain the image of a planet the size of Jupiter in orbit around one of the stars closest to the Sun. Finally we must emphasize the importance of speckle interferometry in the infrared, which is probably the best method of detecting "Brown dwarfs", cold stellar companions, intermediate in size between planets and faint stars.

4. The Interaction of Solar-System Bodies with the Interplanetary Medium

By Michel Blanc

The outer envelopes of the various Solar-System bodies are in direct contact with the interplanetary medium. This chapter describes the different physical processes that take place and govern the exchange of material, energy and momentum between Solar-System bodies and the interplanetary medium.

The interplanetary medium itself consists of four components:

1. dust;
2. cosmic rays, a population of atomic nuclei with very high energies (of the order of, or above 100 MeV), either produced in stellar atmospheres and then accelerated by interplanetary shocks during a long path through the Galaxy (galactic cosmic rays), or directly produced in solar flares (solar cosmic rays);
3. neutral gas of interstellar origin;
4. the solar wind, a plasma primarily consisting of H^+ and He^{++}, produced and accelerated in the solar corona, from which it spreads out into interplanetary space.

After we have discussed the properties of the solar wind, which is the predominant component of the interplanetary medium as far as interactions with the various bodies are concerned (Sect. 4.1), we shall describe the outer gaseous envelopes of the planets (Sect. 4.2). These outer envelopes are directly influenced by solar X-ray and ultraviolet radiation, which cause heating of the neutral atmosphere, excitation of some of the atoms and molecules into higher quantum states and, above all, by ionization, create a plasma envelope (or planetary *ionosphere*) around a planet or a comet.

The interaction of the solar wind with these outer gaseous envelopes is a function of the pressure of these envelopes and of the intensity of the magnetic field of the object under consideration. According to the values of these parameters, this interaction produces very different interaction geometries and involves very different physical mechanisms. The various forms are described in Sect. 4.3. When a planetary magnetic field is sufficiently intense, a magnetic cavity is formed within the solar wind and around the planet. This is known as a *magnetosphere*.

4.1 The Interplanetary Medium

4.1.1 The Interplanetary Plasma: the Solar Wind and Heliosphere

Coronal Expansion

Beyond a few solar radii, the plasma pressure of the solar corona (which has a temperature of the order of 10^5 to 10^6 K) is not balanced by coronal magnetic forces. Some of the solar magnetic field lines are open to interplanetary space, forming flux-tubes along which the coronal plasma can escape in a supersonic flow.

This coronal structure was revealed in detail by photographs in the X-ray region taken by the astronauts on board Skylab. Figure 4.1 is an example. As the X-ray flux is essentially produced only by energetic electrons trapped in closed magnetic loops that are filled with a dense thermal plasma, the luminous regions show the position and the extension into space of the closed magnetic loops. One can see that, to a first approximation, they occur in the equatorial and mid-latitude regions. In contrast, the dark regions, which are seen to extend to the bottom of the corona, correspond to magnetic flux-tubes open to interplanetary space, which are filled with

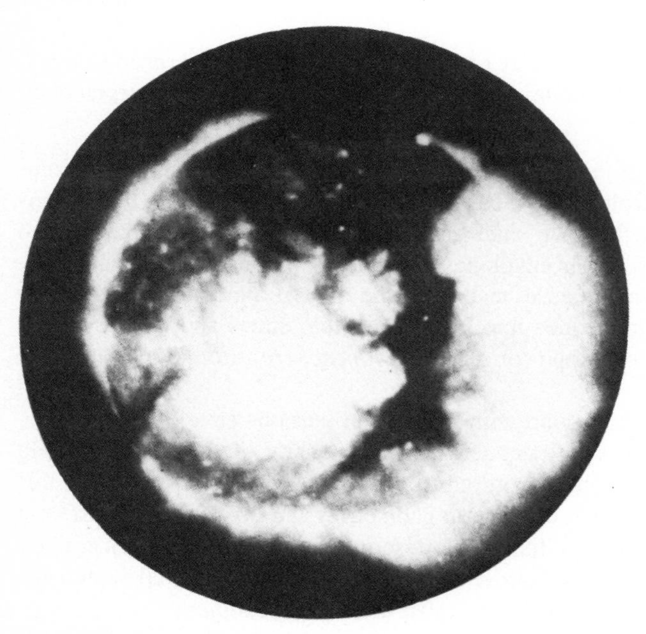

Fig. 4.1. Photographs of the corona in X-rays have shown that between the closed, dense magnetic arches there are less dense regions where the magnetic flux tubes are open to interplanetary space. These are the *coronal holes* (which appear dark in this photograph taken during the Skylab mission), which primarily cover the polar regions but which also extend to equatorial regions at certain longitudes. Coronal holes are probably the source of most of the solar wind, and certainly of the high-speed solar wind. [T.E. Holzer: "The Solar Wind and Related Astrophysical Phenomena", *Solar System Plasma Physics,* Vol 1, ed. by C.F. Kennel, L.J. Lanzerotti, E.N. Parker (North-Holland, 1979). (Reproduced by kind permission of the publisher)]

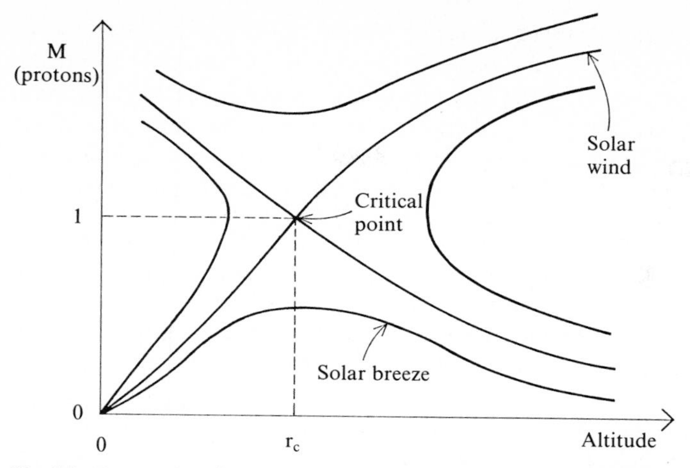

Fig. 4.2. Curves for the vertical profile of coronal expansion velocities (expressed in proton Mach number) for different solutions of the hydrodynamic problem examined by Parker. Only the "solar wind", which crosses the sound barrier at the critical point (a few solar radii above the photosphere) is compatible with the very low pressure exerted by the interstellar medium at very great distances. (By kind permission of J.C. Cerisier)

a plasma that is only about one third as dense as that in the magnetic loops. These regions are the *coronal holes*.

It is the expansion of the coronal plasma into interplanetary space that forms the solar wind. Initial calculations of this flow were made by Parker, who considered only the equations of motion along the magnetic field lines. If the magnetic field is assumed to be radial, the solutions of the equations of motion, coupled with the continuity equation, are represented by the set of curves shown in Fig. 4.2. The only solutions that are physically interesting are, of course, those that connect the surface of the Sun (or a reference level in the lower corona, where the pressure is assumed known and where the flow velocity is negligible) to infinity. It can be seen that there are two types:

1. a family of subsonic solutions, where the flow velocity is nearly zero at the origin, increases to a certain level, and then decreases, tending to zero at infinity: this is the *solar breeze*;
2. a unique solution where the flow velocity increases monotonically, reaching a Mach number of one at a point in the diagram that is known as the critical point, beyond which it becomes supersonic. It is this solution that describes the solar wind that is actually observed. In the case of the solar corona, the critical point is situated at an altitude of between three and ten solar radii, depending on the assumptions made about the divergence of the lines of force and about the values of coronal density and temperature. The velocity of the protons is then 30 km s^{-1}. It increases considerably thereafter, becoming approximately saturated at a speed which is that observed at the Earth (about 400 km s^{-1}), and which remains more or less constant in the outer Solar System.

The fact that the actual flow assumes one or other of these two regimes depends on the boundary conditions set by the pressure at infinity. For the flow to remain subsonic, the pressure at the end of the flux tube must remain higher than a certain critical pressure. If, on the other hand, there is a vacuum at the other end of the tube, only the supersonic-flow solution can be realised. The pressure reigning in the interstellar medium is certainly below the critical pressure, and it thus imposes a supersonic expansion.

The Large-scale Structure of the Heliosphere in the Plane of the Ecliptic

The volume of space occupied by the magnetic lines of force leaving the Sun constitutes a giant magnetosphere: the heliosphere. We shall describe its large-scale structure.

At a level well above the critical point, the magnetic forces acting on the coronal plasma are no longer sufficient for it to be entrained by the solar rotation. On the contrary, it is the coronal plasma, a MHD fluid in which conditions for freezing the magnetic field within the flow are realised to a very close approximation, that entrains the coronal magnetic field in its radial expansion of about 400 km s^{-1}.

As the base of each magnetic line of force remains anchored in the photosphere, it continues to be carried round by the solar rotation. The combination of the uniform rotation of the bases and of the radial motion at each point on a line of force results in the lines of force in interplanetary space becoming twisted. In the equatorial plane of the Sun (and therefore equally in the plane of the planetary orbits, which are not far removed from it) this twisting forms, on average, an Archimedian spiral, the pitch of which is equal to the radial distance covered by one element of the fluid in one solar rotation, i.e. about 10^9 km or six astronomical units (Fig. 4.3).

From these simple remarks, it is quite easy to deduce the radial dependence of the principal parameters of the plasma and magnetic field – at least in the plane of the ecliptic.

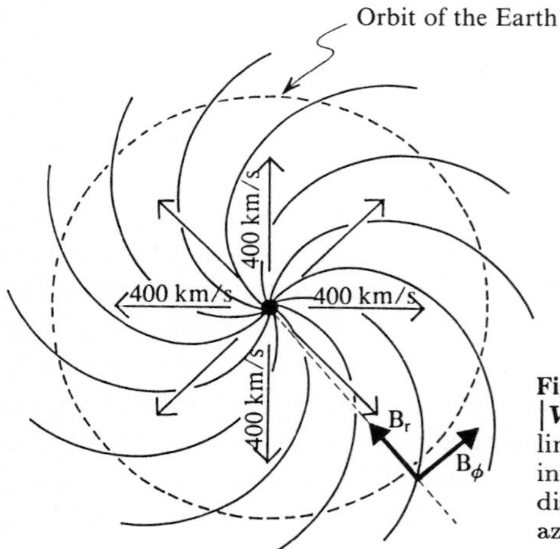

Fig. 4.3. The coronal expansion at a steady rate $|V|$, combined with the solar rotation, twists the lines of force of the interplanetary magnetic field into an Archimedian spiral. As a result, at the distance of the Earth's orbit, the radial B_r and azimuthal B_ϕ components of the interplanetary magnetic field are approximately equal

Conservation of the mass flow results, for a constant radial velocity, in a variation as R^{-2} for the concentration of ions in the solar wind. The value actually observed is:

$$n(\mathrm{cm}^{-3}) \cong 5/R^2 \ (\mathrm{AU}) \quad . \tag{4.1}$$

Conservation of magnetic flux imposes the same inverse-square law on the radial component B_r of the interplanetary magnetic field:

$$B_\mathrm{r} = B_{\mathrm{r}0}/R^2 \ (\mathrm{AU}) \tag{4.2}$$

and the Archimedian-spiral structure then governs the radial dependence of the other component of the magnetic field, the azimuthal component $B\phi$:

$$B_\phi = B_{\phi 0}/R \ (\mathrm{AU}) \quad . \tag{4.3}$$

Comparison with measurements made by various interplanetary missions (Fig. 4.4) shows that these laws are, on average, correct.

At the orbit of the Earth ($R = 1$ AU), $B_{\mathrm{r}0}$ and $B_{\phi 0}$ are more or less equal (the spiral therefore makes an angle of 45 degrees to the direction of the Sun), and are of the order of 3 nT.

The solar wind is a completely ionized gas, the chemical composition of which is close to that of the photosphere, but not as rich in heavy ions. This is doubtless a consequence of the thermal escape mechanism governing the origin of the solar wind, and which favours light ions. As a result, more than 99% are H^+ and He^{++} ions (protons and alpha particles), of which 3.5% to 4.5% are helium. Taking a value of 4 nT as the intensity of the interplanetary field at the orbit of the Earth, we find a radial variation in the Alfvén velocity of the solar wind:

$$V_\mathrm{A}(\mathrm{km/s}) \simeq V_{\mathrm{A}_0} \left[\frac{1}{2} + \frac{1}{2R^2\,(\mathrm{AU})}\right]^{1/2} \tag{4.4}$$

with a value $V_{\mathrm{A}_0} = 40\,\mathrm{km\,s^{-1}}$ at 1 AU.

The azimuthal component B_ϕ of the interplanetary field predominates as one moves farther away from the Sun than the orbit of the Earth. Its radial variation exactly compensates for that of the solar-wind concentration, and the Alfven velocity tends to a value that is uniform, on average, and which is very little different in all the outer Solar System from that measured at the orbit of our planet.

The Three-dimensional Structure of the Heliosphere

When the interplanetary magnetic field in the plane of the ecliptic is examined, it is found that it is organized into sectors of opposite magnetic polarity, the lines of force being either oriented towards the Sun, or away from it. The Pioneer 11 space-probe, which after its fly-by of Jupiter was the first to leave the plane of the ecliptic, has allowed us to study the heliosphere out to a solar latitude of 16 degrees, and to discover that the sector-like structure hid a three-dimensional form that is actually much simpler (Fig. 4.5). In fact, the heliosphere consists of two magnetic hemispheres of opposite polarity, in each of which the lines of force are probably connected to coronal holes of north or south polarity. These magnetic hemispheres

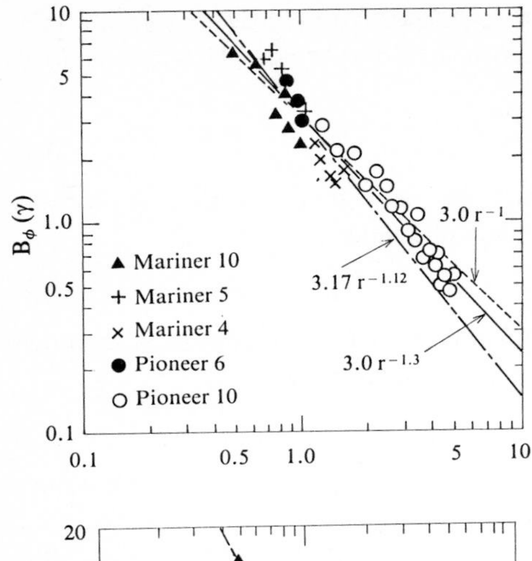

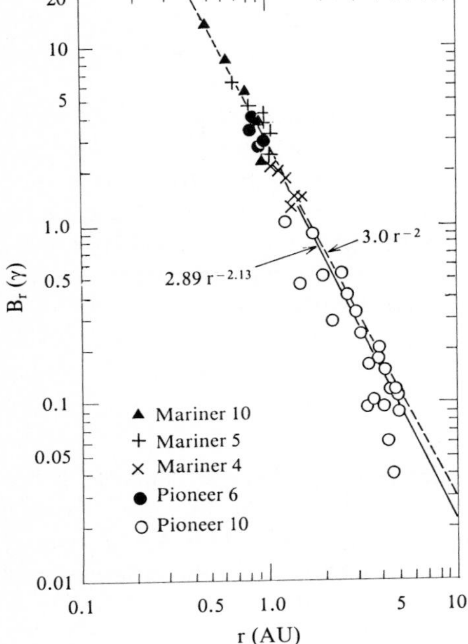

Fig. 4.4. Interplanetary probes have allowed us to determine the radial variation in the interplanetary magnetic field between 0.4 and 30 AU. Plotting the points determined by the Mariner and Pioneer probes shows that the variation in r^{-2} for the radial component B_r and in r^{-1} for the azimuthal component B_ϕ, predicted by Parker's simple spiral model, are in quite good agreement with the observational mean. The two components are equal, and of the order of 3 nT at the Earth's distance. [After K.W. Behannon: "Heliocentric Distance Dependence of the IMF", Reviews of Geophysics and Space Physics **16**, (1978)]

are separated by a neutral sheet, a heliospheric "ballerina's skirt", which rotates with the Sun and undulates from one side of the solar equatorial plane to the other, following the complicated geometry of the photospheric and coronal fields. The sector boundaries are thus simply the intersections of this neutral sheet with the plane of the ecliptic.

The Ulysses interplanetary probe should enable us to explore the highest latitudes in the solar wind in the early 1990s, because it will pass successively over the two poles of the Sun. This mission should enable us to make considerable progress in understanding the relationships between coronal holes and the solar wind.

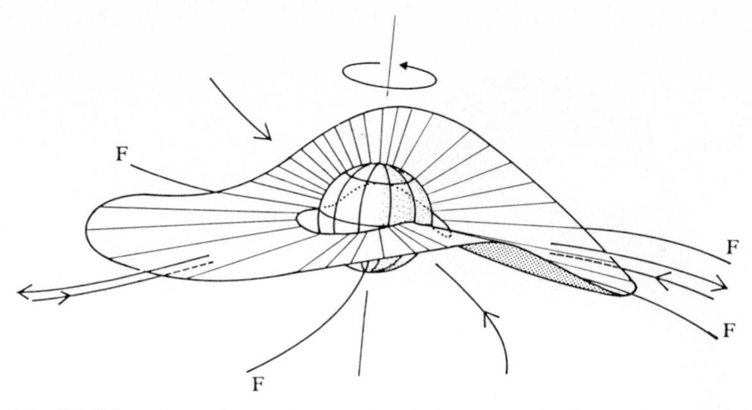

Fig. 4.5. Viewed on a large scale, the heliosphere appears to be divided into two hemispheres of opposite magnetic polarity, one consisting of lines of force directed towards the Sun ("toward hemisphere"), and the other of lines directed away from the Sun ("away hemisphere"). They are separated by a neutral sheet, the position of which undulates around the solar equator. These undulations are the cause of the apparent division of the interplanetary medium into magnetic sectors of opposite polarity, when a section is taken in the ecliptic plane. [After J.L. Steinberg: *Le vent solaire,* Proceedings of the colloquium on "Technologie de l'environment spatial" (Editions CNES 1986). By kind permission of CNES]

Transient Structures in the Solar Wind

Transient, high-amplitude variations are superimposed on the average structure that has just been described. These transient phenomena are the result of deviations from the uniform velocity of the solar wind that permanently exists at its source level in the solar corona. They can be divided into two major categories.

The first type of structure is created by inhomogeneities in the radial-flow velocities in the solar wind. Seen in the plane of the ecliptic, the solar wind is, in fact, divided into longitudinal sectors occupied by high-speed jets in the solar wind (600 km s^{-1} or more), separated by regions where the wind is weaker (about 400 km s^{-1}). It has been possible to establish in a fairly conclusive manner that the high-speed jets are connected by the interplanetary field to equatorial coronal holes. Because of the solar rotation, the slow and fast jets follow one another across any individual radius, leading to the situation shown in Fig. 4.6. The slow-speed solar wind is compressed in front of the high-speed jet, and conversely, rarefied behind it. This perturbation, which co-rotates with the Sun and the sources to which it is linked, may degenerate into an interplanetary MHD shock in front of the compression zone.

The second type of structure is not produced by steady flow, but by coronal events linked with solar activity, such as an eruption, or with some transient coronal phenomenon that leads to localized ejection, at high speed, of a mass of coronal plasma, which expands out into interplanetary space. The different types of discontinuity produced by this mass ejection are shown in Fig. 4.7. This time the disturbance has a finite duration, and affects a longitudinal sector that is of limited extent in the inertial frame of reference.

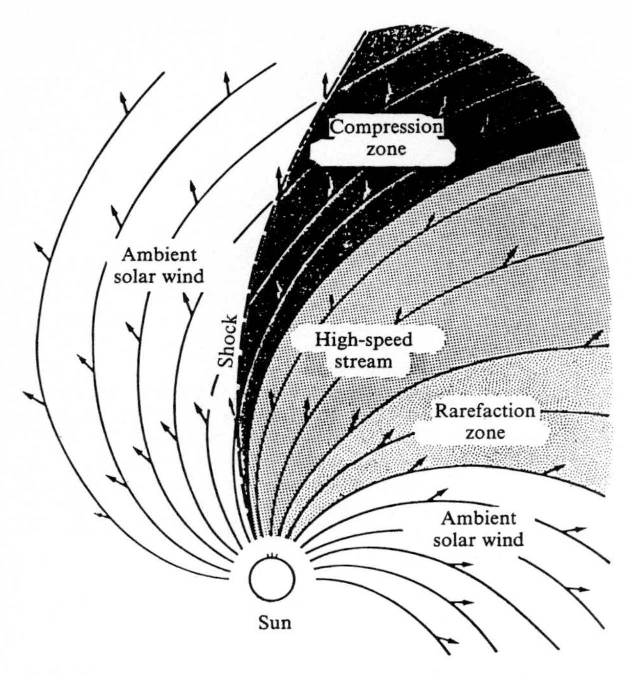

Fig. 4.6. The interaction of a high-speed jet with the ambient slower-speed solar wind – seen here in a reference frame co-rotating with the Sun – creates a plasma compression zone downstream in the ambient solar wind (possibly preceded by an MHD shock), and a rarefaction upstream. The overall structure represented by these sectors rotates rigidly with the Sun, as long as the source of the solar wind remains stationary. [After A.J. Hundhausen: *Coronal Expansion and Solar Wind,* Physics and Chemistry in Space, Vol. 5 (Springer-Verlag, Berlin, Heidelberg 1972)]

The Interaction of the Heliosphere with the Interstellar Medium

The interplanetary medium does, of course, interact with the three principal components of the interstellar medium: neutral interstellar gas, cosmic rays, and magnetized interstellar plasma.

The neutral interstellar gas has a density estimated to be a little less than one particle per cm^3 locally, with a temperature of 10^3 to 10^4 K, and a composition very close to that of the Sun. Its velocity relative to the Sun is 20 $km\,s^{-1}$, and its motion lies almost in the plane of the ecliptic. As it is not influenced by electromagnetic forces, it freely penetrates the heliosphere with this average velocity. Its interactions with the Solar System consist of the deviation imposed by gravity and solar radiation on the trajectory of each particle, and of the exchange of charge between the neutral interstellar particles and the solar-wind ions. This second mechanism, which effectively replaces a rapid ion in the solar wind with a slow ion of interstellar hydrogen, results in a loss of energy and momentum in the solar-wind flow. The effect of this charge-exchange has been revealed by the anisotropy that it causes in the Lyman-α emission from interstellar hydrogen. Effects on the solar wind itself will possibly be detected by the Ulysses mission, which will allow the solar wind to be analyzed in the regions of more rapid flow over the Sun's polar caps.

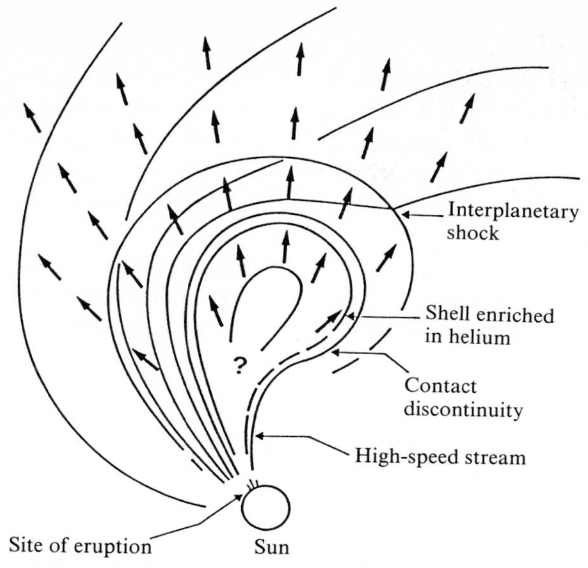

Fig. 4.7. The transient perturbation in the solar wind that arises from an ejection of coronal material following an eruption. The new coronal material that is ejected is separated from the ambient plasma by a contact discontinuity, and preceded by a shock-wave. The external shell of the ejected coronal material is enriched in helium by comparison with the ambient solar wind. [After A.J. Hundhausen: *Coronal Expansion and Solar Wind,* Physics and Chemistry in Space, Vol. 5 (Springer-Verlag, Berlin, Heidelberg 1972)]

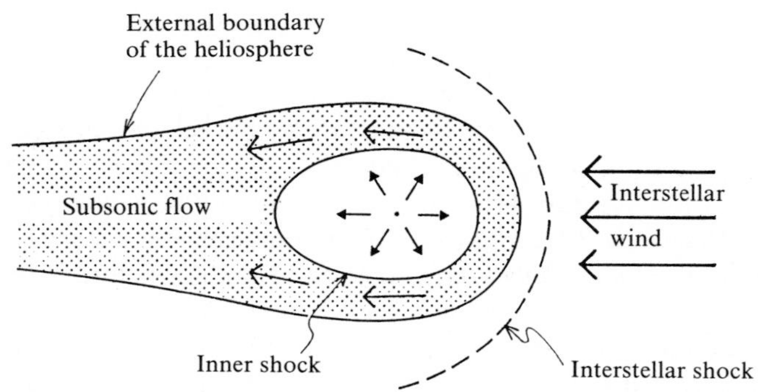

Fig. 4.8. The spatial extent of the heliosphere is defined by the various discontinuities that are thought to be caused by the interaction of the solar wind with the ionized and magnetized component of the interstellar medium. [After W.I. Axford: "Interaction of the Interstellar Medium with the Solar Wind", Space Science Reviews **14**, 582 (1973)]

The characteristics of the interstellar thermal plasma are still poorly understood. It is thought that its composition consists of between 10^{-1} to 10^{-2} particles per cm^3, and that its temperature is between 10^3 and 10^4 K. It is bathed in a magnetic field of close to 0.35 nT. It is the interaction with this magnetized plasma that limits the extent of the heliosphere in space; in effect the expansion of the solar wind is brought to a halt when its total pressure drops to that of the interstellar plasma and magnetic field. The geometry of these outer regions of the heliosphere possibly has the form shown in Fig. 4.8. The interstellar plasma encounters, at a velocity of 20

$km\,s^{-1}$, a contact discontinuity that defines the outer boundary of the heliosphere (unless it has previously been decelerated by a shock-front). Within the heliosphere there is probably another shock across which the solar wind reverts to subsonic flow before being swept back towards the antapex of the flow of interstellar gas (the interstellar wind). Calculation of the equilibrium pressure places the internal shock-front at about 100 to 130 astronomical units from the Sun (about three times the distance of Pluto from the Sun).

Naturally this representation is merely a plausible representation of the boundaries of the heliosphere, which must, in reality, be very variable both in time and space, because of inhomogeneities in the solar wind and in the interstellar medium.

4.2 The Outer Gaseous Envelopes of the Planets: Thermospheres and Ionospheres

The solar electromagnetic UV- and X-radiation, and corpuscular radiation by energetic particles, cause the ionization of the outermost external envelopes of planetary atmospheres. There they give rise to a plasma component, constituting what are known as ionospheric layers.

As Fig. 4.9 shows, these layers of ionospheric plasma play a fundamental role in the dynamics of the ionized envelopes:

1. they are conducting; so much so that each planet is surrounded by a sort of spherical conductor immersed in the upper atmosphere. This conductor carries electrical currents as a result of dynamos produced in the upper atmosphere either by motion of the atmosphere itself (winds), or by the relative motion between the solar wind and the ionospheric conductor;
2. they are sources of plasma. As this plasma of ionospheric origin, which typically has energies of between 0.1 and 1 eV, diffuses outward – particularly along the magnetic field lines – it fills a large portion of the magnetospheric cavity, where it behaves, because of its low temperature, as a MHD fluid "frozen" onto the magnetic field. Its movements obey the equation:

$$\boldsymbol{E} + \boldsymbol{V} \times \boldsymbol{B} = 0 \tag{4.5}$$

and it therefore reflects the "movements" – in the MHD sense – of the magnetic field lines, the determination of which will be one of the principal objects of this chapter;
3. in the regions to which the ionospheric plasma has diffused the lines of magnetic force become strongly conducting. The component parallel to $\boldsymbol{B}$ in (4.5) is simply:

$$\boldsymbol{E} \cdot \boldsymbol{B} = 0 \tag{4.6}$$

because of the high degree of mobility of "cold" ionospheric electrons along $\boldsymbol{B}$. These lines of force therefore act as very efficient electrical connections between the ionospheric conductor and the dynamos that are distributed within

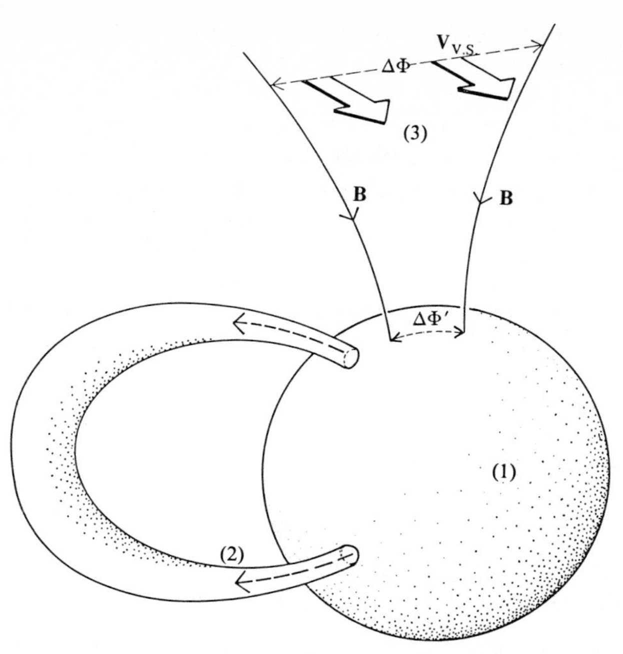

Fig. 4.9. The three principal effects of the ionospheric plasma source on the electrodynamics of ionized planetary atmospheres. (1) The ionospheric layers from a conducting shell surrounding the planet that is immersed in its upper atmosphere. (2) If the planet has a magnetosphere, the plasma created by the ionization of the upper atmosphere diffuses into the magnetic tubes to which it is connected, and thus tends to fill them with a fairly cool plasma (0.1 to 1 eV). (3) The electrons that have diffused in this manner cause the lines of force to be highly conductive and connect the ionospheric conductor to the dynamo set up by the relative motion of the magnetized solar wind and the planet. A fraction $\Delta\Phi'$ of the potential $\Delta\Phi$ induced by the solar wind between two of the planetary magnetic field lines is thus applied to the ionospheric conductor

the magnetospheric cavity or on its interface with the solar wind. The currents generated by these dynamos therefore tend to connect along the lines of force and then across the conducting ionospheric shell.

4.2.1 The Structure of the Neutral Upper Atmosphere

Vertical Structure of the Neutral Atmosphere

The vertical distribution of the constituents of the neutral atmosphere is governed by two phenomena: the chemical and photochemical reactions that exchange different components under the action of the incident solar flux, and each constituent's own dynamic motion, which in an atmosphere horizontally stratified, tends to make it diffuse vertically through other components in order to reach its own hydrostatic equilibrium. According to the importance of molecular diffusion, turbulent mixing within the atmosphere, and chemical reactions, each chemical species present, taken on its own, may be more or less close to its partial hydrostatic equilibrium. But in every case, the atmospheric gas, taken as a whole, obeys the law of global hydrostatic equilibrium (see also Sect. 2.2.1):

$$\frac{d}{dz}(N_n k T_n) = -N_n M g(z) \quad . \tag{4.7}$$

This equation, which describes the vertical variation of N_n, the total concentration (summed over all chemical species) of neutral atmosphere, can be integrated to give:

$$N_n(z) = N_n(0)\frac{T_n(0)}{T_n(z)} \exp - \int_0^z \frac{dz'}{H_n(z')} \qquad \text{where} \tag{4.8}$$

$$H_n(z) = \frac{kT_n(z)}{Mg(z)} \tag{4.9}$$

is the scale-height of the neutral atmosphere (the increase in altitude required for N_n to decrease by a factor of e).

Equation (4.8) can in fact be used only if one knows the vertical profiles of:

1. $T_n(z)$, the temperature of the neutral gas; and
2. $M(z)$, the average molecular mass of the neutral gas.

The determination of these two profiles is in fact a very complex problem, which implies calculating the energy balance of the atmosphere (in order to determine T_n), and knowing the photochemistry of each of the components, in order to be able to calculate the average molecular mass $M(z)$ from the mixing ratio of each constituent. We shall give only a few, more or less intuitive, ideas based on the terrestrial atmosphere as an example. This is represented in Fig. 4.10, which shows the depth to which vertically incident solar radiation penetrates into the terrestrial atmosphere, in the form of the variation as a function of wavelength of the altitude where the optical thickness of the atmosphere equals 1 (see Sect. 2.2.1). Where this curve is more or less horizontal over a wide spectral range, it implies that a specific fraction of incoming solar energy is deposited at a well-defined range of altitudes. Each of these layers corresponds, in the vertical structure of the atmosphere, to a change in the sense in which the temperature varies. This can be seen by comparing Fig. 4.10 and the vertical temperature profile shown in Fig. 4.11. Starting at the ground (where the largest fraction of the solar radiation in the visible and near-infrared regions is absorbed), temperature decreases in the troposphere until it reaches a first minimum at the tropopause (between ten and twenty km high, depending on latitude). Above the tropopause, the stratosphere corresponds to an increase in temperature (as far as the stratopause, at about fifty kilometres altitude), produced by the absorption of ultraviolet radiation in the 2000–3000 Å range by a layer of ozone. (This layer of ozone protects us from direct exposure to solar ultraviolet radiation, and its equilibrium depends on a very high number – about one hundred – of linked photochemical cycles.) Above the stratosphere, the temperature decreases again in the mesosphere. (In the absence of a source of heat, the temperature profile tends to approach the adiabatic gradient, which is a gradient that is marginally stable with respect to the atmosphere's convective instability.) Above the mesopause, at about 90 km altitude, the temperature rises again, and this time to

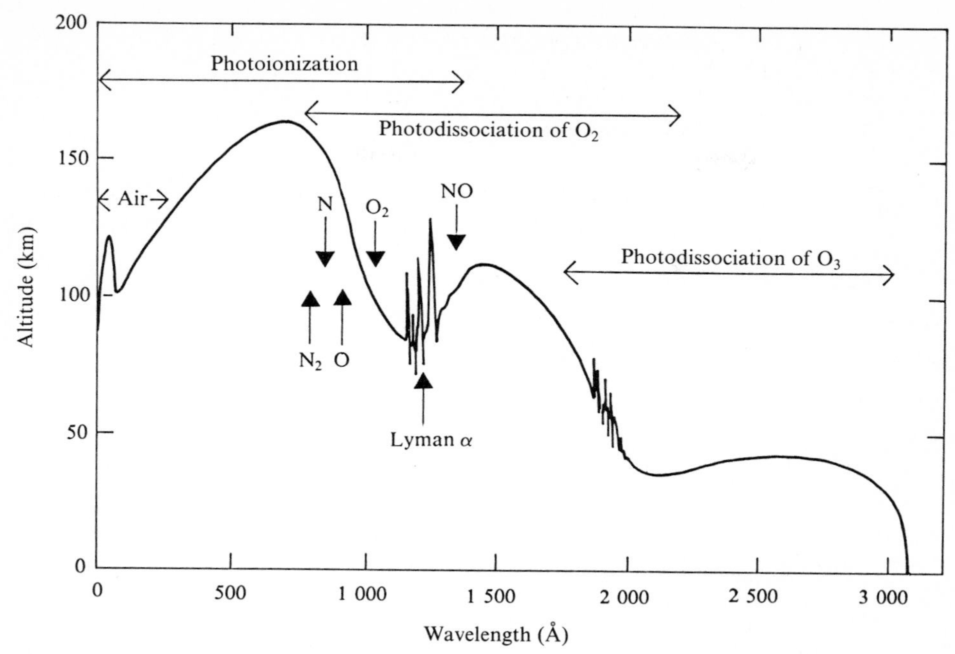

Fig. 4.10. The depth of penetration of vertically incident solar radiation in the ultraviolet region into the terrestrial atmosphere. The different layers of the atmosphere where the radiation deposits its energy can be seen easily. (1) For $\lambda > 3100$ Å, the radiation reaches the ground (or the top of clouds). (2) The 2000–3000 Å band is absorbed by photodissociation of ozone at about 40 km. This is the cause of the increase in temperature in the stratosphere. (3) The band from ~1000 to 1700 Å is absorbed by the photodissociation of O_2 around 100 km altitude. This is the origin of the rise in temperature at the base of the thermosphere. (4) The photoionization of the upper atmosphere above 90 km is produced by the portion of the spectrum with $\lambda < 1400$ Å, approximately

its peak, reaching a very high asymptotic value of 900 K or more, which is known as the exospheric temperature. This is the result of absorption in the EUV (extreme ultraviolet) band by photoionization, and in the ultraviolet band between 1000 and 1700 Å by dissociation of O_2 molecules. The latter comes into play at around 100 km altitude, but is subject to major variations according to diurnal, seasonal and solar cycles, and also as a function of magnetic activity.

Examination of the second curve in Fig. 4.11, that shows the average molecular mass, reveals two other domains. From the ground up to the homopause, situated at about 90 km altitude, the chemical composition of the air remains the same as that at ground-level ($M = 29$). This is essentially owing to turbulent mixing of the air, the principal cause of which is the predominance of the ground in heating the atmosphere. The homosphere is the "convective zone" in a planetary atmosphere.

With the strongly positive temperature gradient found in the thermosphere above 90 km, the atmosphere becomes very stable against convection, and vertical mixing of the components rapidly ceases, becoming negligible above the turbopause (at about 110 km). Here one enters the heterosphere, where a gravitational separation of the different chemical components takes place, leading to a progressive decrease in M with height, the lighter constituents becoming increasingly dominant with

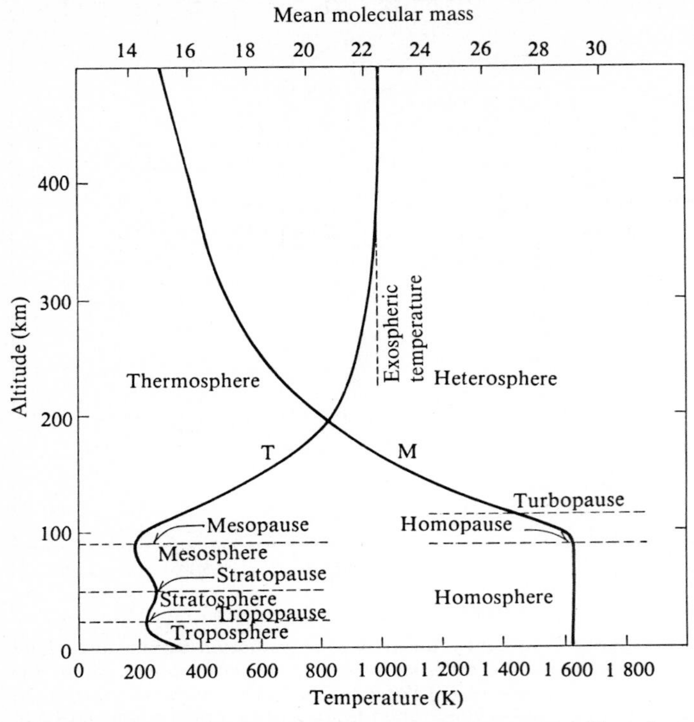

Fig. 4.11. Vertical distribution of the neutral temperature T, and mean molecular mass M, in the terrestrial atmosphere, and the corresponding atmospheric regions

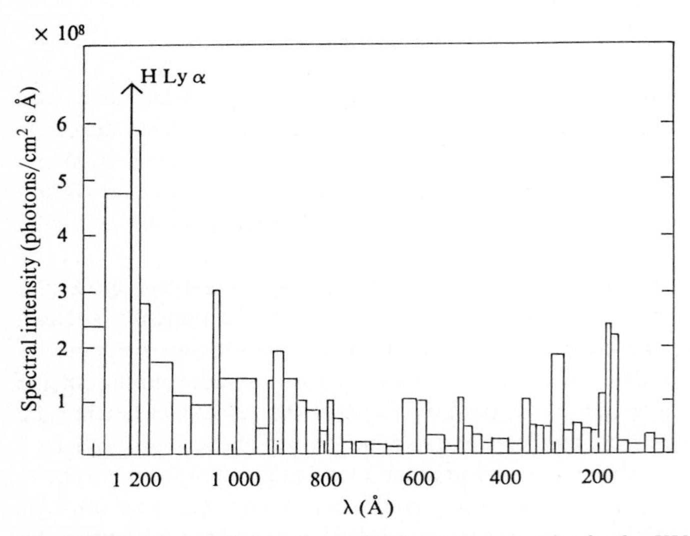

Fig. 4.12. Spectrum of the solar radiation component in the far UV and X-ray regions, which is responsible to the photoionization of the upper atmosphere, and thus for the formation of the normal ionospheric layers. [After K. Takayanagi, Y. Itikawa: "Elementary Processes Involving Ions in the Ionosphere", Space Science Reviews **11**, 380 (1970)]

altitude. It is in the region of the thermosphere and the heterosphere that the basic ionospheric plasma is formed by photoionization driven by solar UV and EUV radiation, a spectrum of which is shown in Fig. 4.12.

Vertical Structure of the Heterosphere/Thermosphere

Above the turbopause, the different atmospheric constituents are no longer mixed by turbulence in an unstable atmosphere. The vertical distribution of each component, taken separately, obeys the equation:

$$(\varrho_i \boldsymbol{g} - \nabla P_i)_z = 0 \quad \text{or} \tag{4.10}$$

$$\frac{\partial}{\partial z}(N_i k T_i) = N_i m_i g \tag{4.11}$$

where N_i, m_i, and T_i are the concentration, molecular mass and temperature of component i, respectively, and P_i its partial pressure. Setting:

$$H_i = \frac{kT_i}{m_i g} \tag{4.12}$$

the equation can be integrated as:

$$\frac{N_i}{N_{i_0}} = \frac{T_{i_0}}{T_i} \exp\left(- \int_{z_0}^{z} \frac{dz'}{H_i(z')} \right) \quad . \tag{4.13}$$

In the absence of chemical reactions, the concentration of each component therefore diminishes exponentially with altitude. The characteristic length of this decrease, H_i, is known as the scale height. It is the altitude gained by a particle possessing thermal energy (kT_i) if its kinetic energy is converted into gravitational potential energy.

The immediate consequence of (4.10) is therefore that – as H_i is inversely proportional to the mass m_i of the component, and proportional to the temperature – the relative concentration of components above the turbopause is a function of altitude, the atmosphere becoming progressively poorer in heavy elements with increasing height.

The effects of this are shown in Fig. 4.13, which depicts, on a logarithmic scale, the vertical density distribution of the principal components for two values of exospheric temperature, still using the terrestrial atmosphere as an example. Following the curves from bottom to top, we see that at first the predominant components are initially N_2 and O_2 (just as at ground-level, because the composition has not changed between the surface and the level of the turbopause). Then atomic oxygen O, produced by photodissociation of O_2 at around 100 km altitude – as shown in Fig. 4.10 – becomes predominant. Finally, with increasing altitude, even lighter elements, molecular and then atomic hydrogen, in turn predominate.

Comparison of the left and right halves of Fig. 4.13 shows the dramatic effects of temperature, which can cause the total atmospheric concentration to vary by an order of magnitude. This effect is well-known to specialists of satellite orbitography. It was, for example, a solar flare that led to NASA's orbital station Skylab reentering the atmosphere several days in advance of the predicted date, because of the

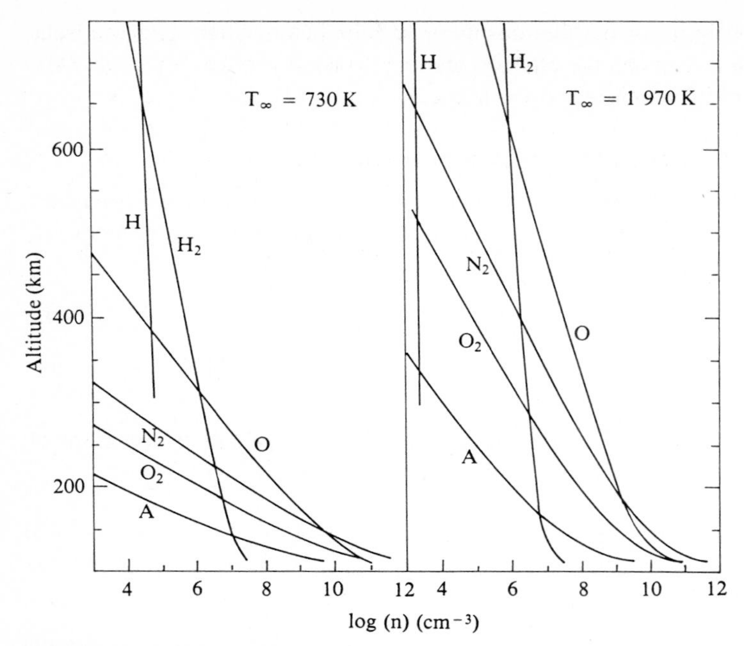

Fig. 4.13. Vertical distribution of thermosphere components for two values of exospheric temperature T_∞. Above the turbopause, the lighter a component is, and the higher the value of T_∞, the more slowly its concentration decreases, thus giving rise to the predominance of the lighter elements (atomic and molecular hydrogen) at the greatest altitudes. [After A. Giraud, M. Petit: Ionospheric Techniques and Phenomena, *Geophysics and Astrophysics Monographs* (D. Reidel Publishing Company 1978)]

increase in the solar ultraviolet flux that accompanied it and the resultant heating of the thermosphere. (The increase in the density and pressure of the outer atmosphere increases the drag exerted by the atmosphere on satellites, and accelerates the "decay" of their orbits.)

The Exosphere

The description of the vertical distribution of atmospheric components by (4.13) is only valid down to a specific lower limit: the turbopause. It also has an upper limit, which is similarly the limit for the validity of the fluid equations, of which (4.7) is essentially the vertical component.

In fact, description as a fluid assumes that the functions describing the velocities for different species do not depart greatly from a maxwellian distribution. This is only true if collisions between particles are sufficiently important. In a heterogeneous medium such as the atmosphere, this local thermodynamic equilibrium (LTE) situation is realized if one has:

$$L_{\text{col}} < L \quad . \tag{4.14}$$

L_{col} (characteristic collision length) is the mean free path, and L is a length characteristic of the spatial variation in the macroscopic parameters of the medium. In our case, we can obviously take $L = H_n$, which, by definition, is the characteristic scale of vertical variation in density. As a result, above an altitude defined by $L_{\mathrm{col}} = H_n$, which is known as the *exobase*, one encounters an atmospheric layer that is no longer maintained in a state close to local thermodynamic equilibrium by collisions. This layer is known as the *exosphere*. In the terrestrial atmosphere the exobase is found at about 500 to 600 km altitude. The exosphere thus consists essentially of just helium and hydrogen.

In calculating the distribution of components in the exosphere, it is no longer possible to resort to hydrostatic equilibrium. It is necessary to:

1. assume the distribution functions for the different species at a reference level, which is considered as being their source (*a priori,* the exobase, where a maxwellian distribution is usually assumed);
2. use the Liouville equation to calculate the distribution function for the various species everywhere else, assuming that all the particles follow ballistic paths, free from collisions.

In fact, this method itself has quite severe, basic limitations. The Liouville equation actually just states that the numerical value of the distribution function $f_i(\boldsymbol{r}, \boldsymbol{v})$ for each species is conserved along each trajectory. Knowing f_i initially only at the exobase, it can only be calculated for points in the $(\boldsymbol{r}, \boldsymbol{v})$ phase-space that are connected to the exobase by some trajectory.

But, as shown by Fig. 4.14, three types of trajectory exist, differing in their degree of connection with the exobase:

1. "ballistic" trajectories, leaving and returning to the exobase. These are trajectories where the initial velocity v of the particle as it leaves the exobase is less than the escape velocity;
2. escape trajectories (hyperbolas or parabolas in the gravitational field of the spherically symmetrical central body) such that $v > v_1$;
3. "satellite" trajectories, so-called because they are orbits that do not touch the exobase.

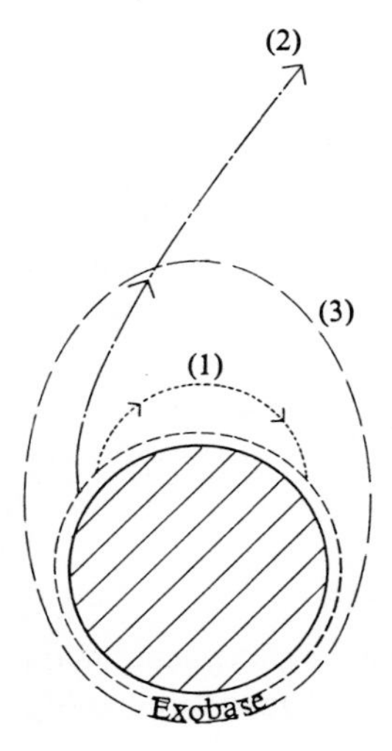

Fig. 4.14. The three types of trajectory for neutral particles in exosphere: (1) "Ballistic" trajectory, from exobase to exobase. (2) Escape trajectory, from the exobase to interplanetary space. (3) "Satellite" trajectory, which does not touch the exobase. (This last type may be elliptical as in the figure or, equally, parabolic or hyperbolic)

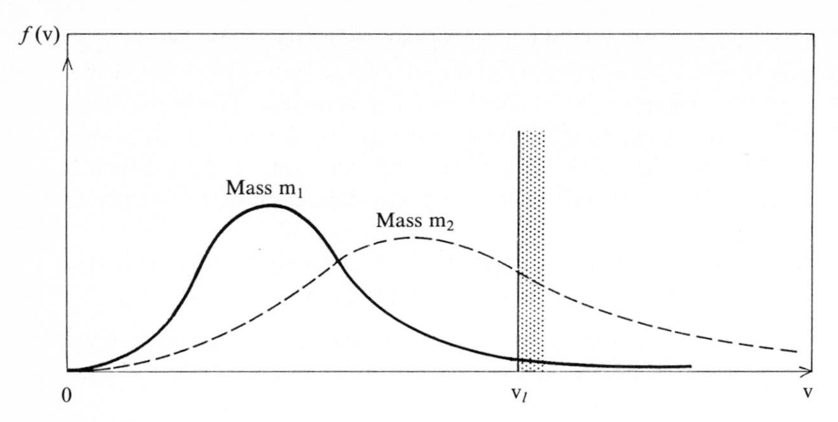

Fig. 4.15. This diagram shows the distribution of velocities $f(v)$ (where $v = |\boldsymbol{v}|$) for two atmospheric species at the exobase of masses m_1 and m_2. If $m_2 < m_1$, at a given temperature T the fraction of particles of mass m_2 that has a velocity higher than the escape velocity v_1 is greater than the fraction of particles of mass m_1 that exceeds that velocity. The escape flux is a very rapidly increasing function of T, and a rapidly decreasing one of m, the position of the peak of the velocity distribution being proportional to the thermal velocity $kT/m^{1/2}$. The "Jeans escape flux" is also a rapidly decreasing function of the planetary mass, which, of course, determines the escape velocity v_1

The kinetic method described earlier allows $f_i(\boldsymbol{r}, \boldsymbol{v})$ to be calculated along trajectories of types (1) and (2). In particular, the total flux of particles on trajectories of type (2) corresponds to a net escape flux from the planetary atmosphere. As is schematically shown in Fig. 4.15, this escape flux becomes more significant the lighter the element considered (because its thermal velocity is inversely proportional to $m_i^{1/2}$), and the weaker the attraction of the central body, which determines the escape velocity v_i.

This explains why the Earth and the inner planets permanently lose hydrogen, and why most of the lighter bodies in the Solar System (the Moon, Mercury, and the small satellites of the giant planets), have not been able to retain any atmosphere.

In contrast, the kinetic method does not allow $f_i(\boldsymbol{r}, \boldsymbol{v})$ to be calculated along "satellite" trajectories of type (3) because, in the absence of collisions, they are not connected to the source formed by the exobase. As far as the kinetic theory is concerned, one should simply advance the hypothesis that $f_i = 0$, and that these trajectories are unpopulated. This conclusion is incorrect, however, because collisions, even if they are very rare in the exosphere, do, when they take place, allow particles initially on trajectories of types (1) and (2) to be transferred to satellite trajectories. This effect, which is weak, but cumulative over a long period of time, tends to re-establish thermodynamic equilibrium even in these rarefied regions. But it is very difficult to construct a conceptually valid model for this situation, half-way between the fluid and the kinetic descriptions.

4.2.2 The Structure and Dynamics of the Ionospheric Layers

The ionospheric plasma, produced by ionization of the upper atmosphere, may, because of its low characteristic energy (0.1 to 1 eV) and its high degree of coupling to the neutral gas – through the production, recombination and collision mechanisms – be represented to a close approximation as a system of $n + 1$ fluids (n being the number of ionic species present), each being coupled to the neutral gas by the collision terms in the Boltzmann equation.

We shall consider the first two moments of this equation in order to deduce the vertical structure of the ionospheric layers, and then their electrodynamic properties.

The Continuity Equation: Birth and Death of the Ionospheric Plasma

For each of the $n + 1$ fluids to be studied (n ions and the electron gas), the first moment of the Boltzmann equation is:

$$\frac{\partial n_j}{\partial t} + \nabla \cdot (n_j \boldsymbol{V}_j) = Q_j - L_j \tag{4.15}$$

where n_j and V_j are the concentration and average velocity for species j, Q_j is the rate of production for that species (in particles per unit time and volume), and L_j is the loss rate for that species.

The conservation of electric charge requires that, with the index e applying to the electron gas, and the indices 1 to n to the ionic species, we have:

$$n_\mathrm{e} = \sum_{i=1}^{n} n_i \quad , \quad Q_\mathrm{e} = \sum Q_i \quad , \quad L_\mathrm{e} = \sum L_i \tag{4.16}$$

thus preserving the neutral charge of the medium, as long as we consider time-scales much greater than the plasma frequency, and spatial scales much larger that the Debye length. It will be seen that, with the electronic equation containing essentially the same information as the ionic equations, it will not generally be directly usable. We can beneficially substitute for it the equation obtained by subtracting the sum of the ionic equations and multiplying the result by e, the elementary charge. This gives us:

$$\nabla \cdot \boldsymbol{j} = 0 \quad \text{where} \tag{4.17}$$

$$\boldsymbol{j} = e\left(\sum_{i=1}^{n} n_i \boldsymbol{V}_i - n_\mathrm{e} \boldsymbol{V}_\mathrm{e}\right) \tag{4.18}$$

is the density of the electric current. This equation, which expresses the fact that the current density field is without divergence, is fundamental in the study of the electrical circuit formed by the ionosphere, the magnetosphere and the solar wind. In practice, one therefore tries to solve the n continuity equations for the ions, and substitutes into the electronic equation (4.18) if one is interested in the distribution of ionospheric currents and electric fields.

The Production Term

The ionization of atoms and molecules in the upper atmosphere has three sources:

1. solar electromagnetic radiation, which may be defined by its spectral energy distribution at the top of the ionosphere $I_0(\lambda)$ (a spectral curve of the sort shown in Fig. 4.12);

2. energetic particles (electrons and ions, and also some energetic neutral particles), the most important example of which are the auroral electrons;
3. meteors, which ionize the air along their tracks.

Meteors are a sporadic, and very localized, source of ionization. Their importance in the global balance of ionospheric plasma production is very small, but they do have the property of introducing metal ions, which the other two sources produce only in very small quantities.

The primary production function Q_j for each ionic species is calculated (as the number of ion-electron pairs per unit volume) for these sources of ionization by modelling each ionizing agent's energy loss along its trajectory through the atmosphere.

When ionization is produced by solar photons, the problem consists of first calculating, for each wavelength, the progressive absorption of the incident solar flux. Not only photoionization is responsible, but also absorption by excitation of atoms and neutral ions. Knowing the absorption, one can subsequently deduce the spectrum of the remaining radiation at any altitude and then the contribution from each individual spectral band to the production function: this is proportional to the product of the spectral intensity and the concentration of the neutral parent species of ion for which one is calculating the production; the coefficient of proportionality being the effective, differential photoionization cross-section.

Integration over the whole spectrum thus enables one to determine the primary production function for each ionic species. The shape of the vertical profile of the function is given in Fig. 4.16 for different values of the solar zenith angle. A very pronounced maximum is noticeable. The production layer typically has a thickness that is two or three times the scale-height. It can be shown easily that when production occurs within a narrow spectral band, the maximum lies at an altitude with optical depth $\tau = 1$ for the given wavelength with vertical solar incidence. It increases to lesser optical depths when the solar zenith angle increases.

When the ionization is produced by energetic particles (ions, electrons, or neutral particles) encountering the atmosphere, means of calculating the production of electron-ion pairs are derived from nuclear-physics methods (absorption of fast particles by a target). The penetration of the target (here the atmospheric layer) by the incident particle is calculated, taking account of elastic collisions (simple deflection of the trajectory by atomic nuclei), and of inelastic collisions, which include, in particular, stripping of outer electrons from the atoms encountered, i.e. ionization. The calculation may be made either by resolving a Fokker-Plank kinetic transport equation (as long as the deflection angles at each collision remain small), or by a Monte-Carlo method, which sums the effect of a large number of incident particles by randomly selecting the parameters of each collision.

For equal incident energy, electrons have the greatest ionizing efficiency; they are principally responsible for the formation of ionospheric layers, and for the luminous auroral emissions themselves. In the terrestrial atmosphere a simple rule describing the ionizing efficiency of auroral electrons is that it requires about 35 eV of kinetic energy in the incident electron to produce an electron-ion pair. So a 1 keV incident auroral electron produces on average 30 electron-ion pairs along its path.

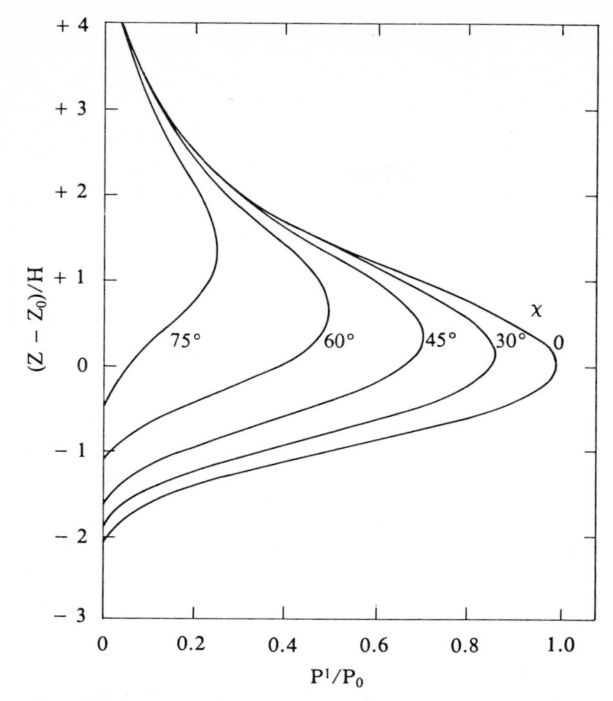

Fig. 4.16. Vertical profile of the electron-ion pair production in the upper atmosphere by solar UV radiation, for various values of the solar zenith angle. (The altitude is normalized to the scale height H, and the production curve to its maximum value P_0 for vertical solar incidence.) [After R.W. Schunk: "The Terrestrial Atmosphere", *Solar-Terrestrial Physics,* ed. by R.L. Carovillano, J.M. Forbes, (D. Riedel Publishing Company 1983)]

Figure 4.17 shows the vertical production profile for an energetic electron incident on the terrestrial atmosphere as a function of its energy. It will be seen that production falls very rapidly below its maximum value, and that, as expected, this maximum is lower the higher the incident energy. The energy values chosen here are characteristic of terrestrial auroral electrons, so it will be seen that the most intense, terrestrial auroral layers occur between 100–150 kilometres.

Following the primary production of ions by ionizing agents, ion-ion or ion-molecular chemical reactions generally produce secondary ionic species. It is obviously necessary to include all these chemical reactions, which are specific to each individual planetary atmosphere, in the function describing the overall production of each ionic species.

The Loss Term

Each ion in the ionosphere has a finite lifetime: either it is destroyed by ionic chemical reactions, or after a certain time it recombines with an electron, returning to the neutral state. For a specific electron-ion pair, the efficiency of recombination mechanisms is essentially governed by the way in which the energy released by the recombination (binding energy of the captured electron) is converted into another form of energy.

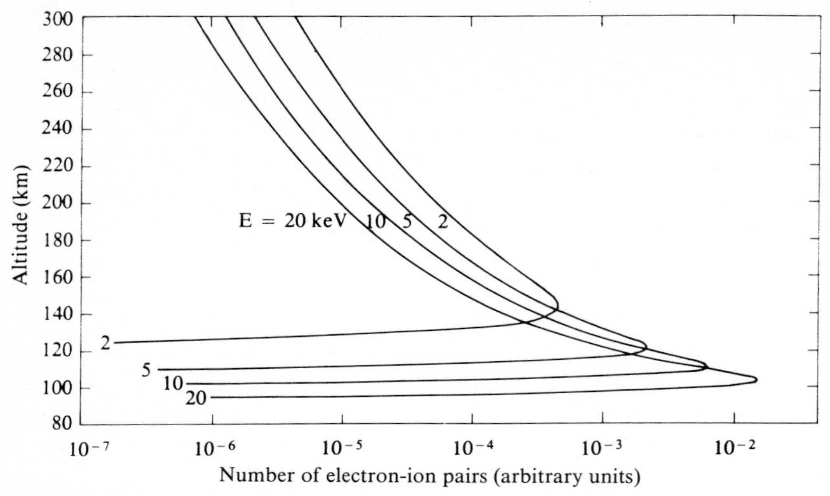

Fig. 4.17. Vertical profile of electron-ion pair production in the upper atmosphere by energetic electrons preciptated along the lines of magnetic force, for various values of the incident electrons' initial energy [After M.J. Berger, S.M. Selzer, K. Maeda: "Energy deposition by auroral electrons in the atmosphere", Journal of Atmospheric and Terrestrial Physics **32**, 1015 (© Pergamon Journals Ltd. 1970)]

So, in general, for ionized atoms (such as atomic oxygen or hydrogen) that require radiative recombination where the binding energy is released by the emission of a photon, direct recombination has a very low efficiency. The recombination chain therefore passes through a preliminary exchange of charge with another ion. For example:

$$H^+ + O \rightarrow O^+ + H \tag{4.19}$$

which is the principal proton-loss reaction. Another chain may occur through a chemical reaction producing a molecular ion, such as

$$O^+ + N_2 \rightarrow NO^+ + N \tag{4.20}$$

which is the principal reaction removing atomic oxygen ions from the terrestrial atmosphere. In both these cases the loss term L_j is a linear function of the concentration n_j.

For molecular ions, recombination with electrons is much faster than it is for atoms. The energy released generally breaks the molecular bond and creates free radicals or excited atoms. Thus

$$O_2^+ + e^- \rightarrow O^* + O \tag{4.21}$$

is the principal reaction removing molecular oxygen from the terrestrial ionosphere. Here the loss term is proportional to the product $n_j n_e$ i.e. to n_e^2 if the ion under consideration predominates.

Plasma Diffusion in the Neutral Gas: the Vertical Ionospheric Structure

The different components of the ionospheric plasma are simultaneously dynamically coupled both to the neutral fluid of the thermosphere, which tends to entrain it with its own motion by means of friction, and also to the magnetic field, which tends to "trap" each charged particle, constraining it to gyrate in a circular orbit around the same line of force. The competition between these two effects entirely determines the dynamic behaviour of ionospheric ions and electrons. It may be characterized, for each charged species j by the dimensionless quantity

$$r_j = \nu_{j\mathrm{n}}/\Omega_j \tag{4.22}$$

the ratio of the collision frequency of the species j with the neutral gas, $\nu_{j\mathrm{n}}$ (in the sense of momentum transfer), to its gyro-frequency Ω_j. When r_j is greater than one, entrainment by the neutral flow is dominant. When it is less than one, trapping by the magnetic field predominates. All the plasma-transport properties are therefore strongly anisotropic.

For every value of r_j, each charged species simultaneously diffuses through the neutral gas and across the lines of force in response to the different forces applied: electric field, pressure gradients, and gravity. The behaviour is described in detail in Appendix 1, where it is shown that the effects of the various external forces on the path of each charged species may be described by a mobility operator, which may be easily expressed in the frame of the centre of mass of the neutral particles and resolved into components acting along the axis of the magnetic field and in the perpendicular plane. For horizontally-stratified ionospheric layers, in the absence of electrical currents, ions and electrons experience the so-called "ambipolar diffusion" (see Appendix 1)

$$V_{jz} = -D_{\mathrm{a}} \sin^2 I \left[\frac{1}{n_{\mathrm{e}}} \frac{dn_{\mathrm{e}}}{dz} + \frac{1}{H_i} + \frac{1}{T_{\mathrm{e}} + T_i} \frac{d(T_{\mathrm{e}} + T_i)}{dz} \right] \tag{4.23}$$

where D_{a} is the coefficient of ambipolar diffusion, inversely proportional to $\nu_{j\mathrm{n}}$, and

$$H_i = \frac{k\,(T_{\mathrm{e}} + T_i)}{m_i g} \tag{4.24}$$

is the plasma scale height. This result will allow us to finally arrive at the vertical structure of the ionospheric layers. We take the continuity equation (4.15) and introduce expression (4.23) for the vertical velocity of the ions. For an equilibrium regime and a layer with horizontal stratification, this reduces to

$$\frac{d\,(n_j V_{jz})}{dz} = Q_j - L_j \quad . \tag{4.25}$$

Let us assume that, as is generally the case, the temperature-gradient term is negligible relative to the concentration gradient. The diffusion term (on the left of the equation) varies vertically as D_{a}, that is, inversely proportional to $\nu_{j\mathrm{n}}$ and therefore to the density of the neutral atmosphere. The loss term L_j behaves very differently. If a molecular species is concerned, it becomes a quadratic function of n_j. As we have seen, if it is an atomic species the loss is not by direct recombination but by production of a molecular ion from a neutral molecule. In every case the loss-rate is either independent of the neutral concentration or a monotonically increasing function of the latter which decreases exponentially with altitude.

Comparison of the loss and diffusion terms reveals the following effect. In the lowest layers of the ionosphere, the diffusion term is small and thus can be neglected relative to L_j. The continuity equation reduces to the local equation $Q_j = L_j$, which simply expresses *photochemical equilibrium*.

In the higher layers, the decrease in the concentration of neutral particles does increase the coefficient of diffusion but also decreases the production and loss rates (the lifetime of an ion at the top of the ionosphere may attain several hours or several tens of hours). The continuity equation reduces, for an equilibrium regime, to cancellation of the diffusion term. We have:

$$n_j(z) = L_j(z_0) \exp\left[-(z - z_0)/H_p\right] \tag{4.26}$$

as shown in Appendix 1. Each ion tends towards hydrostatic equilibrium with a scale height H_p; this is *diffusive equilibrium*.

The existence of these two equilibrium regimes at different heights considerably influences the vertical structure of the layers. Figure 4.18 shows this for the terrestrial ionosphere at middle latitudes.

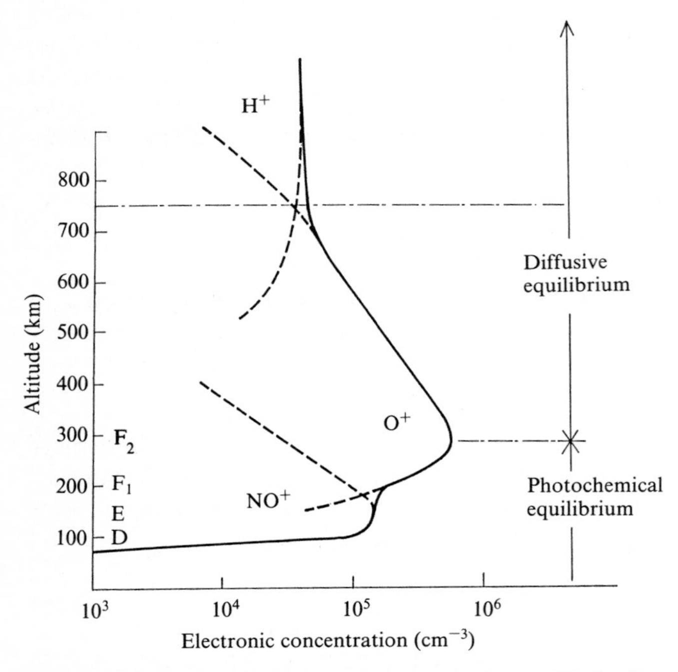

Fig. 4.18. Typical shape of the ion-concentration profile in the terrestrial ionosphere. In the lower ionosphere, the profile of the ionized NO^+ molecule layer is typical of photochemical equilibrium, as is the bottom of the O^+ ion layer (below the maximum). The upper portion of the O^+ ion layer, as well as the layer of protons that "floats" above it are, on the other hand, typical of diffusive equilibrium. [After R.W. Schunk: "The Terrestrial Atmosphere", *Solar-Terrestrial Physics,* ed. by R.L. Carovillano, J.M. Forbes, (D. Reidel Publishing Company 1983)]

The Ionospheric Conductor

As the diffusion rate of ions and electrons through the neutral gas is a function of their respective collision ratios r_i and r_e, and of the sign of their charges, the presence of an electric field $\boldsymbol{E}$, or of an electromotive force $\boldsymbol{V}_n \times \boldsymbol{B}$, caused by the motion $\boldsymbol{V}_n$ of the neutral gas across the lines of magnetic force, induces an electric current in the plasma. The value of this is given by the ionospheric Ohm's law (established in Appendix 1):

$$\boldsymbol{j} = \sigma_{||}(\boldsymbol{E} \cdot \boldsymbol{b}) \cdot \boldsymbol{b} + \sigma_P \boldsymbol{E}'_{\text{perp}} + \sigma_H(\boldsymbol{E}'_{\text{perp}} \times \boldsymbol{b}) \tag{4.27}$$

where

$$\boldsymbol{E}' = \boldsymbol{E} + \boldsymbol{V}_n \times \boldsymbol{B} \tag{4.28}$$

and where $\sigma_{||}$, σ_P and σ_H are, respectively, the parallel, Pedersen and Hall conductivities. The parallel conductivity $\sigma'_{||}$, essentially carried by the electrons, is always much greater than the perpendicular conductivities, which are limited by the mobility of the ions.

In ionospheres without strong magnetic fields (Venus and probably Mars), the conductivity is, of course, isotropic and equal to $\sigma_{||}$. The ionosphere is then a very good conductor that surrounds the planet and (as we shall see in Sect. 4.3), opposes penetration by the interplanetary magnetic field.

In the ionospheres of strongly magnetic planets, the electrodynamic properties of the ionosphere are just as important, as was indicated in Fig. 4.9. Figure 4.19 shows a vertical profile of the typical ionospheric conductivities for the terrestrial daytime ionosphere. It will be seen that

1. the layer of strong conductivities perpendicular to $\boldsymbol{B}$ forms a conducting shell, several times the neutral scale-height in thickness, that surrounds the Earth at the base of the ionosphere. This is the ionospheric *dynamo layer*;
2. this highly conductive layer, transverse to the lines of magnetic force, is connected electrically along these lines of force to regions at very high altitudes – i.e., as we shall see, to the magnetosphere and to the solar wind – by virtue of the very high values of parallel conductivity;
3. the ionosphere is therefore electrodynamically connected to the magnetosphere and the solar wind.

4.3 The Interaction of the Solar Wind with Solar-System Bodies

4.3.1 The Different Types of Interaction

The objects in the Solar System present various types of physical obstacle to the radial flow of the solar wind, which strikes them at a velocity that is simultaneously supersonic and well above the Alfvén velocity, as we have seen in Sect. 4.1. With a velocity of 400 km/s, a concentration of 5 protons/cm^3, and a temperature of 10^5 K, the solar wind has a Mach number and an Alfvén Mach number that are both well above unity.

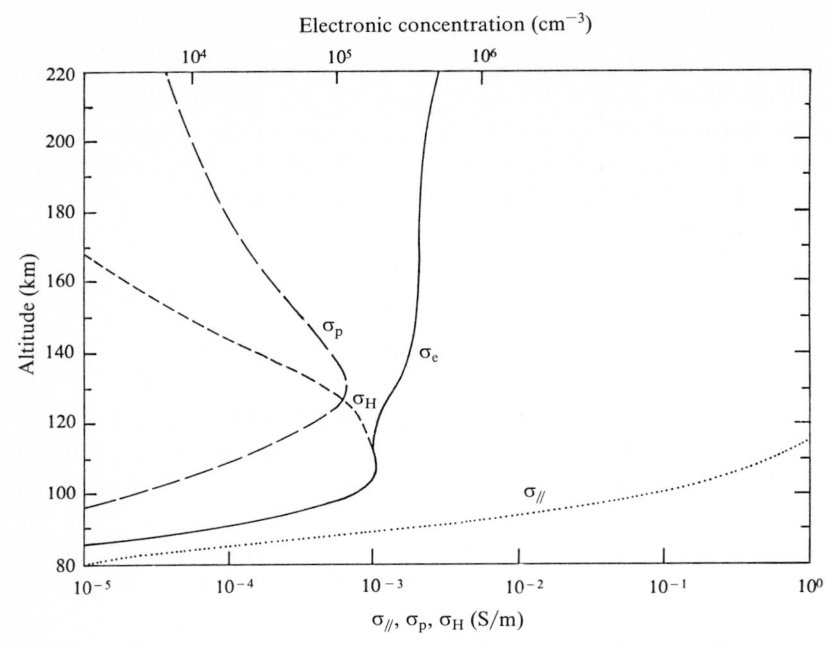

Fig. 4.19. Vertical profiles of the parallel $\sigma_{\|}$, Hall σ_H, and Pederson σ_P conductivities, for typical daytime conditions at middle latitudes. The electronic concentration profile is also shown on a different scale. It will be seen that the parallel conductivity is higher (by several orders of magnitude) than the Hall and Pedersen conductivities at all ionospheric altitudes. The layer where the perpendicular conductivities are significant extends from about 90 to 160 km in height. This is the "dynamo layer" in the terrestrial atmosphere [After A.D. Richmond: "Thermospheric Dynamics and Electrodynamics", *Solar-Terrestrial Physics*, ed. by R.L. Carovillano, J.M. Forbes (D. Reidel Publishing Company 1983)]

The physical nature of the obstacles encountered is very varied. But, briefly, it may be said to depend on three main characteristics:

1. the existence of an atmosphere around the planetary body, and the pressure existing in the ionized component (the ionosphere) that has been created in that atmosphere by the ionizing UV- and X-radiation from the Sun;
2. the magnetic pressure produced in the region surrounding the central body by any intrinsic magnetic field;
3. the conductivity of the central body (liquid or solid). This last characteristic only really becomes significant if the sum of the ionospheric pressure and the magnetic pressure does not suffice to balance the total (i.e. dynamic, thermal and magnetic) pressure of the solar wind. For supercritical flows such as the solar wind, moreover, it is the dynamical pressure term that is predominant, and so the different pressure terms presented by the obstacle should be compared with it.

The four principal types of interaction are shown schematically in Fig. 4.20. In cases (a) and (c), the solar wind is directly incident on the central body. In (b) it encounters the atmosphere, and in (d) the external magnetic field.

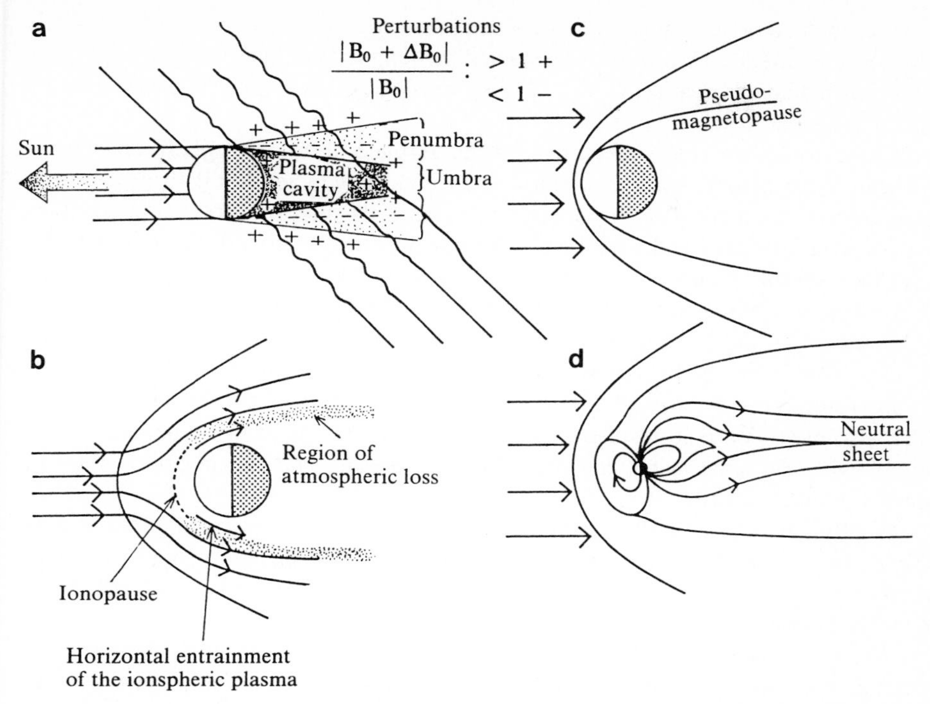

Fig. 4.20. The four principal forms of the interaction between the solar wind and a planetary body. [After R.P. Lepping: "Magnetic Configurations of Planetary Obstacles", *Etude comparative des systèmes magnétosphériques* (CNES/Cepadues 1986)]

The first and third cases arise, of course, in the absence of an atmosphere or an intrinsic magnetic field.

In (a), the central body is an insulator, absorbing the ions of the solar wind. The best-known example is that of the Moon. The lunar surface absorbs ions directly from the solar wind and lunar material overall is such a poor conductor that the interplanetary magnetic field penetrates it by diffusion. Neither the magnetic lines of force nor the solar-wind plasma can therefore accumulate on the obstacle's upstream side, which remains free of any perturbations. Downstream, on the other hand, a cavity that is practically free of plasma is formed. It is progressively filled by diffusion of ions and electrons as one moves farther downstream. As the electrons diffuse more rapidly because of their low mass, a sheath of negative charge tends to form at the boundaries of this conical cavity.

In all the other cases, there is no simultaneous absorption of the solar wind by the obstacle and diffusion of the magnetic lines of force across it, so there is an accumulation of either magnetic flux, or plasma, or generally both, upstream of the obstacle. In flows with Mach numbers well above 1, like the solar wind, this leads to the formation of a stationary shock-wave ahead of the obstacle, across which the plasma is locally compressed and decelerated, passing into a regime of subsonic and sub-Alfvén flow-rates. In this regime it may be diverted to flow around the obstacle. The dynamical pressure ahead of the shock is converted to thermal and magnetic pressures behind it. In the region between the shock and the obstacle, which is known as the *magnetosheath*, the plasma and the magnetic field are compressed, with a resulting adiabatic heating of the plasma and the accumulation of interplanetary lines of magnetic force.

The case shown in (c) may arise, in the absence of an atmosphere and intrinsic field, if the central body is a sufficiently good conductor for it to prevent the interplanetary magnetic flux from penetrating into its interior. The magnetic lines of force would be bent and stretched in the magnetosheath around the central body. Up to the present, none of the Solar-System objects that have been explored seems to show this behaviour, but its existence remains theoretically possible.

In case (b), the central body has a dense atmosphere, but no magnetic field: the solar wind interacts directly with the upper atmosphere. This is the situation that prevails for comets, for Venus, and probably also for Mars. Saturn's satellite Titan, which also has a dense atmosphere, undergoes an interaction of this type, alternately directly with the solar wind and then with the plasma and the magnetic field in Saturn's magnetosphere.

Finally, in case (d), the central body has an intrinsic magnetic field that is sufficiently intense for it to deflect the solar wind. The planetary magnetic field therefore creates a cavity in the solar wind, to which it is essentially confined: this is the magnetosphere. Mercury, Earth, Jupiter, Saturn and Uranus have magnetospheres.

We shall now describe in greater detail cases (b) and (d), which are the most common, but which are also the most interesting from the point of view of the physical mechanisms that they bring into play.

4.3.2 The Interaction of the Solar Wind with Gaseous, Non-magnetized Envelopes (Case b)

Comets

The most spectacular case of the interaction of the solar wind with a Solar-System body is undoubtedly that of comets (see also Sect. 10.5). An outline of this interaction, as derived from astronomical observation and theoretical considerations, is given in Fig. 4.21. As it approaches the Sun, the cometary nucleus, which is a body only a few kilometres in diameter and primarily made of ice and various forms of dust, undergoes sublimation at its surface as a result of solar heating. The gaseous envelope thus created, the coma, expands supersonically in all directions. The neutral gas is subject to two different processes.

1. Being neutral and therefore unaffected by magnetic forces, it partly diffuses into the solar wind itself. Photometric observations have shown that the dimensions of the cloud of neutral gas that diffuses out into the solar wind may even reach 10^6 km for the neutral atomic hydrogen produced by photodissociation of cometary water. In other words, the size of the cometary obstacle is not related to the dimensions of the nucleus. The portion of the neutral cometary gas that diffuses out into the solar wind is gradually ionized by two main mechanisms: charge exchange between the protons in the solar wind and the cometary atoms; and "critical" ionization, a mechanism studied

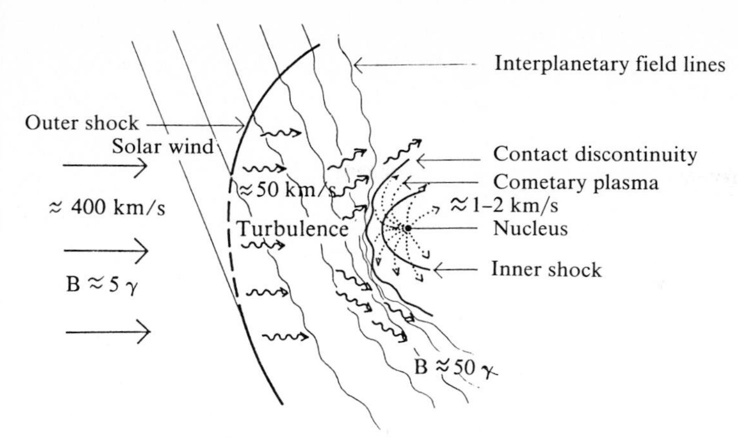

Fig. 4.21. Simplified diagram of the plasma flow and the principal discontinuity surfaces that appear in the interaction between the solar wind and a comet, based solely on observations from a distance and theoretical models. [After J.C.Brandt, D.A. Mendis: "The Interaction of the Solar Wind with Comets", ed. by C.F. Kennel, L.H. Lanzerotti, E.N. Parker, *Solar System Plasma Physics,* Vol. II (North-Holland Publishing Company 1979)]

by Alfvén in 1960. Critical ionization arises when a magnetic plasma and neutral gas interpenetrate at a sufficiently high velocity that the kinetic energy of the neutral atoms, relative to the plasma, becomes greater than, or equal to, their ionization energy. These ionization processes result in the light, but high-kinetic-energy ions of the solar wind being replaced or joined by heavy cometary ions with initially low kinetic energies. As soon as the latter are ionized, they are accelerated by the solar wind's electric field and reach its rate of flow.

The momentum that they thus gain is, of course, extracted from that of the solar wind. The effect is to slow the latter down far ahead of the coma, and to load it with heavy cometary ions. The deceleration is probably insufficiently effective to reduce the flow to sub-Alfvén conditions, which probably accounts for the formation of an outer shock. Inside the shock, the magnetic lines of force, slowed down by the effect of the cometary ions, start to bend round the obstacle.

2. In the coma itself, a fraction of the neutral gas is photoionized directly by the solar EUV- and X-radiation, and then by various ionic chemical reactions. This results in the formation of a highly conducting cometary ionosphere, having a low permeability to the external magnetic field, which is therefore swept up with the neutral gas in the expansion of the coma.

In this scenario, the magnetic plasma of the solar wind (contaminated and loaded with cometary ions) and the unmagnetized plasma of the cometary ionosphere will be separated by a contact discontinuity (the cometary ionopause) that is impermeable to the magnetic field.

The solar magnetic lines of force cannot penetrate the ionospheric conductor and become wrapped round the obstacle. This is undoubtedly the origin of the plasma tail formation, as proposed by Alfvén. Figure 4.22 shows how one can understand

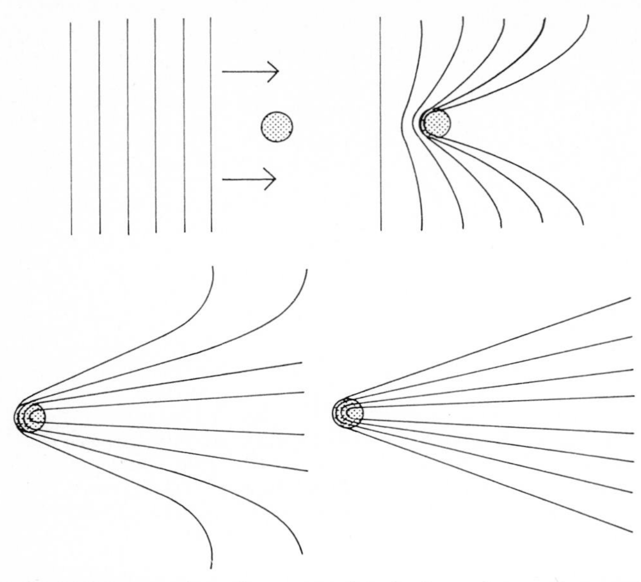

Fig. 4.22. Interaction of a comet with the interplanetary magnetic field explains the formation of the comet's plasma tail and the motion of tail rays. [After H. Alfvén: "On the Theory of Comet Tails", Tellus 9, 92 (1957)]

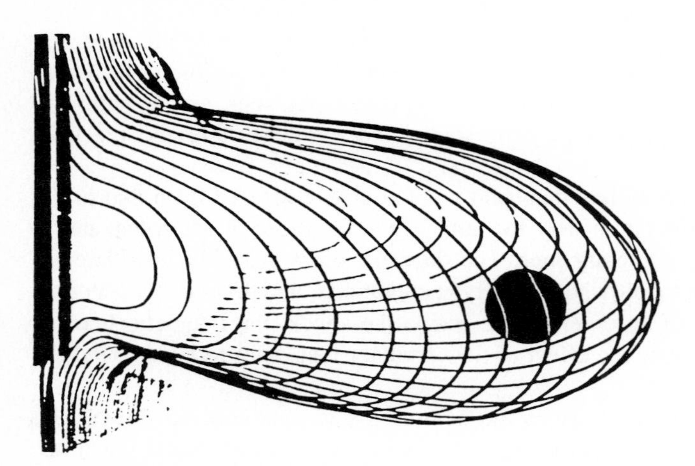

Fig. 4.23. A three-dimensional representation of the lines of magnetic force present in the interaction of the solar wind and a comet, as derived from laboratory simulations. [I.M. Podgorny, E.M. Dubinin, P.L. Israelvich: Moon and Planets **23**, 323 (1980). By courtesy of D. Reidel, publishers]

the formation of the radial structure in cometary tails by this folding mechanism, and the probable three-dimensional structure is shown in Fig. 4.23 (derived from laboratory simulations carried out by Soviet researchers of the interaction between a conductor and a jet of magnetized plasma). Observations do indeed reveal that cometary tails often have a structure consisting of *rays*, radially placed with respect to the nucleus, that are downstream of the comet in the solar wind. These rays often

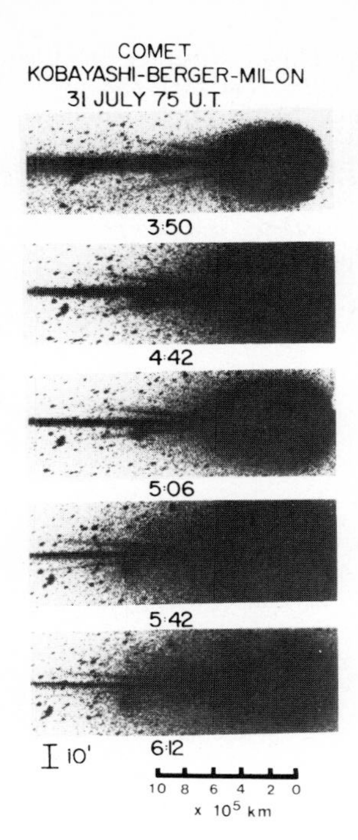

Fig. 4.24. Successive photographs of Comet Kobayashi-Berger-Milon, showing the movement of the cometary rays towards the axis of the tail. (M.B. Niedner: "Magnetic Reconnection in Space and Laboratory Plasmas", Geophysical Monograph **30**, © American Geophysical Union, 1984.)

occur in symmetrical pairs that over the course of time fold back towards the axis of the tail (Fig. 4.24), exactly as if they were tracers of interplanetary magnetic field lines bending round the obstacle (see Sect. 10.5).

In March 1986, five interplanetary probes encountered Comet Halley: the VEGA 1 and 2 Soviet spacecraft, ESA's GIOTTO spacecraft, and the two Japanese interplanetary probes, SAKIGAKE and SUISEI. In September of the previous year, NASA's ICE (International Cometary Explorer) spacecraft had made the first encounter with a cometary environment, crossing the tail of Comet Giacobini-Zinner. These encounters did not show any strong contradictions between in-situ measurements and the predictions by the models just described. But they revealed the complexity and variability of the cometary environment. The outer shock region is certainly much more complicated than was previously anticipated. In the case of Halley, at least three successive discontinuities were identified in the region where interplanetary magnetic-field lines progressively pile up on the obstacle presented by the comet. The contact discontinuity was also very clearly observed by GIOTTO in all the bulk parameters of the flow, and in the magnetic field intensity, which dropped to nearly zero inside it. The structure of this discontinuity seems to result from a complex dynamical and chemical interaction between the plasma, the neutral gas, and the magnetic field, which is still (in 1989) being analyzed.

Venus

The solar wind's interaction with the atmospheres of planets without a magnetic field is essentially known only from the example of Venus, because the case of Mars

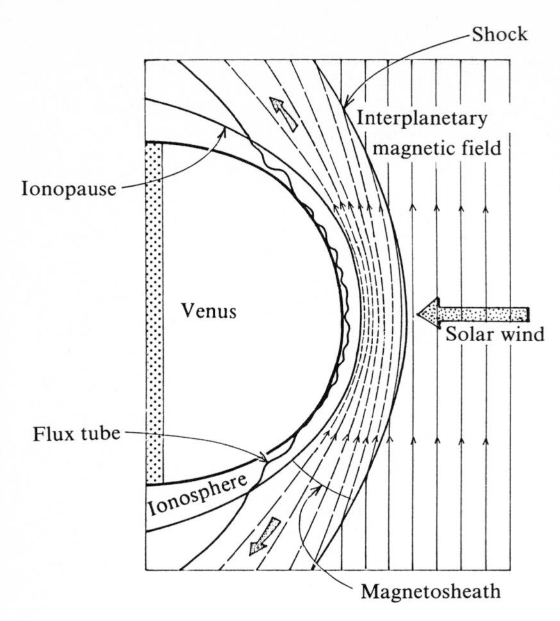

Fig. 4.25. Simplified diagram of the solar-wind interaction with the sunlit side of Venus. [After M.A. Saunders, C.T. Russell, J.G. Luhmann: "Interaction with Planetary Ionospheres and Atmospheres: A review", *Etudes comparative des systèmes magnétosphériques* (CNES/Cepadues 1986)]

is very much more poorly documented even after the Phobos mission, and that of Titan is known only from the single fly-by of Saturn by Voyager 2.

The interaction of Venus with the solar wind has been studied by more than twenty interplanetary missions, in particular by Pioneer Venus Orbiter, which has been in orbit around the planet since 1978. The geometry, shown in Fig. 4.25, is very similar to that of comets: behind the outer shock front, which decelerates the solar wind to a sub-Alfvénic velocity, the lines of force and the solar plasma accumulate on the sunlit side of the planetary ionosphere. They are separated from the ionosphere by a distinct boundary, the *ionopause*, which is poorly permeable to the external magnetic field.

The two interaction mechanisms identified in comets also apply here:

1. there is loading of the solar wind upstream of Venus by diffusion of neutral components from the planet's upper atmosphere (principally consisting of atomic oxygen), although the importance of this effect is weakened by the gravitational attraction linking the atmosphere to the planet, and which causes the concentration to decrease rapidly with altitude;
2. there is direct contact between the planetary ionosphere and the compressed solar wind in the magnetosheath across the ionopause, along which the total pressure (magnetic and particle) of the two plasmas is in equilibrium. Here this pressure equilibrium essentially reduces to equality between the magnetic pressure in the magnetosheath and the pressure of the ionospheric plasma. Figure 4.26, which shows three vertical profiles of the magnetic field and of the ionospheric plasma concentration measured by Pioneer Venus on three passes across the dayside hemisphere of Venus, nicely illustrates the structure of the ionopause and also its considerable variation in altitude and thickness

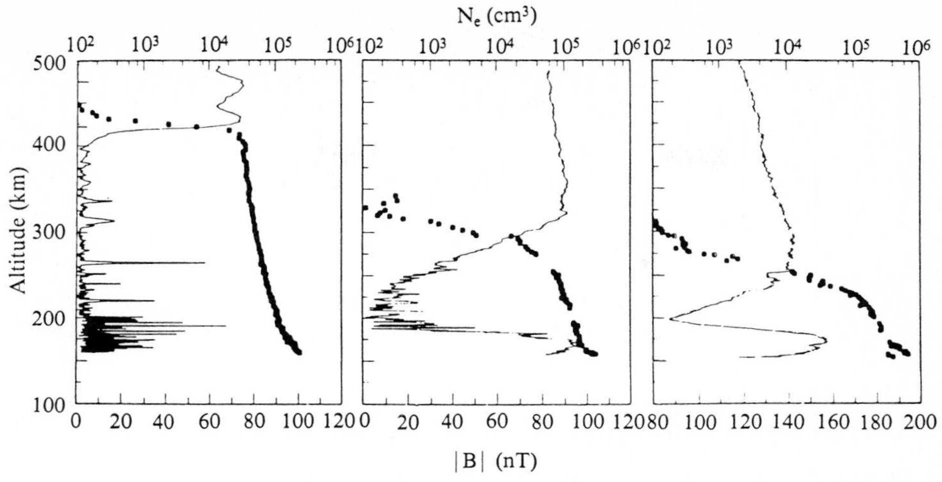

Fig. 4.26. Three altitude profiles of the intensity of the magnetic field (fine lines) and of the electron concentration (black square points) as measured by NASA's Pioneer Venus Orbiter on the illuminated face of Venus, for three different values (increasing from left to right) of the dynamic pressure of the solar wind. [After C.T. Russell, J.G.Luhmann, R.C. Elphic, M.Neugebauer: *Comets*, ed. by L.L. Wilkening, (© University of Arizona Press, Tucson 1982). By courtesy of the publishers]

as a function of the pressure of the solar wind. When the latter increases (from left to right in the figure), the altitude of the ionopause decreases considerably (from 420 to 250 kilometres in this example), and its thickness simultaneously goes from a few kilometres to more than fifty. Current theories still account only poorly for this effect. A possible suggestion, however, is that when the altitude of the ionopause decreases, the effect of collisions with the neutral particles at that altitude increases, which favours diffusion of the plasma and magnetic field in altitude.

Figure 4.26 shows clearly that the decrease in altitude of the ionopause also has a spectacular effect on the permeability of this barrier to the external magnetic field. When the ionopause is located at high altitudes (left-hand panel), only very localized points of the magnetic field (a few kilometres in thickness) persist in the ionospheric layer. These structures are organized spatially like "ropes" of magnetic flux. Although no definitive proof has yet been put forward, the formation of these flux ropes on Venus has been attributed to the development of a Kelvin-Helmholtz instability, excited by the flow of plasma from the magnetosheath onto the ionopause.

When the ionopause is at a low altitude (central and right-hand panels), the external magnetic field appears to penetrate into the ionosphere, and even to accumulate in the lower ionosphere, where a layer forms having a magnetic field comparable in intensity to that of the magnetosheath.

As with comets, the accretion of interplanetary magnetic flux upstream of the planet leads to progressive folding of the magnetosheath's magnetic tubes of force around the planet farther down the flow, as shown in Fig. 4.27. Immediately behind the obstacle, therefore, a cavity is formed that is almost empty of plasma, and which

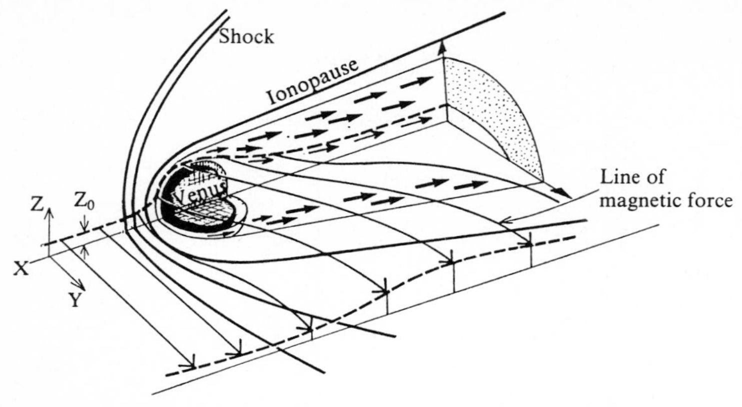

Fig. 4.27. Illustration of the trajectory and deformation of interplanetary lines of magnetic force around Venus. This shows the progressive changes in a single line of magnetic force that passes "above" Venus. [After M.A. Saunders, C.T. Russell, J.G. Luhmann: *Comets,* ed. by L.L. Wilkening (© University of Arizona Press, Tucson 1982). By courtesy of the publishers]

gradually fills farther downstream through the expansion of the tubes of force, and also through diffusion along these force tubes. A few planetary radii downstream of the planet, this complete fold in the magnetic flux tubes leads to the configuration shown in Fig. 4.28. Two magnetic lobes are formed with opposite magnetic polarities, separated by a neutral sheet. We may therefore expect the plane of the neutral sheet to contain the Sun-planet axis and be orthogonal to the plane formed by this direction and that of the interplanetary magnetic field. This has been confirmed by a statistical study of the magnetic-field data returned by the Pioneer Venus mission.

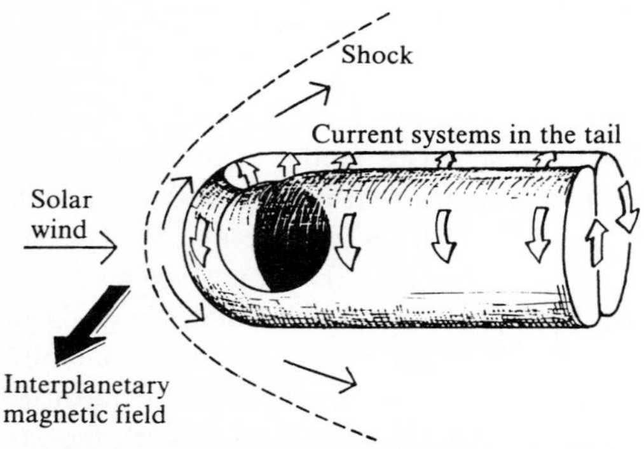

Fig. 4.28. The folding of the lines of magnetic force around the obstacle shown in Fig. 4.27 leads to the formation, well downstream of Venus, of a magnetic tail, consisting of two magnetic lobes of opposite polarities, separated by a neutral sheet. [After K.I. Gringauz: "The Bow Shock and the Magnetosphere of Venus According to Measurements from Venera 9 and 10 Orbiters", *Venus,* ed. by D.M. Hunten, L. Colin, T.M. Donahue, V.I. Moroz (© University of Arizona Press, Tucson 1983)]

4.3.3 The Interaction of the Solar Wind with Planetary Magnetic Fields (Case d)

In the case of planets that possess an intrinsic magnetic field the pressure of which is higher than that of the outermost atmospheric layers (Mercury, Earth, Jupiter, Saturn and Uranus), it is this magnetic field that becomes the obstacle deflecting the solar wind. It makes a magnetic cavity in the solar wind that is, to a first approximation, empty. This is the *magnetosphere*, the shape of which was calculated in the 1960s. Two discontinuity surfaces again appear (Fig. 4.20 d): the *bow shock* and the *magnetopause*, which is a tangential discontinuity separating the decelerated solar wind from the planetary magnetic cavity. The external flow and the form of the boundaries have been reproduced in a very satisfactory fashion by models that impose magnetohydrodynamic equilibrium at the magnetopause, and calculate the external flow by classical hydrodynamic formulae: these models lead to the concept of a closed magnetosphere.

The Closed Magnetosphere

When the permanent existence of the solar wind and of the terrestrial magnetosphere was discovered at the beginning of the 1960s, thanks to the first lunar and planetary spaceprobes, it was immediately obvious that the terrestrial magnetic field, far from decreasing regularly with geocentric distance, ceased at an abrupt discontinuity that marked the limit of its extension out into space: this is the *magnetopause*.

The simplest method of calculating the balance of forces at this discontinuity surface is to consider the solar wind initially as a monokinetic jet of individual nonmagnetic particles that do not collide with one another. Because of the effects of the planetary magnetic field, each particle then undergoes specular reflection as it crosses the magnetopause (Fig. 4.29). This concept of reflection can be used to determine the pressure balance across the boundary. This leads to a very simple condition for magnetostatic equilibrium:

$$2\varrho V_s^2 \cos^2 \chi = \frac{B_T^2}{2\mu_0} \tag{4.29}$$

the derivation of which is given in Appendix 2.

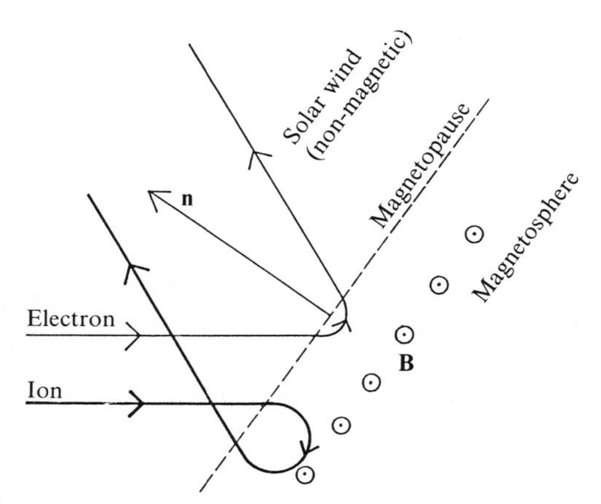

Fig. 4.29. The specular reflection at the magnetopause of a charged particle in the solar wind (which is here treated as a non-magnetic jet of electrical charges) results from the magnetic gyration of the incident particle in the magnetospheric field. This gyration produces net displacements along the magnetopause in opposite directions for ions and electrons, and thus results in an electric current tangent to the magnetopause and orthogonal to the magnetosphere's magnetic field. It is precisely this electric current that isolates the solar wind from the planetary magnetic field

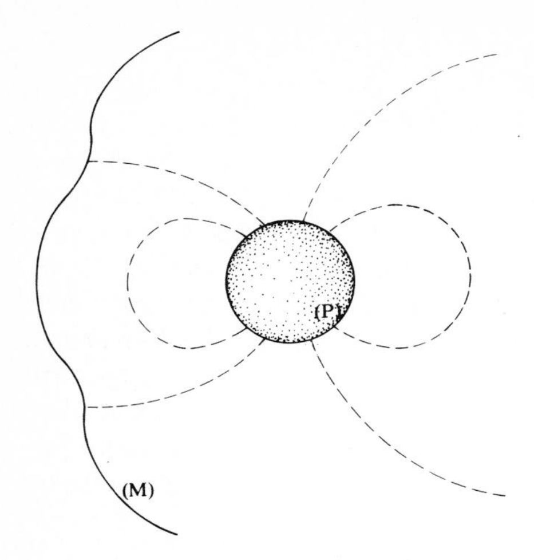

Fig. 4.30. Geometry of the free-boundary problem defining the form of the magnetopause (M) associated with the source of magnetic flux emerging from the surface (P) of the magnetized planetary body (the broken lines represent the magnetic field lines, and also the magnetic equipotentials of the function Φ to be calculated)

In this expression ϱ and V_s are the concentration and velocity of the jet of particles, χ is the angle between the velocity vector and the normal to the magnetopause, and B_T is the tangential component of the planetary magnetic field at the magnetopause.

This expression may be used to calculate the global geometry of the magnetopause and the distribution of the field within the planet's magnetic cavity by solving the following boundary-free problem (the geometry of which is shown in Fig. 4.30):

Find a function Φ that has values within the region (I) separating the planetary surface (P) from a discontinuity surface (M), the magnetopause, such that:

$$\boldsymbol{B} = -\nabla\Phi \quad . \tag{4.30}$$

The effects of electric currents in the magnetic cavity have been neglected except, of course, that of the planet's internal currents responsible for its intrinsic field.

1. B_n, the component of $\boldsymbol{B}$ normal to the surface (P), is set equal to the value given by the internal source of the magnetic field (for example, one can take for B_n the value given by the planetary magnetic dipole); (4.31a)

2. $B^2 = 4\mu_0 \varrho V_s^2 \cos^2\chi$ on (M); (4.31b)

3. $B_n = 0$ on (M) (the magnetopause is impermeable to magnetic flux). (4.31c)

The Chapman-Ferraro Model

In 1940, Chapman and Ferraro obtained a solution (erroneous but approximately correct) to this problem by using the image method. They assumed that the interface between the solar wind and the magnetosphere was plane and perpendicular to the Earth-Sun direction. In order to construct a solution that meets (4.30) and (4.31c) for the magnetospheric side of the interface (M), it is sufficient to superimpose on the dipolar terrestrial magnetic field that of the dipole imaged relative to the interface.

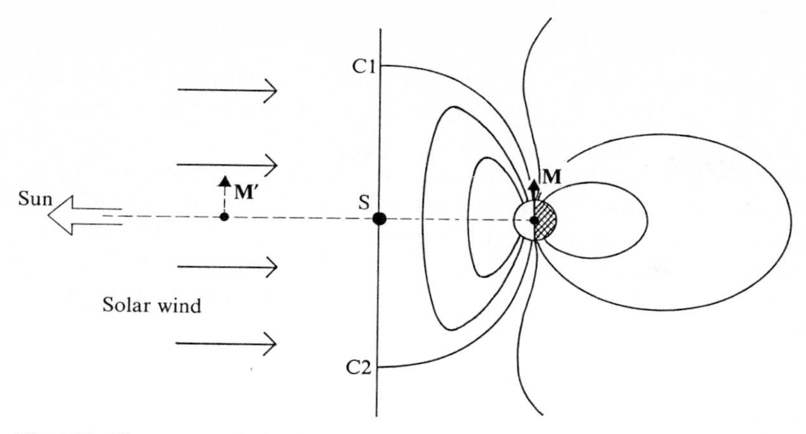

Fig. 4.31. The geometrical configuration of the lines of magnetic force produced by the solar-wind/magnetosphere interaction in the simplified Chapman-Ferraro model. Two neutral points (where the field is cancelled out), appear on the magnetopause in the plane containing the axis of the planetary magnetic dipole and the direction of the Sun. These are the *polar cusps* C1 and C2

The resulting shape of the lines of magnetic force in the plane of the planetary magnetic meridian that contains the Sun-planet line is shown in Fig. 4.31.

For given characteristics of the solar wind, the position of the magnetopause is determined by imposing the condition that the pressures at the subsolar point S, the intersection of the magnetopause with the Sun-planet axis, are in equilibrium. (But this condition is not defined for any other point.) The intensity of the field at S is then twice that of the planetary dipole field itself:

$$B_{\mathrm{m}} = 2B_{\mathrm{e}}/L^3 \tag{4.32}$$

where $L = R/R_{\mathrm{p}}$ is the ratio of the planetocentric distance of S to the planetary radius, and where B_{e} is the intensity of the equatorial field at the planetary surface.

Equation (4.31c) then allows the position L_{M} of the magnetosphere to be determined, expressed as its planetocentric distance (measured in planetary radii):

$$L_{\mathrm{M}} = (B_{\mathrm{e}}^2/\mu_0 \varrho V_{\mathrm{s}}^2)^{1/6} \quad . \tag{4.33}$$

For typical parameters of the solar wind at 1 AU (V_{s} = 400 km/s, N = 5 cm^{-3} in the solar wind), we find that in the case of the Earth, L lies between ten and fifteen radii, which is in good agreement with observations of the terrestrial magnetopause (see Sect. 5.3.4).

The same calculation carried out for Jupiter places its magnetopause at 60 jovian radii from the centre of the planet, whereas measurements of the position of the jovian magnetopause made by the Voyager 1 and 2 spacecraft, placed it instead at about 90 jovian radii (see Sect. 7.3.1). This discrepancy is caused by the deformation of the jovian magnetic field by plasma sources within its own magnetosphere, the pressure of which was not taken into account in the procedure just described.

Three-dimensional Hydrodynamic Models

As there is no analytical solution to the problem in three dimensions, it was computer simulations in the 1960s that allowed the three-dimensional flow of the solar wind around the obstacle formed by the magnetosphere to be determined. This used existing techniques for the calculation of supersonic flows, around aircraft, for example. These calculations used several simplifying assumptions in order to avoid having to provide a general solution to the free-boundary problem that we have just formulated. Taking inspiration from the Chapman-Ferraro solution and from the pressure-balance equation at the magnetopause deduced from the individual-particle model of the solar wind (4.29), it may be assumed:

1. that the pressure of the solar wind at the magnetopause is given by $KV_s^2 \cos^2 \chi$ (K may be set to a value between 1 and 2 in order to simulate various conditions for the reflection of particles by the magnetopause);
2. that the tangential component B_T of the magnetic field at the magnetopause (on the magnetosphere side) is simply equal to twice the tangential component of the dipolar field.

With these two hypotheses, B_T at the magnetopause becomes a function of position and of χ, and the equation

$$K \varrho V_s^2 \cos^2 \chi = \frac{B_T^2}{2\mu_D} \qquad (4.34)$$

may be solved as an ordinary differential equation that determines (via χ) the shape of the magnetopause even before knowing the flow around the obstacle.

This flow is subsequently calculated by hydrodynamic methods, the shape of the obstacle to be bypassed being now determined.

The flow-lines determined by this calculation are shown in Fig. 4.32a for a plane containing the Sun-Earth line. They are drawn as continuous lines. It will be seen that the undisturbed solar-wind flow is bent round the obstacle after the bow shock; close to the Sun-Earth line (the horizontal axis in the figure), the flow becomes subsonic immediately behind the shock. Then, as shown by the lines of equal Mach number (dashed lines) and the supersonic transition line (dotted), in flowing round the obstacle the stream is re-accelerated by the high-pressure zone created behind the shock and close to the Sun-Earth line. At the sides of the magnetosphere, the flow quickly reverts to supersonic velocities behind the bow shock.

Once the flow is known, the distribution of the magnetic field in the region between the bow shock and the magnetopause (known as the *magnetosheath*), may be calculated. Figure 4.32b shows the magnetic-field lines (continuous lines) and the flow lines (broken lines) for two different orientations of the interplanetary magnetic field, which is assumed to be uniform in front of the bow shock. In the two cases shown, there is a concentration of the magnetic-field lines, and therefore of the intensity of the magnetic field, ahead of the magnetopause. The field lines have a tendency to "drape" themselves around the magnetosphere. It should be noted that the accumulated magnetic flux drains away at right angles to the plane of the figure.

Calculation of the magnetic field is made easily by using the equation for the freezing of magnetic flux in the plasma,

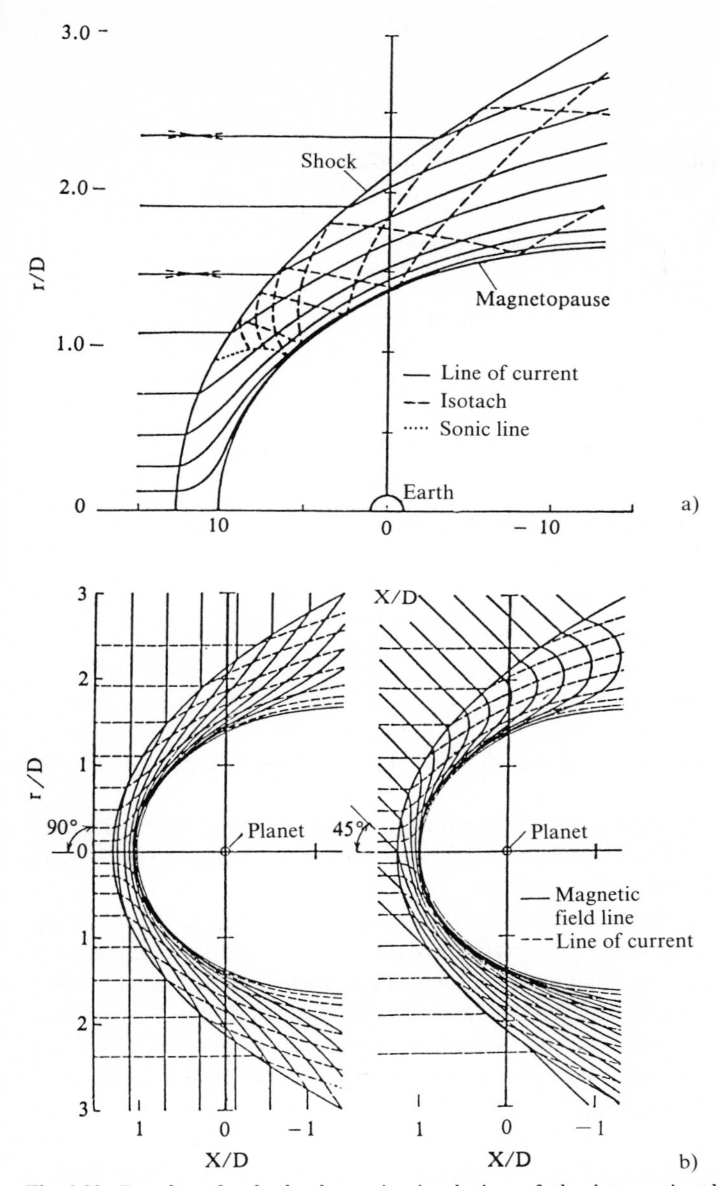

Fig. 4.32. Results of a hydrodynamic simulation of the interaction between the solar wind and the terrestrial magnetosphere, shown in section in the plane of the ecliptic. [After J.R. Spreiter, A.L. Summers, A.Y. Alkne: "Hydrodynamic Flow around the Magnetosphere", Planetary and Space Science **14**, 223 (© Pergamon Journals Ltd. 1966)]

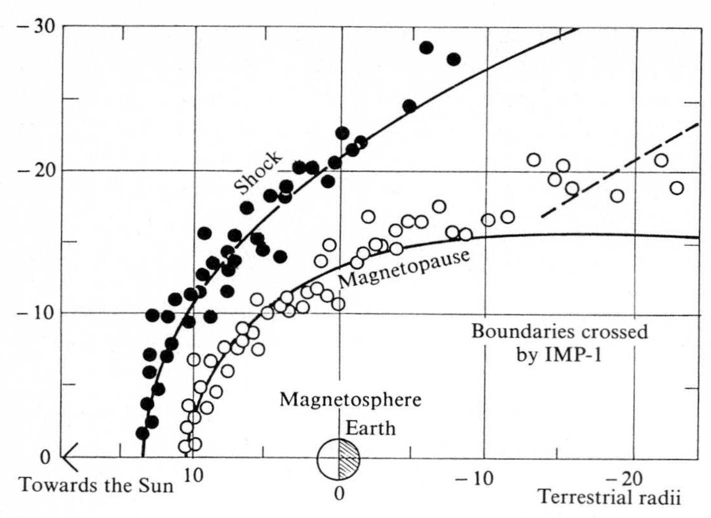

Fig. 4.33. The positions of the bow shock and the magnetopause in the plane of the ecliptic predicted by the hydrodynamic simulation (Fig. 4.32) may be adjusted to give good agreement with the boundaries as shown by satellites (here IMP-1) for similar values of the dynamic pressure of the solar wind. (The observed positions of the bow shock are marked by filled circles, and those of the magnetopause with open circles)

$$\frac{\partial \boldsymbol{B}}{\partial t} + \nabla \times (\boldsymbol{V} \times \boldsymbol{B}) = 0 \tag{4.35}$$

to transport the interplanetary magnetic field of the free-flowing solar wind to the magnetosheath.

Figure 4.33 shows that hydrodynamic simulations correctly reproduce the positions of the bow shock (filled circles in the figure), and of the magnetopause (open circles), as observed by the IMP-1 satellite.

The measurements chosen correspond to a single range of values for the dynamic pressure of the solar wind. The positions of the shock and of the magnetopause predicted for this same pressure value are shown as continuous lines. The agreement is excellent, especially on the Sun-ward side. It should be noted, however, that there is a systematic difference between the theoretical and the observed positions of the magnetopause, which increases in moving away from the Earth on the night side.

This difference, although small, illustrates one of the weaknesses of the hypothesis used in the calculation, in particular the hypothesis of complete freezing of the magnetic field into the flow made within the context of theoretically ideal MHD.

The Open Magnetosphere

Within the framework of theoretically ideal MHD, the magnetospheric cavity is by definition closed to exchange of matter, momentum, energy and magnetic flux with the solar wind. This is naturally contradicted by everyday observation of the influence that the solar wind exerts on auroral and geomagnetic activity.

We now know that a fraction of the solar-wind particle flux (about 1% of the flux, ahead of the magnetosphere, crossing an area equal to the effective cross-section of its day side), actually manages to penetrate the magnetopause. This forms an important source of plasma that feeds aurorae and the radiation belts.

As far as momentum and energy are concerned, satellite measurements in the solar wind and in geostationary orbit have shown that about 10% of the interplanetary electric field is transmitted to the interior of the magnetosphere.

Statistical studies and the analysis of geophysical events, however, show that geomagnetic activity is not equally influenced by every solar-wind parameter. The component of the interplanetary magnetic field that is parallel to the axis of the planetary magnetic dipole, known as the B_z component, appears to play the dominant role, activity being more intense the stronger the orientation of the interplanetary magnetic field towards the south (for the Earth), i.e., in the same direction as the planetary dipole.

Finally, any physical model of the interaction between the solar wind and the magnetosphere has to account for the very specific geometry of the terrestrial auroral zones, as illustrated in Fig. 4.34. This photograph was taken in the ultraviolet by a camera on board Dynamics Explorer 1, at a distance of four Earth radii. Apart from on the sunlit side of the Earth, an almost perfect luminous circle can be seen, particularly in the region of the polar night. Full analysis shows that this circle, the *auroral oval*, is centred about a point that is displaced about five degrees from the magnetic pole towards the night side. Its radius increases with geomagnetic activity and with the intensity of the B_z component of the interplanetary magnetic field; it varies overall between ten and twenty degrees.

The model that gives the best qualitative match with these observational facts is the "open magnetosphere" model proposed by Dungey in 1961. He actually had recourse to the mechanism of magnetic reconnection at the magnetopause, which by definition excludes theoretically ideal MHD. The starting point is as follows: let us take the simplest model of the magnetosphere, that by Chapman and Ferraro

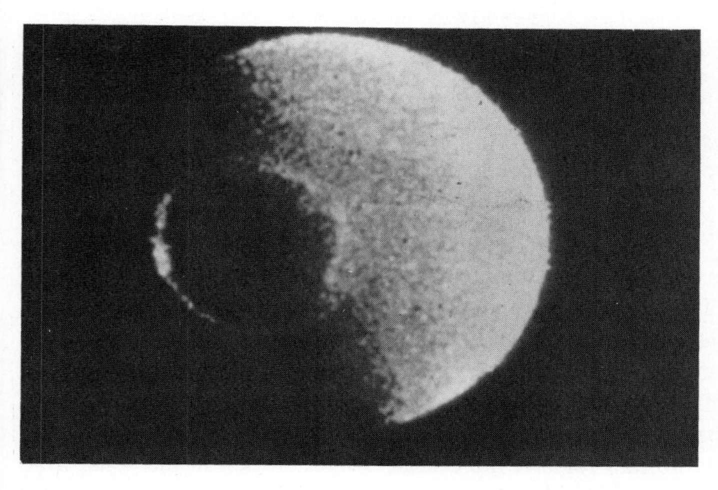

Fig. 4.34. Ultraviolet photograph of the Earth, taken by NASA's Dynamics Explorer-1 satellite, showing the ultraviolet auroral oval, which is particularly distinct on the night side. (By courtesy of NASA)

(Fig. 4.31), and imagine that a vertical magnetic field, directed towards the bottom, exists in the solar wind on the left-hand side of the magnetopause. The presence, on both sides of the magnetopause, of oppositely-directed magnetic fields implies that an intense, and possibly unstable, current sheet is circulating on the magnetopause at right-angles to the plane of the diagram. The development of an instability (known as a "tearing instability" may then lead to reconnection of the solar and planetary magnetic fields. The magnetic topology that results in Dungey's model is shown in Fig. 4.35a, which illustrates the magnetic-field lines in a plane containing the planet-Sun line and the axis of the planetary field (the midday-midnight magnetic meridian).

1. A neutral line is formed in the equatorial plane where the magnetic field has the form of X-shaped neutral points, and where the physical connection of the solar and planetary magnetic fields takes place. This line intersects the plane of the diagram at points N1 and N2.
2. There are three topologically distinct classes of magnetic field lines: class 1 (planetary lines) consists of lines of force entirely contained within the magnetosphere, and connected to the planetary body at both ends; class 2 (interplanetary lines) is formed by lines of force completely contained in the solar wind and the magnetosheath; and class 3 (reconnected lines) is of intermediate type, which specifically arises from reconnection. The lines of class 3 have one end connected to the planetary body and the other extending to infinity in the solar wind.

In the region of the upper atmosphere, the boundary between class 3 lines of force, which form a "tuft" of lines emerging from the polar regions, and class 1 lines, seems to correspond quite accurately with the auroral oval (Fig. 4.34). This is essentially confirmed by satellite measurements of the energy spectrum of electrons precipitating into the upper atmosphere. Whilst the electrons precipitating inside each auroral oval have the same characteristic energy as those found in the solar wind behind the bow shock (in the magnetosheath), those that are found around the auroral ovals and outside them have typical energies that are very much greater (from 1 to 10 keV). The auroral ovals therefore do seem to be close to the projections along the magnetic-field lines and onto the Earth's upper atmosphere of the neutral line that encircles the planet close to the equatorial plane in Dungey's model.

Solar-wind-induced Circulations Inside Magnetospheres

Lines of class 3 have another very important consequence for interactions between the solar wind and planets: they allow the solar wind's momentum to be partially transmitted to the interior of the magnetosphere. Indeed, in Dungey's model magnetic flux is frozen practically everywhere in the material in motion, except in the immediate vicinity of the neutral line. We may therefore determine the motion of the plasma within the magnetosphere by following each line of force of class 3 as it moves. In an equilibrium regime, these class-3 lines, which in the solar wind swing round from the sunlit to the night side, do the same inside the magnetospheric cavity. This circulation takes place as shown by the double arrows in Fig. 4.35a. It is directed towards the night side along open magnetic-field lines (classes 2 and 3).

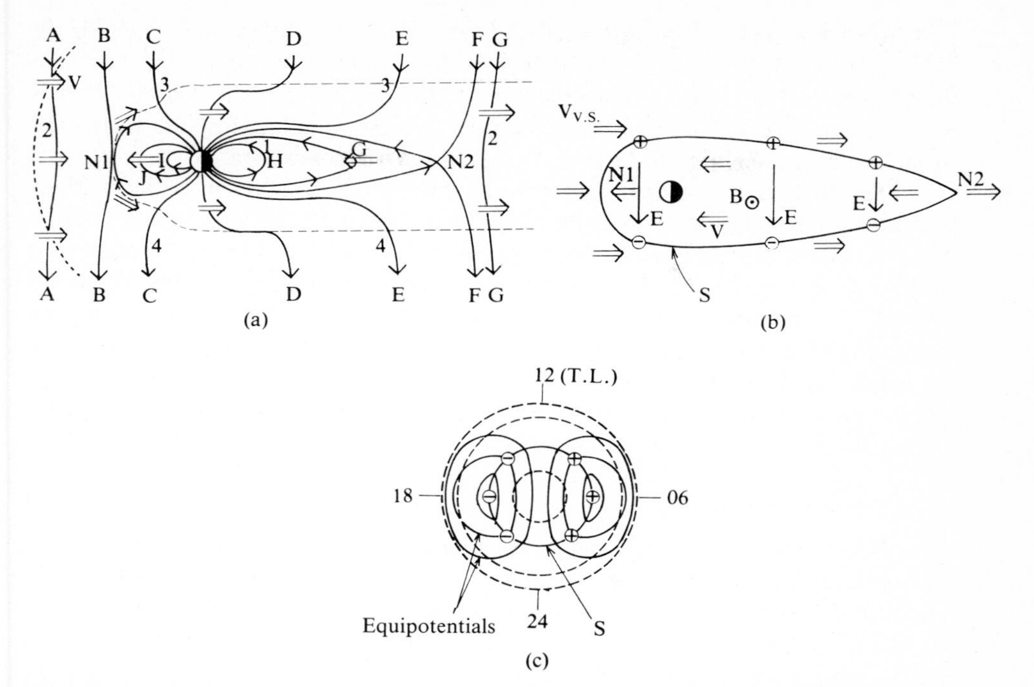

Fig. 4.35a–c. The basic concept and the consequences of Dungey's open magnetosphere model. **(a)** In the plane of the midday-midnight magnetic meridian, the interconnection of the planetary and interplanetary magnetic fields is shown for the simple case of a uniform interplanetary field, pointing towards the south (the bow shock is shown by the dotted line and the magnetopause by the broken line). This interconnection allows the nightward flow of the interplanetary plasma to be transmitted to the lines of force in the polar cusps (3 and 4). Conservation of magnetic flux then implies that the plasma on the closed field lines (1) flows back towards the Sun. **(b)** In the equatorial plane, the continuity of the neutral line (or line of reconnection) (N) becomes apparent, linking the midday reconection point (N1) and the midnight one situated far down the tail (N2). Plasma flows towards the Sun within the neutral line (N), driven by the sunrise-sunset electrical current induced by the solar wind. **(c)** When projected along the field lines onto the ionosphere – circles of constant latitude, which are drawn in broken lines – the curve (N) appears to correspond closely with the polar-cusp/auroral-zone boundary (or auroral oval), along which the most brilliant auroral displays occur. The plasma flow-lines above these regions (which are also electrostatic equipotentials) indicate that the flow is towards the night side within the auroral oval, and flows back towards the midday sector outside the oval. [After L.R. Lyons, D.J. Williams: "Quantitative Aspects of Magnetospheric Physics", *Geophysics and Astrophysics Monographs* (D. Reidel Publishing Company 1984)]

Conservation of magnetic flux then decrees that for the planetary magnetic flux to sustain reconnection on the day side of the magnetosphere, the plasma circulation should be directed towards the Sun on the closed lines of force (class 1). This is exactly what is observed: as shown in Fig. 4.35b, in the equatorial plane the plasma flow is directed towards the Sun everywhere within the neutral line, which, at least on the day side, defines the boundaries of the magnetospheric cavity.

At the level of the planet's ionospheric layers (represented in Fig. 4.35c in polar projection), the movement of the plasma at high altitude (where the collision parameters r_i and r_e are small when compared to unity), can be deduced from the previous diagrams by projection along the field lines. The plasma flows towards the night side within the auroral oval (which is the projection of the neutral line onto the ionosphere), and returns towards the day-time sector outside this circle. It is this twin-vortex geometry that has caused the circulation induced by the solar wind inside the magnetosphere to be called *magnetospheric convection*.

The circulation just described only represents one component of the various systems of magnetic-flux-tube circulation within the magnetosphere. In order to discover the overall circulation, the one induced by the solar wind must be combined with those induced by the dynamos occurring within the magnetosphere. These are produced by the proper motion of the elements of the ionospheric conductor across the planetary magnetic field. Each element is entrained at the local velocity V of the gas in the upper atmosphere at the level of the conducting layer (known as the "ionospheric dynamo layer"), described in Sect. 4.2. This velocity in the planet's inertial frame may be decomposed into:

$$\boldsymbol{V} = \boldsymbol{\Omega} \times \boldsymbol{R} + \boldsymbol{V}_{\mathrm{n}} \quad . \tag{4.36}$$

$\boldsymbol{\Omega} \times \boldsymbol{R}$ – where $\boldsymbol{\Omega}$ is the planetary-rotation velocity vector, and $\boldsymbol{R}$ is the planetocentric radius vector of the ionospheric fluid element – represents the fluid's velocity of entrainment to the planetary rotation, known as the corotation velocity. V_{n}, the second term, is the atmospheric fluid's true velocity relative to the planet, that is, simply the neutral wind. Except in the immediate vicinity of the poles of rotation, this second term is generally negligible and may therefore be ignored, until it is finally added as a correction term, in establishing the global circulation of the magnetospheric plasma in the inertial frame.

The effect of the entrainment of the atmospheric fluid in the dynamo layer by the planet's rotation is simply to superimpose the corotational motion, via the friction exerted by the ionospheric conductor on the feet of each line of force, onto the convection cells induced by the solar wind. As a result, the schemes outlined in Fig. 4.35 have to be modified to represent the overall circulation. At the ionospheric level, scheme (c) remains valid, provided that it is viewed in a frame rotating with the planet, i.e. that of an observer on the surface of the planet. In the equatorial plane, on the other hand, scheme (b) has to be modified as shown in Fig. 4.36. There the frozen plasma flow-lines, which are also electrostatic-field equipotentials for an equilibrium regime, are shown in the plane of the magnetic equator. The overall circulation (right panel) results from the superimposition of the convection towards the Sun that is produced by the solar wind (and associated with an electric field pointing along the morning-evening axis), and corotation, which is itself produced by a radial electric field. It will be seen that two separate circulation regions are produced, separated by a singular flow-line, which passes through a stagnation point lying on the 6 h–18 h axis. Within this boundary, the frozen plasma and the magnetic flux-tubes follow closed trajectories around the planet. These tubes are therefore permanently closed and linked to the source of ionospheric plasma. For the same reason they are never linked to the solar wind, and thus (at least to a first approximation) are not subject to the influence of the interplanetary medium.

In the final analysis therefore, the singular flow-line and the magnetospheric shell that it defines are the true internal limit of the region of the magnetosphere

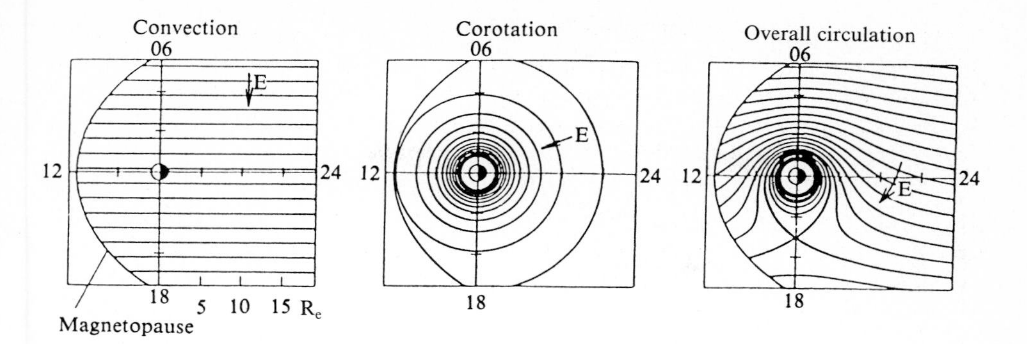

Fig. 4.36. The geometry of the electrostatic equipotentials (which are also current lines in the cold plasma), in a magnetosphere's equatorial plane. The overall electric field (the equipotentials of which are shown *right*), in fact result from the superimposition of two systems of induced electric fields, one caused by interaction with the solar wind (*left*), the "convection field", shown here for simplicity as being uniform, and the other by the diurnal rotation of the planet's ionospheric conductor (*centre*), the "corotation field". [After L.R. Lyons et al.: "Quantitative Aspects of Magnetospheric Physics", *Geophysics and Astrophysics Monographs* (D. Reidel Publishing Company 1984)]

governed by the interplanetary medium. As seen in Fig. 4.36, circulation outside this boundary is from the magnetotail towards the sunlit hemisphere, flowing round the corotating region either on the evening or morning side. Again we find the movement of the closed field lines that was seen in (a) in Fig. 4.35. If we look at that diagram again it reminds us that the closed field lines, after having been transported from the tail towards the magnetopause, become open to interplanetary space in crossing that boundary, and that they return towards the magnetotail above the two polar caps as two open field lines.

These facts will allow us to understand the extent of the region within the magnetosphere that is filled by ionospheric plasma. Figure 4.37, which shows a section of the magnetospheric cavity in the midday-midnight meridian, indicates the principal features. The shaded zone represents the interior of the magnetic shell defined by the singular flow-line. It is the zone where the lines of force remain closed and are connected to the source of ionospheric plasma, while being entrained into corotation. This source is always present on the illuminated side of the planet (because of the effect of solar ultraviolet radiation). The tubes of force therefore fill with ionospheric plasma until their pressure becomes equivalent to that of the source. As a result the concentration of cold plasma is always high within this region, which is known as the *plasmasphere*. Outside its boundary, which is known as the *plasmapause*, the flux tubes are periodically open to interstellar space as they are transported across the poles. There they lose most of their ionospheric plasma, which in the case of the Earth is mainly H^+ ions, and which escapes from the planet's gravitational field into interplanetary space through a jet effect, similar to the source of the solar wind. This is why this loss of plasma at supersonic speeds, which has been observed over the Earth's polar caps, is known as the *polar wind*.

This recurrent loss maintains the concentration of cold plasma outside the plasmasphere, in the region of the magnetosphere that is strongly coupled to the interplan-

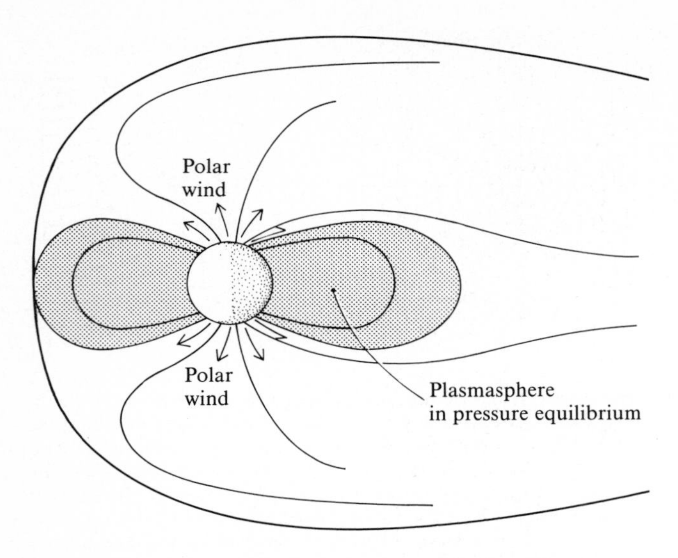

Fig. 4.37. The superimposition of convection and corotation of the magnetic flux-tubes in a planetary magnetosphere produces three distinct regions, populated to different extents by the ionospheric plasma. [After P.M. Banks, A.F. Nagy, W.I. Axford: "Dynamical Behaviour of Thermal Protons in the Midlatitude Ionosphere and Magnetosphere", Planetary and Space Science **19**, 1053 (© Pergamon Journals Ltd. 1971)]

etary medium, at levels that are at least one order of magnitude less than those in the plasmasphere.

At every opening to interplanetary space, on the other hand, the flux tubes outside the plasmapause fill with plasma from the solar wind that diffuses across the magnetopause along the magnetic field lines. This plasma is transported by magnetospheric convection to the equatorial plane of the magnetotail, where it accumulates in a plasma reservoir surrounding the equatorial plane, and which is known as the *plasma sheet*. With the closure of the flux-tubes across the night-side neutral line, this plasma is pulled towards the Earth (Fig. 4.35a). The adiabatic compression that it undergoes during this motion, as a result of the rapid decrease in the volume of the flux-tubes, raises it to temperatures above the keV level. It is the electrons from this hot plasma that, precipitated by various mechanisms in the upper atmosphere, give rise to the aurora borealis and aurora australis (Fig. 4.38), whilst the ions are partly fed into the outer regions of the radiation belts.

The size of the plasmasphere varies greatly from one magnetosphere to another. It may be calculated quite simply, on the basis of the diagrams in Fig. 4.36, by determining the position of the stagnation point, which is where the convection and corotation electric fields are equal and opposite. With the terrestrial magnetosphere, the relatively slow planetary rotation allows a major portion of the magnetic cavity to be coupled to the solar wind. Both calculation and observation show that the plasmapause lies at a geocentric distance that typically varies between four and seven planetary radii. We may speak of the magnetosphere as being dominated by convection.

Fig. 4.38. Aurorae, which illuminate the night sky in the Earth's polar regions, are the most spectacular signs of the influence exerted by the interplanetary medium on the upper atmosphere. Aurorae were also observed on Jupiter and Uranus by the Voyager 2 spaceprobe. (By courtesy of the Max Planck-Institut für extraterrestrische Physik, Garching)

In the case of Jupiter, on the other hand, the same calculation places the plasmapause beyond the magnetopause. This means that the entire magnetospheric cavity takes part in the planet's rapid rotation (once in ten hours). The effects of the interplanetary medium are therefore doubtless limited to the immediate vicinity of the magnetopause and the magnetotail. The regions of Jupiter's magnetosphere with closed magnetic topology are therefore dominated by the internal sources of plasma (the planet's ionosphere and the satellites).

The case of Uranus is at present probably unique in the Solar System. At the time of the Voyager 2 encounter, this planet's rotation axis was pointing nearly exactly towards the Sun. But its magnetic dipole axis lies at an angle of 60° to the rotation axis. With such a configuration, the convection and corotation flows do not combine at all, in the way illustrated in Fig. 4.36, because they are essentially orthogonal to one another. As noted by Vasylunias ("The convection-dominated magnetosphere of Uranus", Geophys. Res. Lett. **13**, 621–3, 1986, AGU publications), in that situation the convection flow is stationary in the corotating frame of reference, rather than in the inertial frame of reference. Thus one can predict that in the corotating frame the convection flow penetrates freely to the inner magnetosphere, and is nowhere opposed by corotation. Uranus' magnetosphere is probably entirely dominated by convection at the present time, and will return to a situation that is more like that of the Earth as, in the course of its orbital motion, its rotation axis turns away from the direction of the Sun.

4.3.4 "Auroral" Radio Emissions of Planetary Magnetospheres

It has been known since 1955 that Jupiter is a powerful radio source at decametric wavelengths, and it is only since the space age that we know that our own planet Earth is a radio source. Thanks to the Voyager mission, it is now recognized that it is a common property of planetary ionosphere/magnetosphere systems that they generate non-thermal radio emissions. Among all observed emissions, a rather broad class seems to be generated basically by the same mechanisms and in the same context in all known magnetospheres: these are the so-called "auroral" emissions, which seem to be a direct consequence of electron-precipitation phenomena taking place in regions of strong electrodynamic coupling between the ionosphere and the magnetosphere. With the Earth, these regions are the northern and southern auroral zones, which are traversed by magnetic field lines carrying particularly intense currents aligned with the magnetic-field and connecting the solar wind, the magnetosphere, and the ionospheric conductor.

Figure 4.39 shows records of dynamic spectra of the known "auroral" radio emissions of the four planets studied by the Planetary Radio Astronomy experiment on board Voyager: the terrestrial AKR (Auroral Kilometric Radiation), the jovian DAM (decametric) and HOM (hectometric) radiations, Saturn's kilometric Radiation (SKR) and Uranus' Kilometric Radiation (UKR). Each diagram (in which a grey scale represents the intensity of emission in a time-frequency domain) shows one planetary rotation (represented by 24 hours for the Earth and 360 degrees of longitude for the other planets). Each emission is actually modulated by the planetary rotation, but one can see in the Figure that they all display a considerable degree of variability and many small-scale structures in the time-frequency domain. The complexity made it a particularly difficult task to establish the basic morphology of each emission, and to locate its sources. For instance, in the case of Jupiter's DAM, it happens to have sources both in a region close to the foot of the magnetic field line linking Jupiter's ionosphere with its satellite Io, and in a more extended region corresponding more or less to Jupiter's auroral zones. Both regions are probably characterized by intense electron-precipitation fluxes into Jupiter's upper atmosphere, and by the presence of strong magnetic-field-aligned currents connecting an external generator (the solar-wind/magnetopause interface, or the e.m.f. induced by the motion of Io through the magnetic field lines of Jupiter) to the jovian magnetospheric conductor.

Table 4.1 shows the main characteristics of the emissions. Taken as a whole, they share the following characteristics:

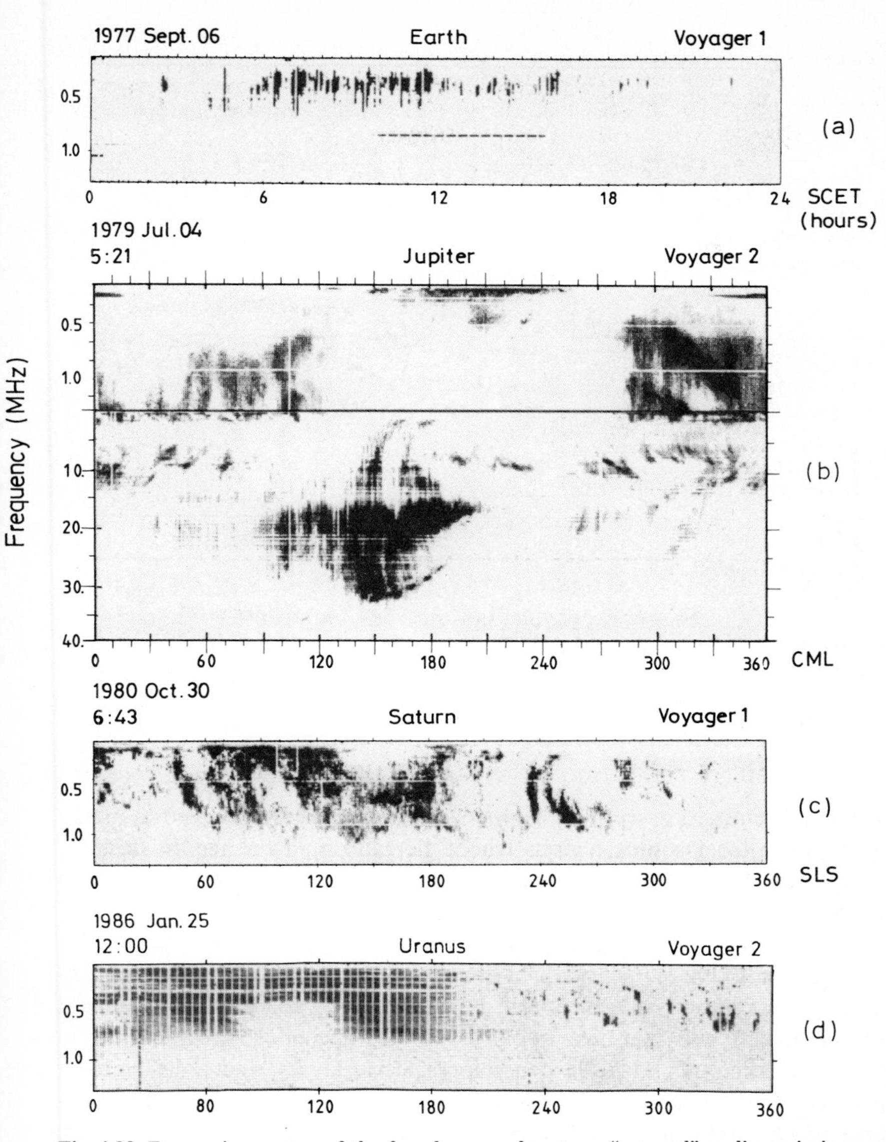

Fig. 4.39. Dynamic spectra of the four known planetary "auroral" radio emissions, as observed by the Voyager Planetary Radio Astronomy experiment. From top to bottom: Earth's Auroral Kilometric Radiation (AKR); Jupiter's Decametric (DAM) and Hectrometric (HOM) radio emissions; Saturn's Kilometric Radiation (SKR); and Uranus' Kilometric Radiation (UKR). Each diagram shows one planetary rotation. [After F. Genova: "Les émissions aurorales des planêtes", Ann. Phys. Fr. **12** , 57–107 (1987)]

1. They are all extremely intense. For instance, the brightness temperature of the jovian emission is larger than 10^{12} K even if one assumes the source to be uniformly distributed over the planetary disk;
2. all emissions seem to originate from the source nearly perpendicularly to the magnetic field, at a frequency close to the local electron gyrofrequency f_c and to the cut-off frequency of the x mode of electromagnetic propagation in plasmas;

Table 4.1. Main characteristics of planetary "auroral" radio emissions

Planet	Emission	Mean radiated power at 1 AU ($W \cdot m^{-2} Hz^{-1}$)	Peak frequency	Probable source location
Earth	AKR Auroral Kilometric Radiation	$\sim 10^{-23}$	200 kHz	$60° < \Lambda < 78°$ from 1.3 to 3.5 R_T Max. occurrence in evening sector, but exists at all local times
Jupiter	DAM Jovian Decametric Radiation HOM Jovian Hectometric Radiation	$\sim 10^{-19}$	10 MHz	Io-controlled emission: close to the foot of the Io flux tube Non-Io-controlled: high latitudes
Saturn	SKR Saturn Kilometric Radiation	$\sim 10^{-21}$	200 kHz	$70° < \Lambda < 80°$ 1 to 6 R_S Noon sector limited in longitude
Uranus	UKR Uranus Kilometric Radiation	$\sim 4 \cdot 10^{-22}$	200 kHz	Magnetic polar regions

3. actually, the emissions seem to belong to the right-hand-polarized x mode; they originate from source regions where the ratio f_P/f_C of the plasma frequency to the electron gyrofrequency is much smaller than unity, and of the order of 0.2 to 0.4 at most.
4. the source regions always correspond to particular regions of intense electron precipitation into the planetary upper atmosphere, in which electron energization and precipitation is driven, for a large part, by electrostatic potential drops of several kilovolts developing along the magnetic field lines, in the high-altitude regions above the dense ionospheric layers where the ambient ionospheric plasma density drops to very low values. These precipitating electrons are likely to be the carries of the free energy feeding the radio emissions. The conversion efficiency from electron energy to radiated wave energy can be as large as 1% in some cases.
5. finally, the source regions seem to have a small extent across magnetic field lines, of the order of 100 km in most cases.

Taken as a whole, this set of common features is an efficient constraint on identification of the mechanism responsible for these emissions. There is now a rather broad consensus that the most likely generating mechanism is *Maser Synchrotron Instability*. Basically, this instability results from the gyroresonant interaction between energetic electrons and the right-hand-polarized x-mode close to its cut-off. There is a resonance of this kind when an electron gyrating in the magnetic field as it drifts alone the field line "sees" an electric field rotating at exactly the same frequency in its own frame of reference. In such a situation, there is a net exchange of energy between the electron and the field, leading both to a change (growth or damping) of the wave energy, and to a diffusion of the electron in its velocity space. The resonance condition can be written simply as

$$\omega - k_{\|} v_{\|} = \omega_c / \Gamma \quad . \tag{4.37}$$

If ω and $k_{\|}$ are the frequency and parallel wave vector of the wave, the left-hand side of the equation is simply the apparent frequency of the wave (modified by the Doppler effect) in the frame of reference moving with the velocity $v_{\|}$ of the electron along the magnetic-field line. The right-hand side is the gyrofrequency of the electron, taking into account the relativistic correction by the factor $\Gamma - (1 - v^2/c^2)^{-1/2}$, which happens to be important for this particular mechanism.

The tendency for a wave of particular $(\omega, \boldsymbol{k})$ characteristics to grow or decay in its interaction with a collection of energetic electrons depends upon the velocity distribution of this electron population (which must contain some free energy for the wave to grow). The theory of Maser Synchrotron Instability shows that the type of free energy source need is an electron distribution function $f(v_{\perp}, v_{\|})$ (where $v_{\|}$ and $v_{\perp}$ refer to electron velocity components parallel and perpendicular to the local magnetic field, respectively) in which there is a positive slope $(\partial f / \partial v_{\perp} > 0)$ in the electron distribution with respect to the perpendicular velocity.

In 1986, the Swedish VIKING satellite studied in detail the structure of the auroral acceleration regions in which auroral electrons are energized before being precipitated. It provided simultaneous measurements of the AKR emissions within their source regions, of the plasma characteristics of these source regions, and of the energetic-electron distribution functions. An example of a distribution function measured directly inside an AKR source region is shown in Fig. 4.40. This 3-D representation shows the level of the distribution function f along the z coordinate, as a function of $v_{\perp}$ and $v_{\|}$. This distribution function displays two spectacular characteristics, found within every AKR source region.

1. There is an empty "loss cone" along the $v_{\|}$ axis in the ascending direction, which is produced directly by the fact that all electrons that might have filled this region of velocity space have been precipitated into the atmosphere at the end of their descending trajectory,
2. and the rest of the function shows a "ring"-type of distribution, in which the distribution function around the central peak remains small, showing a lack of medium-energy particles, and then abruptly increases in essentially all directions around a certain energy "threshold". One can show that this type of distribution function is produced by the passage of a Maxwellian electron population through a region along a magnetic-field line where it is accelerated by an electrostatic potential drop of a few kilovolts accelerating each electron downwards. While the potential drop produces the trough in intermediate-energy electrons, the mirror effect produced by the variation of the magnetic-field intensity along the filed lines converts total energy into perpendicular energy at constant total energy, and produces the "ring" distribution observed.

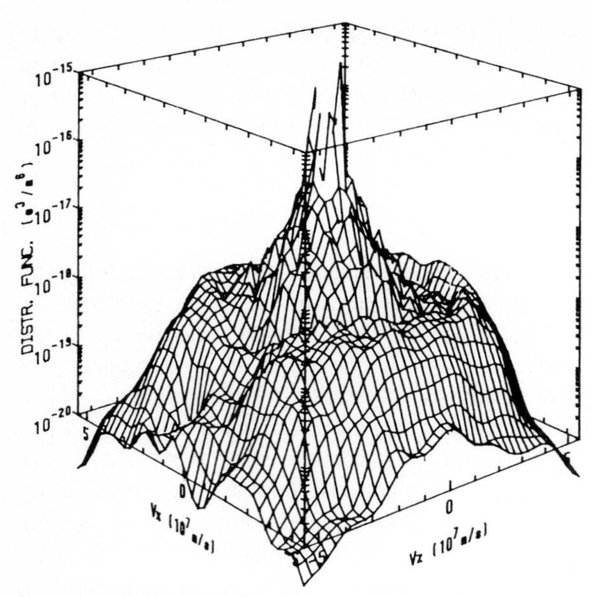

Fig. 4.40. Three-dimensional representation of the electron distribution function measured by the Swedish VIKING satellite inside a source region of the Auroral Kilometric Radiation. Coordinates in the horizontal plane of the representation correspond to velocities orthogonal to the local magnetic field (V_x) and parallel to the field (V_z).Two spectacular characteristics of the AKR source region are visible: the presence of an empty "loss cone", seen as a strong decrease in the distribution function around $V_x = 0$ for negative V_z, and a "ring" type of distribution in the rest of the velocity plane (with kind permission of M. André and L. Madsen, Institute of Space Physics, University of Umeå, Sweden)

It is this distribution, with strongly positive $\partial f / \partial v_\perp$, which has been identified as the free-energy source for the AKR.

In-situ spacecraft measurements of the AKR phenomenon have also clarified many other functions. They have shown that the emission sites are cavities aligned with the magnetic field, within which the density of the cold plasma of ionospheric origin drops to a very low level, while the hot plasma becomes dominant, thus modifying significantly the propagation characteristics of the x mode. It has been suggested that these cavities might partly "trap" the x-mode radiation, forcing it to pass several times through the amplifying medium constituted by the energetic electron population, and produce a gain in the resulting total amplification by a mechanism similar to what happens in a laser cavity. It was also confirmed that these cavities are directly connected, along magnetic-field lines, to the bright, visual auroral forms, showing that it is indeed the same precipitating electrons that produce the radio emissions and the visual emissions, which form the bright auroral forms at a lower altitude.

5. The Inner Planets and Their Satellites

The four planets closest to the Sun: Mercury, Venus, Earth, and Mars, to which one can add the Moon, have many similarities that justify their being considered together. These planets are sometimes called the *terrestrial planets*, Earth serving as a reference. It does not seem to us, however, that in dealing with astrophysics the Earth should be put in a privileged category, so we prefer the description "inner planets", emphasizing that it is only their proximity to the Sun that is responsible for the primary characteristics that these bodies have in common.

The inner planets have high densities and possess atmospheres that form only an extremely small fraction of their total masses. The atmosphere of Venus, which is by far the most massive, amounts to only one hundred-millionth of the planet's mass. The situation is the opposite for the giant planets, where most of the mass is in the form of gas. A corollary of this situation is that the actual atmospheres of the inner planets, whatever their density, are not primaeval: they have developed their current properties through planetary evolution.

There is another similarity: these objects have undergone periods of planetary activity that have profoundly modified their internal structure and surface features. The mineralogical differentiation of these various objects has, however, not produced similar interiors, which means, in particular, that the magnetic fields are of different intensity. The duration of the periods of activity, their extent and the effects visible on the surface are all specific to individual planets, and their origins are still poorly known.

Meteoritic impacts were predominant throughout the first few hundred million years, in particular by their creation of enormous craters hundreds of kilometres in diameter. Subsequently, accumulation of heat from the radioactive decay of long-period isotopes (mainly U, Th and K) caused the internal temperature to rise steadily, and eventually became sufficiently high to produce volcanic activity with magma rising to the surface. This is how the lunar maria were formed: dark basalt flooded some of the impact basins that had been created earlier. As it derives from isotopic decay, the internal heat naturally decreases with time. The less massive bodies cooled quicker: as the amount of energy radiated away being proportional to the surface area, and the amount of energy stored within a body being proportional to the volume, the central temperature increases as the square of the radius, all other things being equal. The Moon and Mercury therefore became extinct slightly more than one thousand million years after their formation. Mars, being more massive, remained active for at least another two thousand million years: imposing volcanoes have modified its surface, and gigantic floods have left an extensive network of gorges. As for Venus, its dense atmosphere has so far not allowed any definite evidence for volcanism – even extinct – to be detected (see also Chap. 12).

In this context, the Earth appears very notable for its level of activity, and this is particularly shown by its plate tectonics, which has spectacular effects. Its atmosphere, where O_2 and N_2 predominate, is equally distinct from those of Mars and Venus, where CO_2 is the major component. Finally, and this is by no means the least of the differences, chemical evolution on the Earth led to the appearance of living species more than 3 thousand million years ago.

The advent of interplanetary space missions over the past twenty years has allowed astronomical observation to be linked with in-situ measurements for all the inner planets. Our ideas about these objects have undergone amazing changes as a result. In particular it is now possible to study them comparatively, which opens up new ways of interpreting the phenomena observed on each of them, and, in particular, on the Earth.

Each of the inner planets discussed in this chapter is the subject of a separate section.

5.1 Mercury

When compared with the other planets of the inner Solar System, Mercury is the least well-known. This is mainly because of its location: of all the planets it is the closest to the Sun, from which it is never more than 28° distant to a terrestrial observer, and this does not favour astronomical observation, especially as Mercury is the smallest of the inner planets. Until 1965, for example, its period of rotation was estimated to be 80 days, i.e. the same as its orbital period. As a result, the presence of an atmosphere around the planet had to be assumed as otherwise the night-time temperature would approach absolute zero, which did not agree with observation. It was only towards the end of the 1960s that the rotation period was corrected, not by just a couple of hours, but by more than twenty days.

Only one space mission has had Mercury as one of its objectives, which were primarily concerned with photographic observation. The surface that it revealed, covered in impact craters, seems to indicate that there has been no internal activity for thousands of millions of years. The daytime temperature of Mercury is very high, which, together with a low escape velocity, has prevented it from retaining a dense atmosphere.

Mercury therefore appears to be extinct, and the large number of differences from the Earth and, conversely, of similarities to the Moon have meant that it has not been favoured in past programmes, and that the various space agencies have no plans for it in the next twenty years. It is nevertheless probable that Mercury will play a part in comparative planetology in the next century.

5.1.1 Orbital Parameters and Macroscopic Properties

The orbit of Mercury has a semi-major axis of 0.466 AU (69.7 million km), and is strongly eccentric: $e = 0.206$. It is the most elliptical of the planetary orbits, if we exclude Pluto, where $e = 0.25$. Its perihelion distance is 55 million km, and its aphelion distance 84 million km. Another feature of Mercury is its orbital inclination,

7°, which again puts it in second place after Pluto with 17.2°. The sidereal orbital period is 88.97 days, with a mean orbital velocity of 48 km/s.

Mercury appears to be a sphere, with essentially no flattening, and with an equatorial radius of 2439 km. It is the smallest of the four inner planets, not much larger than the Moon. Its mean density of 5.44, on the other hand, is close to that of the Earth (5.52) and of Venus (5.25), and sets it apart from Mars (3.94) and the Moon (3.33). It leads to a very low escape velocity, 4.3 km/s, compared with 11.2 km/s for the Earth, 10.3 km/s for Venus, 5.0 km/s for Mars, and 2.4 km/s for the Moon. Finally, Mercury's rotation is direct, about an axis exactly at right-angles to the plane of its orbit, and takes 58.65 Earth days. Note that its rotation period is equal to 2/3 of its orbital period. The fact that these two periods are commensurable, reflecting dynamic coupling between the rotational and the sidereal orbital periods, may be interpreted as arising from the braking of the planet's rotation, either during or at the end of its accretion, under the gravitational influence of the Sun. Finally, this feature of Mercury's dynamical behaviour leads to the mercurian day being equal to 176 terrestrial ones.

The absence of a dense atmosphere gives rise to surface temperatures that may range from very high values during the day – up to 700 K at the subsolar point at perihelion – to very low values at night – 100 K in the equatorial regions.

5.1.2 Observation of Mercury

Until the beginning of the 1970s, Mercury was known only from telescopic photometry, polarimetry and spectroscopy. These showed it to be without an atmosphere, covered with a surface layer of dust, whose low albedo resembled that of lunar soil. In addition, spectral signatures in the near infrared, despite being faint, showed a band at about 0.95 μm. By comparison with what was then known about the lunar surface and terrestrial silicate samples, it was possible to identify this band as being that of iron in the Fe^{++} state, characteristic of the presence of pyroxene basalt. The overall shape of the spectrum could be interpreted as being that of a surface dominated by the presence of glasses, arising from meteoritic-impact vitrification of a surface layer rich in silicates and ilmenite (an iron (FeO) and titanium (TiO_2) oxide, with the formula $FeTiO_3$). But the poor spatial resolution ($\sim$500 km) did not allow contrasts in albedo, similar to those between the lunar maria and highlands, to be observed.

On 1973 November 3, NASA launched the Mariner 10 probe, which, for the first time in the history of space exploration, had a double objective: to fly past Venus and then, using the gravitational field of the planet, to change course towards Mercury. After having first photographed and then swung round Venus on 1974 February 5, Mariner 10 reached Mercury on 1974 March 29 and approached to within 756 km. The chosen trajectory then placed Mariner 10 into solar orbit with exactly twice the planet's orbital period, so it made further approaches to Mercury, which was in the same relative position to the Sun on each occasion. Moreover, as Mercury rotates thrice on its axis in the course of two orbits of the Sun, it showed the same regions for observation, with the same conditions of lighting. Only the minimum-approach distances were different, giving different resolution of surface features: 48 069 and

Fig. 5.1. Mercury as seen by Mariner 10 on 1974 March 29. A mosaic built up from eighteen images obtained at a distance of 210 000 km. (By courtesy of NASA)

327 km for the next two passes, which took place on 1974 September 21 and 1975 March 16 respectively.

The probe mainly carried a camera and Mariner 10 produced maps of about 45 % of the surface with a resolution of the order of twenty km (Fig. 5.1). Only a small fraction is known with a resolution better than one km. There was no instrumentation to carry out chemical or mineralogical mapping of the surface. A set of four filters in front of the camera gave some indication of albedo, polarization and "colour difference", on a scale of 1 to 2 km. The absence of any strong polarization in the UV indicates that there have been no recent basalt flows but that the surface is instead covered with fine dust.

Albedo measurements confirm (at kilometre scales), the overall observations of Mercury made from the Earth. More particularly, there are no albedo gradients similar to those between the maria and highlands, or encountered in the crater rays, on the Moon. Any possible basalt flows on Mercury have not, apparently, modified the surface mineralogy. This may be interpreted as indicating a considerable thicker crust than on the Moon. Finally, the average albedo of Mercury, which is higher than that of the darkest lunar maria, as well as the colour differences (between orange and ultraviolet), might be explained by the absence of *ilmenite*, which is a relatively important component (up to 7 %) of lunar maria.

Measurements of thermal flux on the first and third fly-bys enabled the surface temperatures that have been mentioned earlier to be determined.

Mariner 10 carried a magnetometer, intended to measure the intensity of the interplanetary magnetic field and to study the interaction of the interplanetary plasma with the planet. The instrument revealed the features expected for a planet without any atmosphere and with an internal magnetic field: a bow shock and a magnetopause. The maximum field measured was about 100γ ($1\gamma = 10^{-5}$ gauss), which should be compared with the value of about 20γ for the interplanetary magnetic field at the orbit of Mercury. Such a magnetic field, although weak, was not really expected: although the average density of Mercury had strongly suggested the presence of a core rich in iron, its small mass seemed to indicate that it should be solid; the slow axial rotation of the planet, moreover, might well have prevented it from generating an intense magnetic field. In the direction in which the solar wind is incident, the magnetopause develops at about 1 mercurian radius above the surface. This should be sufficient to deflect most of the particles in the solar wind, at least when the Sun is quiet. But at times of major solar activity, on the other hand, the full force of the solar wind probably reaches the surface of Mercury. If we take into account the close proximity of the Sun, and thus the much greater flux, the overall number of ionized particles (mainly protons and helium nuclei) striking the surface must be essentially the same as that found on the Moon. We may therefore expect similar radiation effects to be present. In particular, it is those particles that, after implantation into the surface grains, diffuse out to form its very tenuous atmosphere. The partial pressure of helium, derived from UV-spectroscopy measurement at the terminator, is of the order of 5×10^{-12} mbar. An upper limit for the total atmospheric pressure is 2×10^{-9} mbar.

5.1.3 The Surface Topography

The photographic coverage obtained in Mariner 10's three fly-bys has enabled geological maps of about 35 % of the surface to be constructed: about 50 % was in shadow, and 15 % was photographed at too oblique an angle. It is this documentation that is used by planetary geologists in their attempts to make out traces of the internal and surface activity that have occurred since the formation of the planet.

The structures that dominate the surface are impact craters of all sizes, separated by intercrater plains. The craters' topography mainly depends upon their size: the smallest are hemispherical, but they appear less and less deep relative to their diameter as the latter increases, becoming flat when the diameters reach about twenty km or more. A central peak is then seen, together with clearly identifiable ejecta. Although the number of craters decreases as the size increases, reflecting the mass-distribution of interplanetary objects, there is a large population of craters over 200 km in diameter, which are known as *basins*; the largest, 1300 km in diameter, is the *Caloris Basin*.

The absence of any erosion other than that from impacts allows the different types of terrain to be dated by the density of craters, their overlapping, and their structure (eroded sides, darkened ejecta, etc.). Systematic study leads to the following conclusion: the frequency of impacts was very high during the first few hundred million years, but rapidly declined to a low, and stable level, some 3.8 thousand million years ago. This first period of intense bombardment corresponds to the *accretion tail* described in Chap. 11 that was responsible for the formation of the large impact craters visible on the Moon (in particular). It seems that the planet's internal

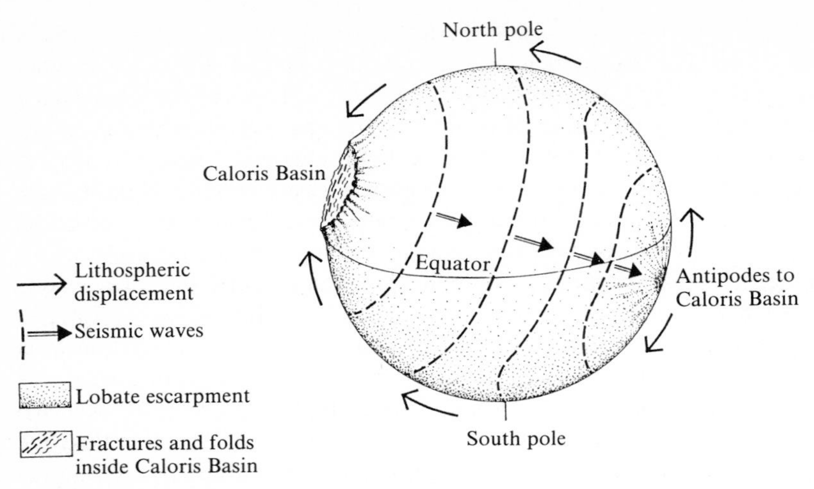

Fig. 5.2. The tectonic consequences of the Caloris Basin impact. The tectonic events that followed the formation of the Caloris Basin are of two very distinct types, which may be summarized as follows: at the moment of impact, seismic waves produced by the shock were focussed on the region at the basin's antipodes; following the impact, the lithosphere, which was in compression as a result of the cooling of the planet's core, moved slightly towards the centre of the basin; this movement led to the formation of ridges and fractures inside the basin as well as of peripheral faults, and dictated the orientation of lobate scarps [after P.H. Schultz, D.E. Gault (1975), and L. Fleitout, P.G. Thomas (1982). [P. Masson: "Mercury", in *Le grand Atlas Universalis de l'Astronomie* (Encyclopedia Universalis 1983). By courtesy of the publisher]

activity, stimulated by these violent impacts, also ceased shortly after the formation of Mercury's largest basin, Caloris, which has an age of about 3.8 thousand million years. The formation of this basin gave rise to a last bout of tectonic activity, in particular, to volcanism that can be seen in the presence of the smooth plains that fill the basin. In addition, a certain number of structures (ridges and fractures) can be seen around the Caloris Basin, as well as at the basin's exact antipodean point, and these give information about the large-scale properties of the *lithosphere* (the surface layer of the planet, above the outer mantle), at the time of impact. In particular, it is thought that the direction of the surface's movement may be explained by the lithosphere then being in a state of compression, because of the contraction of the central metallic core, which was in the course of cooling (Fig. 5.2).

Mercury does thus show some structures that reveal tectonic activity triggered by impact. There are, however, no signs of global activity similar to the plate tectonics known on Earth. This tends to exclude the existence of internal convection, unless such tectonics is prevented by too thick a lithosphere. The intercrater plains may be of volcanic origin. In any case, it is obvious that all geological activity ceased on this planet at least 3.6 thousand million years ago, and that the present surface retains the traces of the first thousand million years of its evolution, dominated by its interaction with the interplanetary medium.

5.2 Venus

Its proximity to the Sun and its high albedo cause Venus to be the brightest planet in the sky: it has always been a beacon to shepherds, whose "star" it has been since ancient times. But that same atmosphere that makes it so bright allows nothing of the surface to be seen. Even now, its properties are poorly known, despite having been the target of more than twenty American and Soviet spaceprobes. NASA and Intercosmos are already planning new programmes for the exploration of this planet over the next ten years. It is not that Venus is primarily interesting as regards exobiology: its temperature and surface pressure do not even allow the strongest automatic probes to last for more than a few tens of minutes; let alone the composition of its atmosphere, where the clouds consist of sulphuric, nitric and hydrofluoric acids. Its main attraction lies in the fact that its mass and density, and thus its composition, are very similar to those of the Earth, but that it shows striking differences with regard to its present-day atmosphere. Has there been, or is there still volcanic or tectonic activity on Venus? What are the factors governing the widely different evolution of the two planets?

5.2.1 The Observation of Venus

Study of Venus in just the visible sunlight that it reflects gives practically no information. But we should note that Galileo, thanks to his astronomical telescope, first observed the *phases of Venus*. He realised that they could not be explained within the framework of Ptolemy's geocentric world-system. Towards the middle of the 17th century, features were seen on the disk, and their variable nature later led to their being regarded as arising in an atmosphere that completely shrouded the planet. Around 1930, observations in the ultraviolet clearly showed identifiable Y-shaped absorptions, which allowed the atmospheric rotational period to be determined as 4.2 days. Spectral analysis of reflected sunlight gave the first indication of the composition of the outermost atmospheric layers. This established that carbon dioxide CO_2 is the principal component, CO and H_2O being detected in trace amounts. But on the other hand, a mistake of nearly a factor of two was made in the (absolute) temperature, and one of about a hundred in the pressure until the end of the 1950s, when the first spectral analyses in the radio wavelength region were carried out. It then became clear that the temperature had to be close to 680 K (and not 400 K); the pressure was calculated from the amount of carbon dioxide required to produce a greenhouse effect that would give such a temperature, and this led to a value close to the actual value. But the characteristics of the surface, even the diameter and its speed of rotation, still remained unknown.

The strategies for the exploration of Venus that have been adopted by the Soviet Union and the United States are different. The Americans decided to carry out fly-bys of the planet (Mariner 2, Mariner 5, Mariner 10) and then observe from orbit with Pioneer Venus 1 and 2 (Fig. 5.3); the last mission dropping probes onto the surface. The Soviets devoted their efforts to placing descent modules on the surface itself (the series Venera 7 to 14, and Vega 1 and 2), information being relayed to probes in orbit. The first photographs of the surface in black-and-white were taken

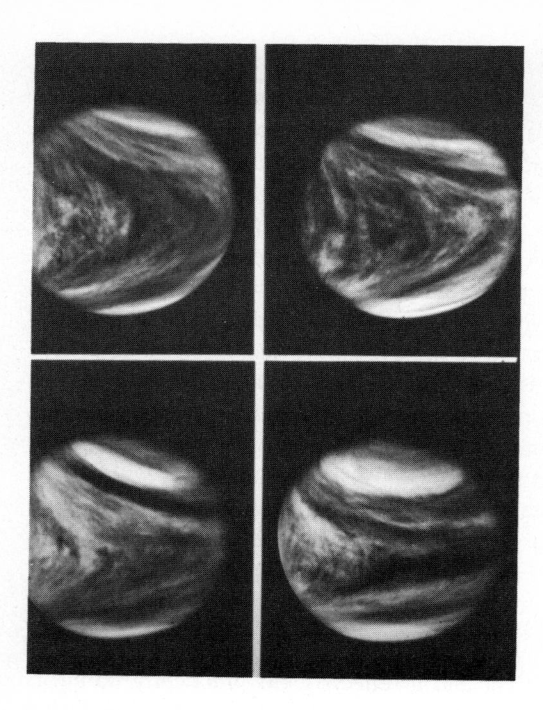

Fig. 5.3. Changes in the super-rotating atmosphere of Venus, as photographed by the Pioneer Venus probe on 1979 February 10, 11, 19 and 20. (By courtesy of NASA)

by Veneras 9 and 10 (in 1975 October), and then in colour by Veneras 13 and 14 in 1982 March (Fig. 5.4). These probes were also the first to determine the composition of the landing sites. The Pioneer Venus missions have provided information about the chemical and isotopic composition of the atmosphere and, above all, radar mapping of the surface with a resolution of about 50 km in horizontal extent, and about 200 m in altitude. The Soviet Venera 15 and Venera 16 probes carried out high-resolution (1 to 2 km) radar mapping close to the north pole. NASA launched a probe (Magellan) to Venus in 1989, with the primary goal of complete radar cartography of the surface, with a resolution of a few hundred metres.

5.2.2 Orbital Parameters and Overall Properties

Tables 1.1 and 1.3 in Chap. 1 summarize the principal orbital data about Venus, as well as its basic physical properties.

Its average distance from the Sun, which is 108 million km, means that it receives about twice as much solar energy as the Earth. This amount hardly varies over the course of the year, because of the very low eccentricity of the orbit. The angle between the rotational axis and the perpendicular to the plane of the orbit is only 2.2°, which does not give rise to seasons on Venus. The planet's rotation is, however, retrograde, so the inclination is 177.8°.

Precise determination of the planet's diameter, and thus its density, could only be made when the surface could be detected, i.e. by radar. The measurements are made in the following manner: a radar signal is emitted by the satellite (Pioneer Venus) at right-angles to the surface of Venus. The time taken to receive the echo

Fig. 5.4. The surface of Venus photographed by the Venera 13 (*top*) and Venera 14 (*bottom*) probes in 1982 March. (Academy of Sciences of the USSR)

gives the altitude of the satellite with reference to the surface. The distance of the satellite from Venus' centre of mass is given by the satellite's orbit. By subtraction one then obtains the planet's radius assuming a perfect sphere. This procedure gives a radius of 6051.0 km for zero flattening. The mass of the planet is also determined by the satellite's orbit, leading to a mean density of 5.25 g/cm^3.

As far as the dynamical properties of the planet are concerned, there are considerable differences from the values obtained before observations were made from space-probes. What was thought to be the rotation of the planet was actually that of the atmosphere. We now know that the latter is super-rotating, and the origin of this remains one of the main points to be explained. It was generally assumed that the planet had synchronous rotation, always presenting the same face to the Sun and therefore having a period of 224 days. In fact the period is −243.01 days, the minus sign indicating that the rotation is retrograde; the reason for this is still the subject of controversy. Taking account of the orbital period of Venus around the Sun, the day on Venus lasts 117 terrestrial days.

Venus is therefore the planet that has by far the slowest speed of rotation on its axis. With the exception of Mercury, which has a period of 59 days, all the other rotation periods are less than 25 hours. This indicates that the rotation of Venus (like that of Mercury) has been braked during the course of the planet's evolution, probably because of the tidal effects caused by its proximity to the Sun. This rotation period of Venus has one consequence that has intrigued many people: Venus presents (almost exactly) the same face towards the Earth at every conjunction. This led to the idea that there might have been a *synodic resonance*, in other words synchronization of the movements of Venus and the Earth. As part of this hypothesis, the idea has even been advanced that Venus had once been captured by the Earth, which is very difficult to accept. In fact, the exact resonance period is 243.16 days and the small difference (0.15 days) is sufficient to suggest that it is just a coincidence.

5.2.3 The Surface of Venus

Topography

The first radar images of the surface of Venus were obtained in 1974–5 by the Goldstone observatory in the United States and by Arecibo in Puerto Rico. Both operated at wavelengths of 12.6 cm, corresponding to the S band at 2.32 GHz. The spatial resolution of these observations is good and may reach several km. But the great distance of the planet, on the other hand, does not allow differences in altitude to be differentiated, only differences in albedo, which essentially means in roughness. This is how numerous circular formations were attributed to being impact craters.

The Pioneer Venus mission completely changed our ideas about the surface of Venus, with its almost complete coverage of the planet at a moderate resolution and with a remarkably high precision in altitude (200 m). It appears that most of the surface (70 %) consists of a series of plains, whose altitude is close to the reference level, above which there are a few high regions, corresponding to about 10 % of the surface. The remaining 20 % of the surface lies below the reference surface (Fig. 5.5).

The plains are slightly undulating and they contain numerous circular structures, which appear dark on terrestrial radar images. These structures might be of volcanic origin or the results of meteoritic impact. Venera 15 and Venera 16 seem to have shown numerous small-sized volcanoes in the high-latitude regions that they explored (Fig. 5.6).

Major relief primarily consists of two plateaus: *Ishtar Terra* and *Aphrodite Terra*. The former contains in its eastern section *Maxwell Mons*, the highest point on Venus, which reaches up to about 12 km. Each vast plateau is the size of a terrestrial continent, and dominates the surrounding plains with high cliffs. Smaller in size, *Beta Regio* has a summit 4000 m high. It may be a region of recent volcanic activity.

The depressed regions generally have roughly circular outlines, and are reminiscent of impact craters.

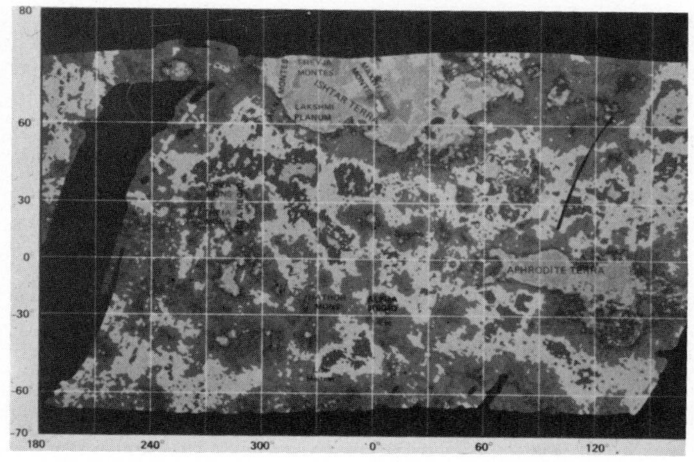

Fig. 5.5. Topographic map of Venus obtained by radar from the Pioneer Venus Orbiter in 1979. (By courtesy of NASA)

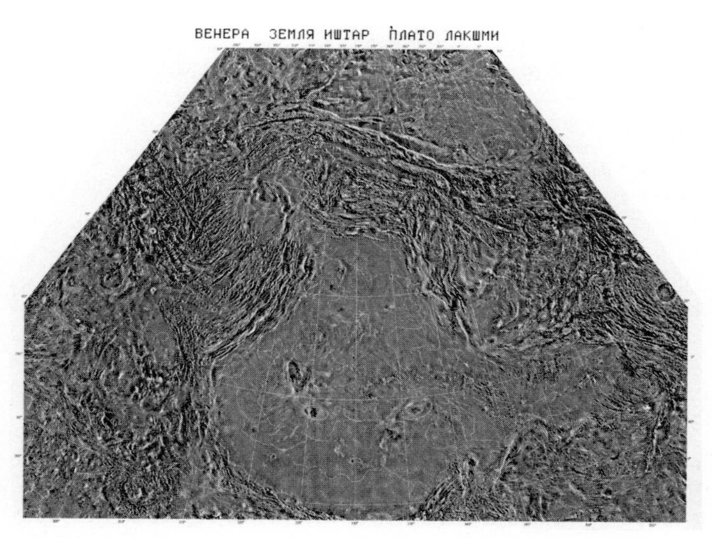

Fig. 5.6. High-resolution (1.5 km) radar map of the Lakshmi Planum region, obtained by the Soviet Venera 15 and 16 probes. (Academy of Sciences of the USSR)

At the resolution given by the Pioneer Venus imagery, the geological formations that have been discovered do not seem to indicate plate tectonics. But there are, on the other hand, several signs of intense activity, which could be the results of single-plate tectonics having caused the crust or the lithosphere of Venus to deform. It is possible to make out a network of fractures and collapse structures, grabens and ridges, especially in the Beta Regio area. The most striking characteristic may be the presence of extremely large volcanoes. Overall, the activity on Venus appears to be closer to that of Mars than to that on the present-day Earth.

The Composition of the Surface

The first series of measurements was carried out by Venera 8 (in 1972), Venera 9 and Venera 10 (in 1975). Their lander capsules had γ-ray spectrometers, which allowed the amount of the radioactive elements U, Th and K to be determined. The three sets of measurements revealed great differences in the abundances of these elements (Table 5.1); the soil analyzed by Venera 8, which was systematically the richest, resembles a sample of granite, unlike the two others, which have compositions close to those of basalts.

Table 5.1. Radioactive element content of surface rocks of Venus analyzed by Soviet automatic probes

Rocks at landing sites and terrestrial rocks	Content of radioactive elements		
	potassium (%)	uranium (10^{-4} %)	thorium (10^{-4} %)
Venera 8	4.0 ± 1.2	2.2 ± 0.7	6.5 ± 0.2
Venera 9	0.47 ± 0.08	0.60 ± 0.16	3.65 ± 0.42
Venera 10	0.30 ± 0.16	0.46 ± 0.26	0.70 ± 0.34
Venera 13	4.0 ± 0.6	—	—
Venera 14	0.2 ± 0.1	—	—
Basalts	0.76	0.86	2.1
Granites	3.24	9.04	21.9

(By courtesy of Y. A. Surkhov.)

The most precise results were obtained by the Venera 13 and 14 probes, which landed on Venus in 1982 March, and the Vega 2 probe, which dropped a landing module in 1985 June. They each carried an instrument that was able to measure the X-ray fluorescence of soil samples introduced into a vacuum chamber at 300 K. The fluorescence was excited by irradiation from two radioactive sources: ^{55}Fe and ^{238}Pu. Venera 13 obtained 38 spectra, over an integrated counting time of 384 s, whilst Venera 14 measured 20 spectra in 192 s. Despite the extremely short experimental times, imposed by the probes' own short lifetimes, the statistical information was sufficiently good to allow the chemical abundances for the two sites to be determined. Expressed as the percentage oxide abundances, these measurements are given in the second and third columns in Table 5.2.

It will be noted that, apart from the quality of the results (which have accuracies comparable with those from the Viking experiments on Mars, which had the benefit of much longer integration times), the samples analyzed at the two sites show considerable differences, the most important being the potassium abundance. Venera 13 landed at a site at a moderate altitude (1.5 km) slightly higher (by about 1000 m) than Venera 14.

Table 5.2. Chemical composition of surface rocks of Venus analyzed by the Venera 13 and 14 probes, compared with that of a terrestrial basalt

Elements	Venera 13	Venera 14	Tholeitic basalt
MgO	11.4 ± 6.2	8.1 ± 3.3	6.3
Al_2O_3	15.8 ± 3.0	17.9 ± 2.6	14.1
SiO_2	45.1 ± 3.0	48.7 ± 3.6	50.8
K_2O	4.0 ± 0.63	0.2 ± 0.07	0.8
CaO	7.1 ± 0.96	10.3 ± 1.2	10.4
TiO_2	1.59 ± 0.45	1.25 ± 0.41	2.0
MnO	0.2 ± 0.1	0.16 ± 0.08	0.2
FeO	9.3 ± 2.2	8.8 ± 1.8	9.1

(By courtesy of Y. A. Surkhov.)

If these compositions are compared with those of sites on Earth, the samples analyzed by Venera 13 may be likened to the alkaline basalts rich in potassium of the Earth's crust, which are only found occasionally and mainly on oceanic islands and rift zones in the Mediterranean region. In contrast, the samples analyzed by Venera 14 resemble basalts found in regions of the oceanic floor. These similarities in composition may correspond to provinces with the same geomorphology and petrographic characteristics.

Geochemists are currently working on possible models of the interaction between the surface and the chemically very active atmosphere, together with possible tectonic activity, in order to account for the presence of minerals that have already been identified.

5.2.4 The Atmosphere of Venus

The atmosphere of Venus is very massive: the surface pressure is about 1.03×10^6 kg/m^2, and therefore about 100 times the terrestrial value. The average temperature is also very high: 730 K. Profiles of the temperature and pressure to an altitude of 100 km also reveal important differences when compared with the Earth's atmosphere (Fig. 5.7).

The main difference is the following: on Earth, absorption of solar ultraviolet radiation by oxygen and ozone causes heating of the atmosphere, which has a temperature maximum at about 45 km altitude. There is therefore a temperature inversion at about 12 km altitude, and the appearance of a *stratosphere* between about 12 and 45 km, above which extends the *mesosphere*. On Venus, the extremely high surface temperature caused by a strong greenhouse effect (see Sect. 5.3.5), as well as a lack of oxygen and ozone, prevents this structure from occurring: temperature declines continuously with height until about fifty km altitude, remaining approximately constant beyond that point until about 100 km. The *troposphere* is directly beneath the *thermosphere*. The latter name, which indicates that the temperature increases with height, only partially corresponds to the truth: although the temperature does increase to about 300 K on the day side, it decreases to about 100 K on the night side, which has sometimes led to it being called the *cryosphere*. This decrease in temperature

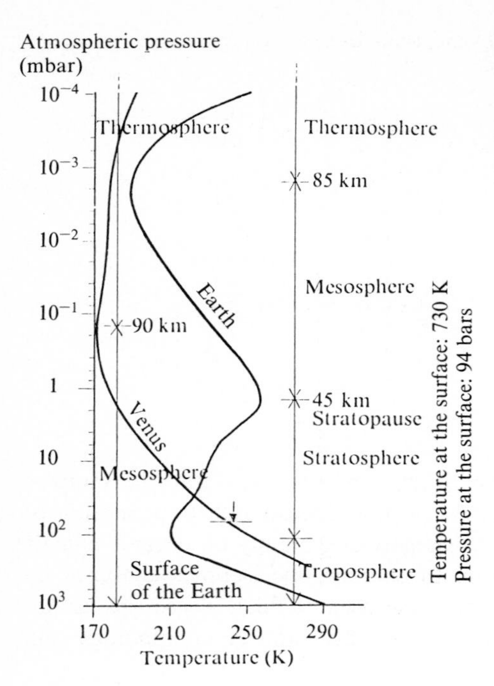

Fig. 5.7. Temperature profile of the atmosphere of Venus compared with that of the Earth. [After G. Israël: "Venus", in *Le grand Atlas Universalis de l'Astronomie* (Encyclopedia Universalis 1983). By courtesy of the publisher]

is much more important than it is for Mars, where a similar profile is observed. (In fact, the physics of the atmosphere of Venus is still very poorly known.) Above the thermosphere the *exosphere* is found, at an altitude of 135 km on Venus and about 700 km on Earth.

The composition of the lower atmosphere of Venus (the troposphere) is such that CO_2 (96.5 %) and N_2 (3.5 %) make up more than 99.9 % of the whole. Next in importance are SO_2 (0.02 %), A (0.007 %) and Ne (0.001 %). Water has been observed at a possible concentration of about 0.01 %. Among traces of other rare components detected we may mention CS and H_2S. At higher altitudes, where H_2O might be more abundant, CO, O_2 and O have been found. These arise from photodissociation of CO_2 by solar radiation, which also causes photoionization above about 150 km. O_2^+ and then O^+ are the most abundant ions.

One important characteristic of Venus' atmosphere is the presence of clouds and hazes at altitudes between 30 and 80 km. The vertical stratification of the clouds appears to be extremely stable. Below 30 km, the sky is always "clear". Above that altitude there is a thin haze. The base of the cloud layer itself is continuous except for some possible "breaks in the clouds". This layer ends at about 64 km, and is overlain by yet another haze layer. The cloud layer is itself subdivided into three layers as a result of differences in concentration and composition.

The particles comprising the clouds, or *aerosols*, are generally very small, being below 10^{-2} mm. In the upper layer they consist mainly of sulphuric acid H_2SO_4 in an aqueous solution. The mechanism by which this is formed is probably similar to terrestrial acid-rain pollution, a reaction of SO_2 with H_2O under the action of solar radiation. These droplets form a fine rain and evaporate at the lowest altitudes,

H_2SO_4 being then turned back into SO_2 and H_2O. At the base of the haze layer, the solar ultraviolet flux is insufficient for acid-synthesis to take place. Ultraviolet radiation with a wavelength of less than 200 nm actually penetrates down to a level corresponding to a column density of about 10^{18} cm^{-2}. This is found at about 60 km. Convection within the troposphere carries water vapour and sulphur anhydride up to the top of the cloud layer, and the cycle continues. It has two consequences: the maximum abundance of water is at cloud level (around 50 km); and most of the components in the clouds found in the lower layer are not sulphides but chlorides.

The presence of H_2O and SO_2 in the clouds adds considerably to the greenhouse effect on Venus, and this even though H_2O and SO_2 are relatively minor components by comparison with CO_2. The effect is very important, because the amount of solar flux available for heating the planet is less than 5 % of that incident at the top of the atmosphere, when allowance is made for the very high albedo of the gas and clouds (which absorb, reflect or scatter more than 95 % of incident photons). Despite its preponderance, carbon dioxide alone cannot account for the greenhouse effect, because the absorption bands of CO_2 in the planet's thermal spectrum leave very wide "windows" through which thermal radiation could escape. The role of H_2O and SO_2 is to fill in the atmospheric absorption spectrum.

The low concentration of water in the atmosphere of Venus is explained by photodissociation of H_2O, followed by gravitational escape of the hydrogen. The rate of loss is estimated at 10^8 atoms of hydrogen per cm^2 per second. This rate of loss, assuming that it has been constant throughout the planet's lifetime, leads to a very low estimate of the original water content, so that Venus would always have been an extremely dry planet. The major problem with such a model is that it does not take the very high temperature of the present surface into account. If current abundances of CO_2, H_2O and SO_2 are sufficient to maintain high temperatures, it still remains to be explained how the greenhouse effect started in the first place. Somehow, starting at a much lower temperature – because the original solar luminosity was probably much lower – large quantities of CO_2 have to be extracted from the rocks. One possibility would be to assume that Venus once had a considerable amount of water, in a quantity similar to that of the Earth. An initial violent degassing, for example by volcanism, could have injected this water into the atmosphere, at rates of the order of 10^{10} H_2O $cm^{-2}\,s^{-1}$. This could have led to an increase in the temperature of the surface, accelerating the CO_2 degassing, which would in its turn have increased the greenhouse effect. This could have led to a chain-reaction, which finally reached an equilibrium dominated by a much slower loss of water, corresponding to present conditions. According to this model, Venus passed through a stage in its evolution when its surface may have had water in the liquid state. So one cannot exclude the possibility that significant organic or even biochemical activity may have taken place, only to disappear as a result of later heating. In this case, high-resolution images of the surface ought to be able to confirm the presence of earlier water-flow or dry basins.

Among the tests of atmospheric evolution, the ratio between the two hydrogen isotopes D and H is one of the most sensitive. In the theory of violent degassing by volcanism, no fractionation is to be expected as the two isotopes show similar behaviour under such conditions. But this is not the case with gravitational escape following photoionization. This process favours the escape of the lightest elements, and one may therefore expect a D/H ratio that is considerably enriched over the original value. The measurements that have been made (D/H $= 10^{-2}$) appear to indicate that there has been such enrichment by a factor of about 100. These measurements should be confirmed in the next series of space-probe experiments, and it should be possible to carry out calculations following the atmospheric evolution of Venus over the last four thousand million years.

5.3 The Earth

Study of the Earth involves such a wide range of disciplines that it would be useless to attempt to describe everything within the limits of this book. Just to take a single example, the evolution of the Earth involves various specialized aspects of geology and geophysics, as well as the physics of solid bodies, nuclear physics, hydrodynamics, plasma physics, chemistry and biology. Our knowledge of the way in which the Earth evolves and its past history has therefore been helped immeasurably by the amazing advances that have been made in many areas of science. Our view of the Earth's place in the universe, like that of the uniqueness of the appearance of living organisms, has changed very rapidly over the course of the last few decades.

A decisive step was taken when Man acquired the capability of actually going into orbit around the Earth, and then of "leaving" its gravitational field by acquiring a velocity higher than escape velocity, thus beginning the era of interplanetary travel. The Earth suddenly became a planet in front of one's eyes, rather than just the ground beneath one's feet. Moreover, the Earth came to be seen as a perfectly ordinary planet. More recently, the fact that no biological activity, not even fossil, has been found on the surface of any other planet in the Solar System has slowly

Fig. 5.8. The Earth photographed from the Meteosat satellite. (By courtesy of the European Space Agency)

changed that view, and emphasized the very characteristic physical features of this planet: the composition and dynamical behaviour of its atmosphere, and the presence of plate tectonics are just examples. At present, despite the sophisticated experiments that are being carried out to try to understand the processes underlying these specific features, many questions remain unanswered. The aim of this chapter is to summarize the knowledge that we have gained purely from the point of planetology.

5.3.1 Orbital Characteristics

The Earth is the third planet in heliocentric distance, and the most massive of the inner planets. Tables 1.1 and 1.3 in Chap. 1 give its orbital data and its essential overall physical characteristics.

The orbit of the Earth around the Sun, which defines the plane of the *ecliptic*, is not strictly circular. The eccentricity of the ellipse, although small, leads to a difference between the minimum heliocentric distance (at perihelion) and the maximum (at aphelion) of 5 million km, that is 3.3 %. The amount of solar energy received at the Earth's distance, which depends on the square of the heliocentric distance, therefore varies by more than 6.5 % between one and the other. However, this is not the reason for the seasons, which are instead governed by the considerable inclination ($23°27'$) of the rotational axis to the normal to the ecliptic.

It is generally accepted that the cause of this obliquity arose from the impact of protoplanetary bodies (meteorites) that took place during the process of the Earth's formation, or in the few hundred million years that immediately followed it. Nothing indicates that it has changed significantly since then. This is not the case with the period of the Earth's rotation, i.e, with the length of the day. The Moon's tidal effect, discussed in the next section, causes a simultaneous braking of the Earth's rotation and a corresponding increase in the Moon's distance. The uninterrupted increase in the length of the day has been shown by measuring the variation with geological age of the thickness of layers that laid down each day by certain micro-organisms. Over the course of the last 400 million years, is seems that the day has increased on average by about ten to twenty millionths of a second per year. As tidal effects appear unable to influence the length of the year, it is therefore the number of days in a year that has decreased by about one day every ten million years.

The Earth's rotation is like that of a gyroscope: the direction of the axis of rotation is not fixed but describes a cone, tracing an angle of $23°27'$ (the obliquity) around the normal to the ecliptic. This movement of *precession* is called the precession of the equinoxes. The line of the equinoxes slowly rotates around the sky, with an angular velocity of $50.3''$ per year, which corresponds to a period of 26 000 years. For a terrestrial observer the position of the stars on the sky with respect to the Earth's north pole therefore varies with time, and this effect was discovered by Hipparchos in the second century B.C. by comparing his observations with those of Timocharis made 150 years earlier.

The precession of the equinoxes is produced by the gravitational attraction of the Sun and the Moon on a body that is not perfectly spherical and which is rotating about an axis inclined to the ecliptic. The point at which the forces are applied are not precisely the same as the Earth's centre of mass. A torque therefore results, which attempts to pull the equatorial bulge down into the plane of the ecliptic.

Other variations with time in the orbital parameters are produced by the gravitational perturbations of the other planets in the Solar System, mainly that of Jupiter. We may mention the variation in the eccentricity of the orbit, which has a period of 100 000 years, the rotation of the line of apsides (the ends of the major axis of the ellipse) or *precession of perihelion*, and the variation in the obliquity, which varies from 21.5° to 24.5° with a period of 41 000 years.

Overall, the effects on the climate of the annual variation in the Earth's insolation (more than 6 %) depend on latitude and evolve with time with a number of periodicities that are now well-established. At present, the Earth is at perihelion on January 3: the Sun is closest during the northern hemisphere's winter. We may therefore expect seasonal climatic variations to be less in the northern hemisphere than in the southern, where the Sun is closer in the summer and farther away in the winter. Because of the precession of the equinoxes, this situation will be reversed in about ten thousand years. Naturally, accurate climatology must take other factors into account, in particular the effect of the oceans which show a strong north-south asymmetry. However, astronomical effects are the basis for modern theories about climatic variations, and which, in particular, account for the periodicity in the ice ages.

5.3.2 Internal Structure

The structure of the Earth has been deduced from several sources of information. The first indication is that the average density of the Earth (5.5) differs from that of the silicates that form most of the surface rocks (around 3). This suggests that the composition is differentiated, with denser materials in the interior. However the deepest samples that have had their composition analyzed, and have been brought to the surface by volcanic eruptions, only come from the outermost few hundred km, where silicates greatly predominate.

Deeper probing can only be carried out rather indirectly, by *seismology*. In earthquakes, and artificial explosions underground, pressure waves propagate through the Earth. They are detected at different seismological stations on the surface of the Earth. The analysis of the variation with time of the amplitude of the waves enables their velocity to be calculated, and this is dependent on the density and the elastic properties of the regions that have been crossed. The two principal components, transverse (the *S wave*) and longitudinal (the *P wave*) of the seismic waves behave differently in media with different properties. An S wave can only be propagated in a solid medium, unlike a P wave which can travel through either a solid or liquid medium. Analysis of seismic records does not allow the density of the different zones to be determined, but does give information about changes in density. By making many measurements, from many stations, a model of the Earth's internal structure is obtained. This shows several distinct layers (Fig. 5.9): a central *core*, which has a radius of 3400 km, surrounded by a *mantle* with a thickness of 2900 km, itself surrounded by a thin *crust*, only 10 to 40 km thick. The discontinuity between the crust and the mantle is known as the *Moho* – or, more correctly, the *Mohorovičic*

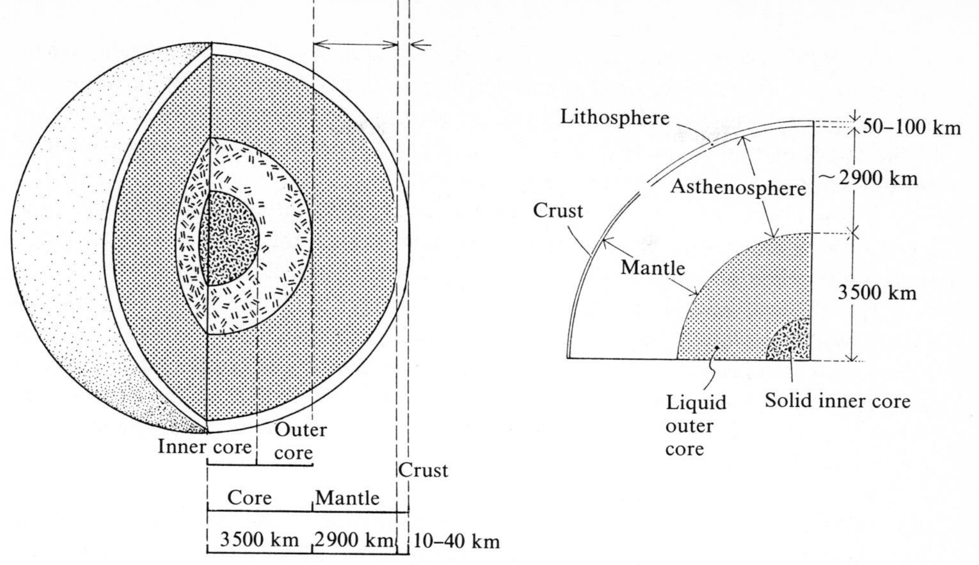

Fig. 5.9. Schematic section of the Earth, showing how the structure may be regarded as consisting of either core, mantle and crust, or core, asthenosphere and lithosphere

discontinuity – named after the Yugoslav geophysicist Andrija Mohorovičic (1857–1936) who discovered it in 1909.

Core, mantle and crust differ in their mineralogical composition. The crust, consisting of aluminium, calcium, potassium and sodium silicates, is directly accessible for analysis, as is the outer mantle. The composition of the inner mantle and the core are deduced from models and from laboratory data concerning the density and mechanical properties of different materials as a function of their temperature and pressure. As a result it is thought that the mantle consists of ferro-magnesian silicates, i.e. mainly olivines. One important consequence of the difference in composition between the crust and the mantle is that the density of the mantle is greater than that of the crust. The main constituent of the core must be iron, probably in the form of a nickel alloy, with a mixture of lighter elements such as oxygen and sulphur.

This division into core, mantle and crust only takes the differences in the Earth's physical properties partly into account. First, the Earth's crust consists of two types of landform that are very different, the continents and the oceans. The continents are generally of granitic composition whilst the ocean floors consist of basaltic rocks. From a geological point of view, the distinction between oceans and continents does not correspond to common usage, which simply describes the distribution of the Earth's water masses. At present, water covers not only the ocean floors, but also a large part of the continents. If the water were to disappear, a variety of different forms of terrain would become visible, distinguished by their age, origin, composition and physical properties.

Beneath the terrestrial crust, the mantle, rigid at the top, becomes viscous (*magma*), below a depth of about 100 km. This "softer" layer is known as the *asthenosphere*. Everything above it is the *lithosphere*, which therefore includes part of the mantle and the crust (Fig. 5.9). What distinguishes the upper mantle, which is part of the lithosphere, from the lower mantle is not chemical composition, but mineralogical state, caused by the increase in pressure towards the interior. As far as the upper mantle is concerned, it is generally considered that the rock consists of peridotites, mixed with pyroxenes and olivines.

Below about 2900 km, S-waves are no longer transmitted, indicating the existence of a liquid phase. For this reason the core is thought to consist of an inner, rigid region with a radius of about 1250 km, and an outer, liquid zone.

The thermal profile of the Earth, in accordance with this structure, may be as follows: the temperature at the surface, 288 K, is explained by heating by solar radiation, which is absorbed in the uppermost few metres. The temperature increases progressively towards the interior, reaching more than 3000 K at the mantle/core interface, and perhaps 5000 K at the centre of the core, where the pressure would reach several megabars.

As thus described, models of the internal structure deduced from seismic data show spherical symmetry. In reality, closer examination shows important deviations from this symmetry, notably as far as the lithosphere is concerned. The Earth's crust is lower in density than the underlying mantle. In particular, the continents "float" on the outer mantle, like blocks of ice in water, in other words in accordance with Archimedes' principle. When applied to solid materials, this principle is known as *isostasy*: each element of the surface of a particular horizontal plane in the mantle supports an equal mass of material. As the continents have, in addition, less density than the oceanic floors, the Moho, which is located at a depth of just a few kilometres below the oceans, may attain depths of anything from ten to nearly one hundred kilometres beneath the continents. This depth changes with time: erosion of the surface of the continents leads to their being elevated, rather like a ship that rises as its cargo is discharged.

The geographical distribution of the outer layers of the Earth can be interpreted in terms of the modern theory of plate tectonics.

5.3.3 Plate Tectonics

It was Wegener (1880–1930) who was responsible for providing a firm basis, between 1910 and 1930, for the idea that the continents "drift", i.e. that they move relative to one another at rates that are of the order of a few cm per year. Africa and South America, for example, are drifting apart, giving rise to the Atlantic Ocean.

Nowadays there are abundant experimental proofs showing the relative movements of the various continents, and indicating that they formed a single land-mass about 250 million years ago. First of all one can cite the similarities between different coastlines on both sides of seas and oceans: Brazil and the Gulf of Guinea, the Arabian peninsula and the Horn of Africa, etc. Second, geological formations can be traced on the opposite sides of the oceans: the change from crystalline basement to sedimentary terrain agree perfectly between South America and Africa if the pieces are "fitted together" on a map from which the Atlantic has been removed. In the same way, identical species of rare fossils can be found in southern Africa and Brazil, which enables one to assume that they were once in contact: a general spread via the ocean would have led to a far greater geographical range of the species concerned.

Additional indications come from palaeomagnetism. Magnetic minerals lose their magnetization when their temperature rises above the Curie point. In a volcanic eruption, incandescent lava is "demagnetized". In cooling, magnetic minerals (such as magnetite, Fe_3O_4) acquire the characteristics of the magnetic field at the time and place that they crystallize. Analysis of the magnetic grains trapped in sedimentary strata, which can be dated by independent methods, allows one to follow the changes with time of the orientation and intensity of the magnetic field. As we are dealing with a dipolar field, these characteristics are a function of latitude. Consequently, a variation with time of the orientation of the field at a given place indicates how that point has moved, thus allowing the drift of the various continents to be traced over time.

Reconstruction of the drift allows us to go back 250 million years. It seems that there was then a single continent, Pangaea (from the Greek $\pi\alpha\nu\gamma\eta$) surrounded by a giant ocean, Panthalassa ($\pi\alpha\nu\theta\alpha\lambda\alpha\sigma\sigma\alpha$). Pangaea first broke into two more or less equal portions: Eurasia and North America, linked by Scandinavia and Greenland, slowly separated from South America, which was joined to Africa, Antarctica, and Australia. Then the Indian and Arctic Oceans, and the Mediterranean, formed. Africa separated from America, and India, moving northward, collided with Asia, creating the greatest continental folding along the Himalayan chain. This movement continues today, at a rate of about 5 cm per year.

Continental drift is one sign, one consequence, of a more widespread activity by the terrestrial lithosphere. First, it is not just the continents that shift. Between the continents the sea floors show the same sort of movement. The sea floor is continuously created from mantle material that has been heated and has risen towards the surface by convection. It emerges in the zones known as "ridges", where its cooling leads to the formation of new sea floor. The oceanic ridges, which have a total length of more than 50 000 km, are important relief features that may attain heights of several thousand metres. They are mainly below the sea, but sometimes develop on dry land, as in the cases of Iceland and in the Afar region of Africa.

Production of new sea floor does not increase the Earth's surface area. It is accompanied by the disappearance of an equal quantity of sea floor, which returns to the mantle in what are known as *subduction zones*, which form oceanic trenches (Fig. 5.10). Overall, the lithosphere is divided into about twelve large *plates*, separated by ridges on one side and trenches on the other (Fig. 5.11). Although these plates are rigid, they are in perpetual movement, at rates of 2 to 10 cm per year. This motion can take place because of the elastic properties of the asthenosphere on which they rest.

In moving, the lithospheric plates carry along any continents that may lie on top of them. There is, however, a fundamental difference between the motion of the sea floor and that of the continents: whilst the sea floors, created at the ridges, return to the mantle by subduction at the trenches, the continents, with lower densities, are not carried down into the subduction zones. They cannot be submerged.

The oceanic floors therefore consist mainly of young material, the period of time between their creation at a ridge and their sinking at a subduction zone does

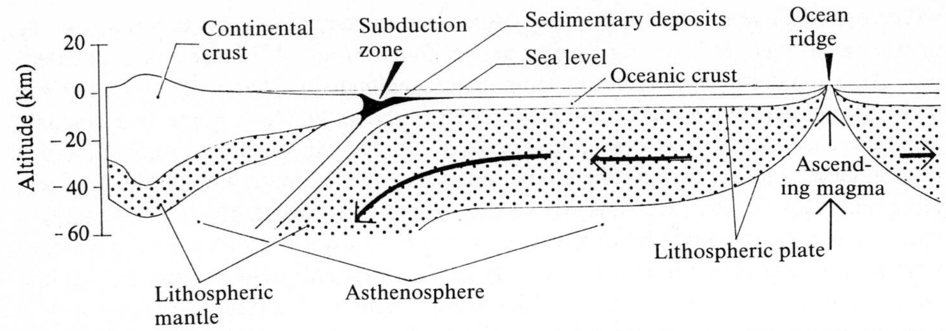

Fig. 5.10. A schematic diagram, in section, of a part of the Earth's lithosphere. The vertical scale is deliberately exaggerated. [After B.W. Jones: *The Solar System* (© Pergamon Books Ltd. 1984)]

not exceed 100 to 200 million years. This contrasts with the age of the continents, which may be in excess of three thousand million years, as is the case with certain areas of Greenland, which are probably the oldest rocks on Earth.

A direct effect of plate motion is *orogenesis*, or mountain-building. This mainly occurs at subduction zones, where two plates moving in opposite directions collide. Two specific cases may be mentioned. The first is when only one of the two colliding plates carries a continent. When this reaches the trench, its motion is halted and this results in the appearance of a mountain chain parallel to the trench. A typical example is that of South America, which is in contact with a subduction zone, the Peru-Chile Trench, on its western side. This prevents it from travelling farther to the west and has caused the emergence of the mountain chain forming the Cordillera of the Andes.

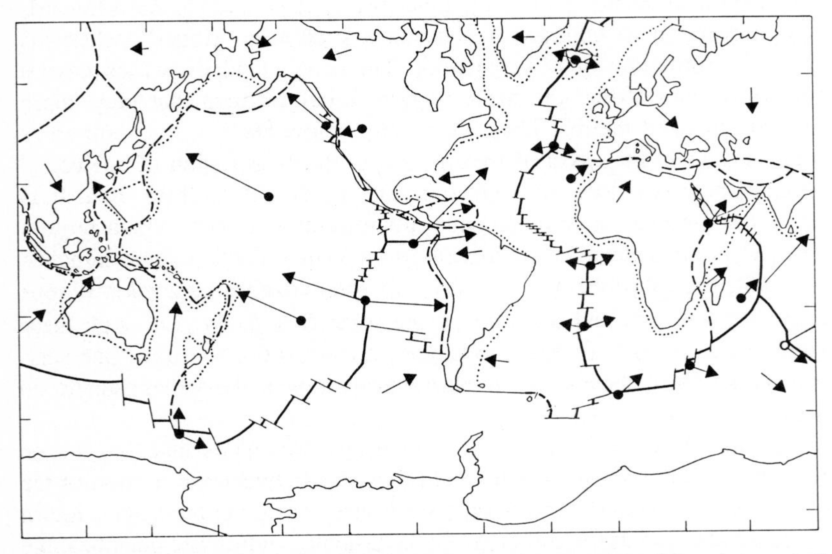

Fig. 5.11. A map by Jason Morgan, showing the *absolute motion* of the plates relative to a fixed reference formed by the system of "hot spots". The length of each arrow is proportional to the absolute velocity of the plate. [C. Allègre: *L'écume de la Terre* (Fayard 1983). By courtesy of Editions Fayard]

This is unlike Africa, for example, which is situated between two ridge-systems, the Mid-Atlantic and the Indian (Fig. 5.12).

The second case is where the collision between plates is that between two continents, which have both arrived at the same subduction zone. In this way the collision between the African plate, which is moving north-east, and the European plate has resulted in the formation of the Alps. In the same fashion, India, which separated from Africa about 120 million years ago, drifted northwards and collided with Asia along a subduction zone lying along the edge of what is now Tibet. The "welding together" of India and Asia was accompanied by the formation of the Himalayan chain.

The energy to drive the plate motion, which is particularly evident in orogenesis, is provided by the mantle, and perhaps also the core. In the mantle, the source is probably the radioactive disintegration of long-lived isotopes, that is of uranium (^{235}U and ^{238}U), of thorium (^{232}Th), and of potassium (^{40}K), which have half-lives of 7.1×10^8 and 4.51×10^9, 1.41×10^{10}, and 1.28×10^9 years, respectively. In the core, energy may be liberated by progressive solidification of the outer, liquid core. Although these heat sources are evenly spread throughout the volume, they lead to a temperature profile that shows a continuous decrease in temperature T with radius R. The energy can only be lost through radiation at the surface. The thermal regime may, a priori, be either one of conduction or convection, according to the value of the absolute temperature gradient dT/dr. This can only be measured within the crust, where it is about 25 K/km. Under these conditions the transfer of heat towards the exterior takes place by conduction. In the asthenosphere on the other hand, it is thought that the transfer is by convection. The hot material, with lower density rises through the cooler material. The ascending material mainly reaches the surface beneath the ridges. Descent of the colder and thus denser material, by contrast, begins at the subduction zones. It appears that these mantle convection currents occur at rates of a few metres per year at the most: we are not dealing with a liquid medium but with a solid, and it is only on a time-scale of hundreds of thousands of years that the mantle behaves as a convective fluid.

Other signs of plate motion are seismic activity and volcanism. The plates move without deformation and so it is at the boundaries between plates that most of the internal energy is released. These boundaries are of three types: ridge, trench and fault. The complex motion of the plates relative to one another basically takes place along "transform faults", which are where most of the Earth's surface seismic activity occurs. Various forms of fault are recognized: those that cut the ridges as found along the Mid-Atlantic Ridge; faults that split or connect trenches, like the Mediterranean faults along which the great seismic activity of Turkey and Yugoslavia occurs; and the faults that connect ridges to trenches, as with the San Andreas Fault, which is responsible for the considerable seismic activity in California. The earthquakes occurring at the base of the subduction zones arise from release of extreme mechanical stresses accumulated in the fold zone where one plate is forced underneath another. The steady movement of the plates, which is measured in cm per year, stresses the plates at their subducting boundaries. Sinking of the plates into the mantle takes place in fits and starts, as soon as the stored energy passes a certain critical level. Several metres of plate then disappear as they sink, which causes a

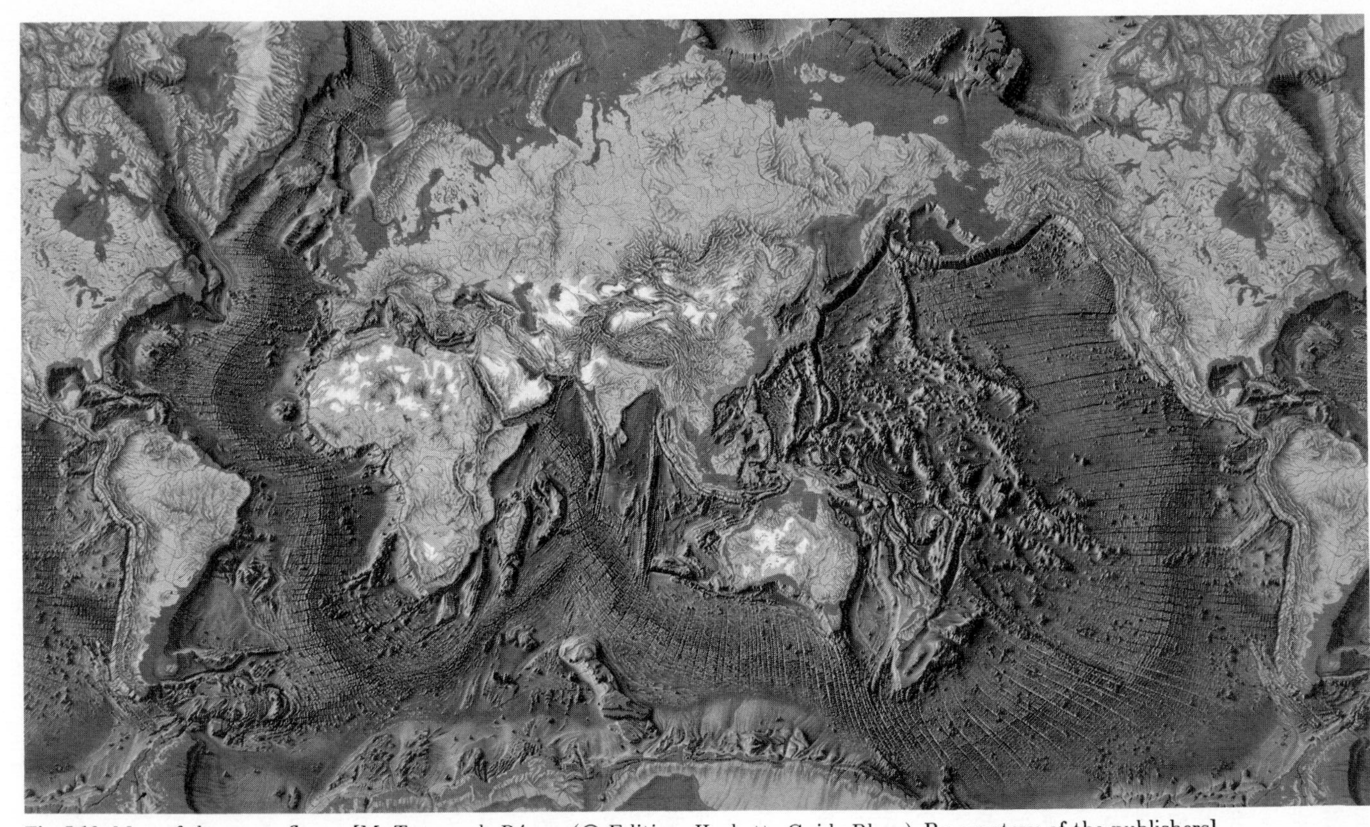

Fig. 5.12. Map of the ocean floors. [M. Tanguy de Rémur (© Editions Hachette Guide Bleus). By courtesy of the publishers]

violent "earthquake". When this is mainly beneath the sea, it is accompanied by a "tidal" wave – a tsunami – as happens in Japanese seismic events.

The oceanic ridges and the subduction zones are also the sites of considerable volcanic activity. The former therefore produce submarine volcanism, except in Iceland. Volcanoes in Japan, the Sunda Isles, the Philippines, Central and Latin America are along subduction zones. It will be seen that volcanic regions, lying on plate boundaries, undergo very considerable activity that is both seismic and volcanic. Transform faults, on the other hand, see considerable seismic activity, but no volcanism (Turkey, California, Mexico, etc.).

Although it had difficulty in becoming accepted, plate tectonic theory is now able to account for most geological activity. Nevertheless it should be mentioned that certain observations suggest a more complex interpretation, without, however, calling into question the basic principles of plate tectonics. For example, the statement that geological activity is concentrated along the boundaries between plates is contradicted by a number of facts: the presence of volcanoes in the middle of a plate, like the Hawaiian volcanoes and the very numerous submarine volcanoes in the Pacific; the intense seismic activity in China; the formation of the Pyrenees, etc. To explain the origin of intraplate volcanoes it has been proposed that there are sources of heat deep beneath the lithosphere, known as "hot spots" that create plumes of ascending magma, which rise to the surface not just at the ridges but also at various, fixed, points around the world. Sporadic eruption of this magma through a moving lithosphere, would account for the existence of chains of oceanic islands (Figs. 5.13 and 5.14), rather like a perforated paper tape. As for intracontinental orogenic or seismic activity, that would arise at points, far from the margins, where continents had been weakened during the course of their travels and collisions, and would represent the release of forces generated at the plate boundaries.

So over the course of their lifetime, the continents have grown, collided, broken up, been welded together again, etc. The mosaic of present-day plates, as we know, has continued to evolve in the last 250 million years. It is now possible to follow the history of the destruction of Pangaea and the geological evolution that was a result. On the other hand, we have hardly any information about the continents that existed before that phase when there was just one: did Pangaea appear as a result of the incessant movement of the continents, shortly before the break-up, or had it existed as a stable entity for several thousand million years? From what we now know of the causes of continental drift, the first possibility seems the more likely.

5.3.4 The Earth's Magnetic Field

The presence of a zone containing a metallic liquid, the outer core, is the origin of the Earth magnetic field. Rotation of this fluid, either as a result of the Earth's rotation, or because of thermal convection, gives rise to electrical currents along closed field lines: the resulting magnetic field is dipolar, in other words, at distances that are large compared with the size of the loops of current, the shape of the lines of force, independent of the lines of current, is that of a magnetic dipole located at the Earth's magnetic centre. By analogy with an electrical dynamo, which converts

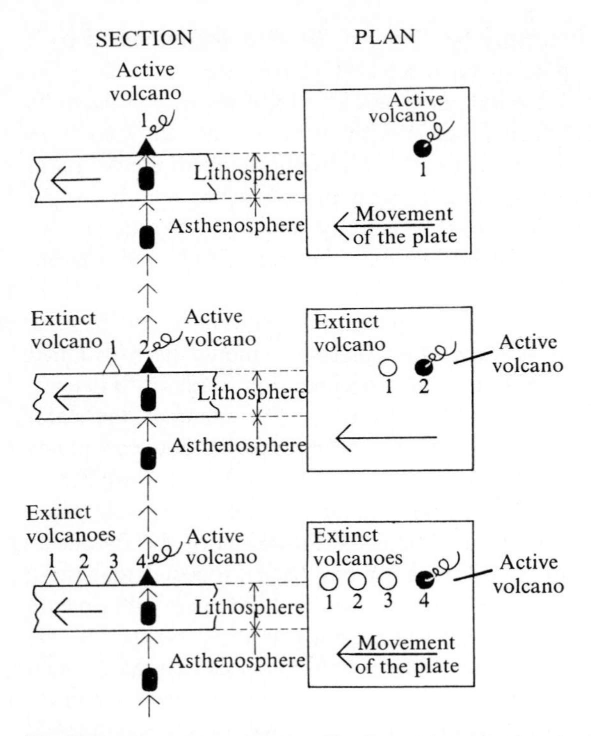

Fig. 5.13. Diagram explaining the formation of volcanic chains. A fixed point under the lithosphere generates "bubbles" of magma. The lithosphere moves across the hot spot and is "pierced" by the bubbles of magma, producing a series of blisters on the surface. The result is an alignment of the structures formed over the perforations. The model is shown in both section and in plan. [C. Allègre: *L'écume de la Terre* (Fayard 1983). By courtesy of Editions Fayard]

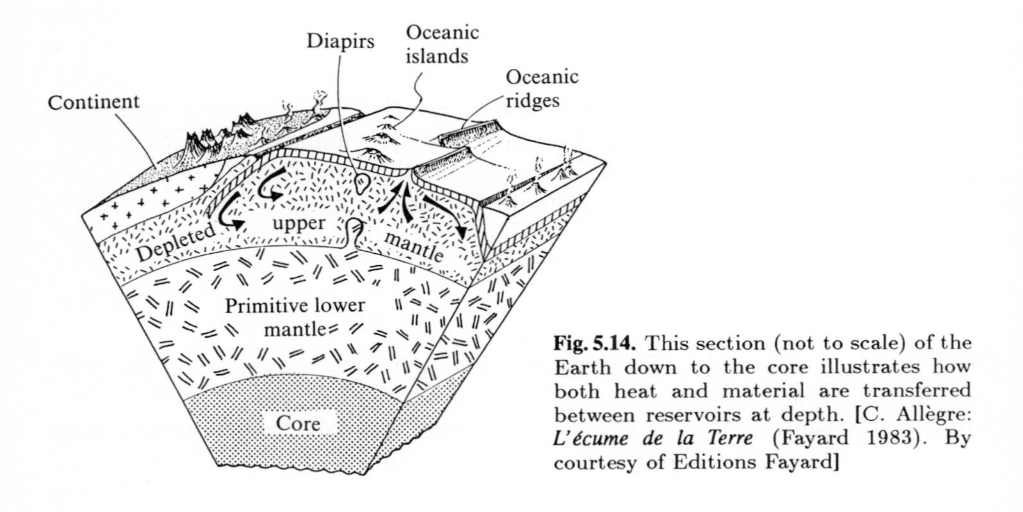

Fig. 5.14. This section (not to scale) of the Earth down to the core illustrates how both heat and material are transferred between reservoirs at depth. [C. Allègre: *L'écume de la Terre* (Fayard 1983). By courtesy of Editions Fayard]

kinetic energy into electrical energy, the theory explaining the origin of the Earth's magnetic field is known as the *dynamo theory* (see Sect. 4.3.3).

As reconstructed from its characteristics at the surface of the Earth, the magnetic field has an axis of symmetry, the magnetic axis, slightly tilted (by 11°) with respect to the Earth's rotational axis. The Earth's magnetic centre, the intersection of the magnetic axis and equator, is therefore different from the Earth's centre of mass. At any point on the surface the geomagnetic field is characterized by its intensity and its direction, defined by the angle of declination between the directions of geographic and magnetic north, and the angle of inclination (or dip) between the magnetic field and the local horizontal plane.

The Earth's magnetic dipolar moment is 8×10^{15} Tm^3. The intensity of the field is weaker at the equator than at the poles. Present values are respectively 30 000 and 60 000 gammas (0.3 and 0.6 gauss). Moreover charts giving the distribution of magnetic-field intensity at the surface of the globe clearly show that there are important differences from a perfect dipolar field. These are *magnetic anomalies*, which may be positive or negative, and which may be small-scale (local or regional) or of continental scale, such as the positive anomaly in Siberia, and the negative anomaly in Brazil. These differences are attributed to the influence of the material in the Earth's surface rocks.

Slow variation with time has been established for the magnetic inclination, which drifts towards the west at about 0.15° per year, and for the magnetic-field intensity, which decreases by about 20 gammas per year. These variations, the origins of which are poorly understood, are generally thought to reflect changes in velocity and flow deep in the Earth's outer core.

Even more intriguing are the reversals of the direction of the Earth's magnetic field, which have been clearly established for the last few million years. As yet we do not know if there is a continuous rotation of the magnetic axis, carrying the north pole southward and vice versa, or if we are dealing with a variation in the intensity of the magnetic field, which declines to zero before re-establishing itself with reversed polarity. These reversals allow one to define "epochs" corresponding to the field's present, or reversed, orientation, with durations typically of a million years, which have succeeded one another over the last few million years. However, within these major epochs with a single overall field polarity, shorter periods of time "events" are observed during which the field is reversed. No simple law has so far been discovered to account for the observed reversals.

These changes in the properties of the Earth's magnetic field are found trapped in magnetic minerals that crystallized at different, but known, periods. This is how palaeomagnetism allows continental drift and sea-floor expansion to be followed, and gives the strongest evidence for the theory of lithospheric plate-tectonics.

The magnetic-field lines above the Earth's surface are not those of a field in a vacuum. Their interaction with the interplanetary magnetic field frozen into the plasma forming the solar wind produces a characteristic shape, shown in Fig. 5.15.

The solar wind's supersonic flow results in the formation of a stationary shock-wave, ahead of a *magnetopause*, which is the equilibrium surface between the terrestrial and interplanetary magnetic pressures (of the order of a few gammas). In the solar direction, the magnetopause lies at about ten Earth radii. In the antisolar

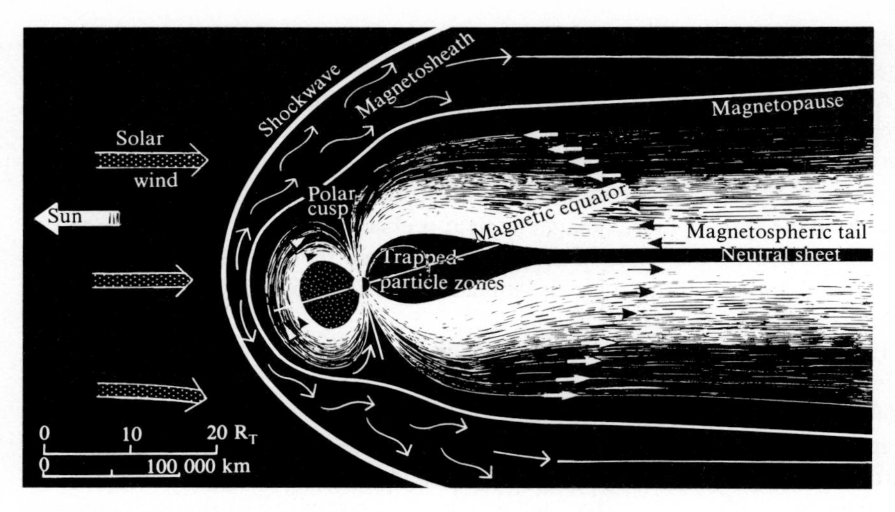

Fig. 5.15. This representation of the terrestrial magnetosphere shows the principal characteristics of the interaction between the solar wind and the geomagnetic field

direction, a *geomagnetic tail* develops, several tens of Earth radii in diameter, which extends for more than 80 Earth radii (500 000 km). So the Moon, for example, in its orbit around the Earth, passes several days each month in this magnetotail.

5.3.5 The Present-day Atmosphere

Atmospheric pressure at sea-level has an average value of 10 360 kg/m^2. The rapid decrease in pressure with altitude (90 % of the atmosphere's mass lies between 0 and 16 km altitude), means that its overall composition is essentially the same as that in the immediate vicinity of the surface (Table 5.3). The main outstanding feature of the terrestrial atmosphere is undoubtedly its extremely high concentration of molecular oxygen. This is directly connected to the existence of living organisms, which build organic tissue from carbon dioxide and water and liberate oxygen O_2. This synthesis is powered by solar radiation (photosynthesis). Contrary to the case in other planetary atmospheres, CO_2 and H_2O are only minor components. They are, however, by no means absent from the surface of the Earth, but are found in non-gaseous forms: water is mainly liquid and solid (oceans and ice), whilst carbon dioxide is trapped in marine sediments in the form of carbonates. Overall, they are the most abundant of terrestrial "volatile" components, but are stored in reservoirs on the surface of the globe itself.

The temperature profile of the atmosphere, shown in Fig. 5.16, clearly indicates the different zones that are distinguished by their thermal gradients, and which have different overall properties: the troposphere, stratosphere, mesosphere, thermosphere and exosphere (see also Sect. 4.2.1).

The *troposphere*, or lower atmosphere, has a net decrease in temperature with altitude. The thermal gradient depends on the water content of the air: 10 K/km for dry air, it drops to 5 K/km for air saturated with water. These values mean that a convection regime is easily established. The troposphere is heated by direct contact

Table 5.3. Composition of the Earth's lower atmosphere

Gas		Proportion (by volume)	Isotopic composition (%)
Nitrogen	N_2	78.084 ± 0.004 %	^{14}N : 99.63 ; ^{15}N : 0.37
Oxygen	O_2	20.946 ± 0.002 %	^{16}O : 99.759 ; ^{17}O : 0.037 ; ^{18}O : 0.204
Argon	A	0.934 ± 0.001 %	^{36}A : 0.337 ; ^{38}A : 0.006 3 ; ^{40}A : 99.60
Carbon dioxide	CO_2	0.035 ± 0.001 %	^{12}C : 98.89 ; ^{13}C : 1.11
Neon	Ne	18.18 ± 0.04 ppmv	^{20}Ne : 90.92 ; ^{21}Ne : 0.257 ; ^{22}Ne : 8.82
Helium	He	5.24 ± 0.004 ppmv	^{4}He : 100 ; ^{3}He : 0.000 13
Krypton	Kr	1.14 ± 0.01 ppmv	
Xenon	Xe	0.087 ± 0.001 ppmv	
Hydrogen	H_2	0.5 ppmv	^{1}H : 99.985 ; ^{2}H : 0.015
Methane	CH_4	2.0 ppmv	
Propane	C_3H_8	2.0 ppmv	
Nitrous oxides	N_2O etc.	0.5 ± 0.1 ppmv	
Ozone	O_3	0.04 ppmv	
Various aerosols		0.001 à 0.01 ppmv	
Water	H_2O	5 300 ppmv	(average)

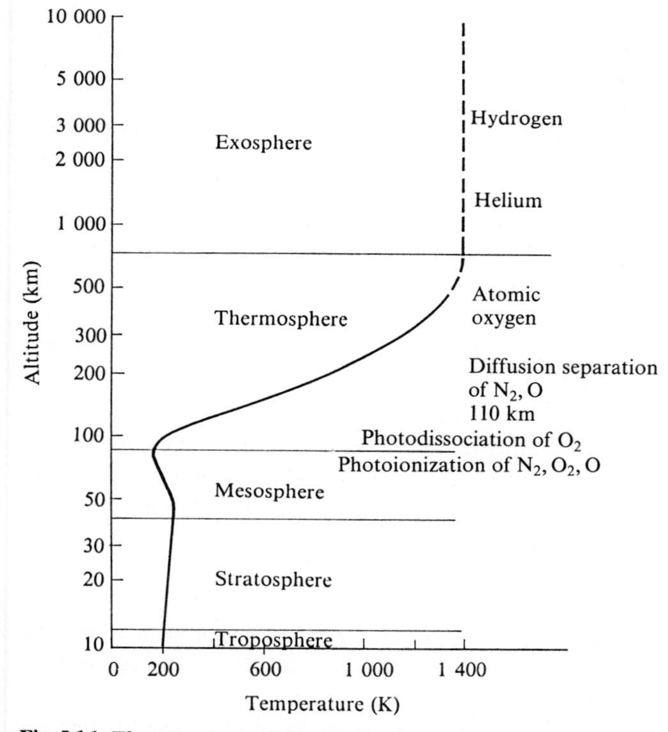

Fig. 5.16. The structure of the atmosphere

with the ground, which absorbs solar radiation (primarily visible) at its surface. The (thermal) radiation from the ground, whose maximum lies in the infrared, is absorbed by asymmetrical molecules in the atmosphere (those that have dipolar moments), i.e. H_2O and CO_2. In fact these molecules have intense vibration-rotation modes in the spectral region close to the maximum emission of the Earth's surface ($10\,\mu$m). So the atmosphere is not transparent to radiation from the ground. A fraction of the energy absorbed by the gas is re-radiated towards the ground, and this leads to an average equilibrium temperature of 288 K, different from the value of 255 K that would be found for a surface, receiving the same amount of solar radiation, but in contact with an atmosphere transparent in the infrared. This heating of the atmosphere by part of the ground's thermal radiation is known as the *greenhouse effect*. It was an indispensable factor in biological evolution, particularly in allowing water to remain in liquid form.

The top of the troposphere, or *tropopause*, is defined by the altitude where the temperature ceases to decline: this altitude varies from ten to twenty kilometres depending on latitude; it is higher over the equator than over the poles. The temperature in this region is around 200 K. Above this is the *stratosphere*, bounded at the top by the *stratopause*. At the base of the stratosphere the temperature changes little with altitude, then it climbs slowly up to an altitude of about fifty kilometres. The stratosphere is in radiative equilibrium. The energy input to the stratosphere comes partly from absorption by water vapour of thermal (infrared) radiation from the surface of the Earth, and partly from absorption by oxygen and ozone of solar ultraviolet radiation. Ozone (O_3) is produced by a series of chemical reactions, initiated by the photodissociation of O_2 by energetic solar radiation. It is this gas that by absorbing most of the solar radiation at short wavelengths (< 3000 Å) protects the troposphere from a photodissociative effect which would, in particular, prevent all molecular biochemistry.

The *mesosphere* stretches above the stratosphere up to the *mesopause* at an altitude of about 80 km. In it the temperature decreases with altitude. This arises from the fact that the competition between the increase in the solar flux with altitude and the decrease in the density of oxygen, and therefore of ozone, swings in favour of density at the stratopause. Above it there are less and less molecules able to heat the surrounding gas by absorption.

In the *thermosphere*, the temperature again increases with altitude, attaining more than 1000 K at about 500 km. In this region the heating comes from absorption of solar radiation at yet shorter wavelengths (< 1000 Å), which is not only capable of photodissociation, but also of photoionizing molecules and atoms, particularly O and N_2, and this again increases with altitude.

At an altitude of 500 km the density becomes of the order of $10^6\,\mathrm{cm}^{-3}$. The mean free path $(n\sigma)^{-1}$, where σ is the effective collision cross-section, becomes of the same order as the atmospheric scale-height. There is therefore a transition between a regime where the atmosphere is retained by collisions to one where it is solely governed by ballistic properties: if a particle has a vector directed outwards and a modulus that is greater than the escape velocity it is no longer bound to the Earth. At a temperature of 1000 K, the mean-square velocity of an atom of atomic number A is $5/A^{1/2}$ km/s. This value is distinctly lower for most atoms and molecules than the escape velocity (10.4 km/s at 1000 km altitude). Only hydrogen and helium can escape, even though only slowly, with a characteristic time that is measured in millions of years (see Sect. 4.2.1).

The absorption of ultraviolet radiation at very high altitudes does not just contribute to raising the temperature of the atmosphere, but also to ionizing it. When the density is too low for collisions and recombinations to dominate, we have a plasma regime. This defines the *ionosphere*, which therefore exists in the outer thermosphere. Overall, the medium is electrically neutral, the density of electrons being equal to that of the positive ions that are present, mainly O^+ and O_2^+. UV radiation is not the only source of ionization: X-rays and energetic particles (from the solar wind and cosmic rays) also contribute. These flux sources are dominant during the day, so much so that ionic concentrations decrease sharply at night.

Ionospheric components have three possible sources of origin. They may be: atmospheric components carried to high altitudes and subjected to solar radiation; particles of interplanetary plasma, which predominate between the bow shock and the magnetopause (a region known as the *magnetosheath*); products of the ablation of extraterrestrial meteor and cometary particles in the atmosphere's outer layers. This last source mainly produces Al, Mg, Si and Fe ions.

The action of the geomagnetic field on the charged particles in the ionosphere produces electrical currents. These in turn create magnetic fields (*magnetic storms*) which are superimposed on the terrestrial dipolar field. Variations in solar activity with time lead to perturbations in the terrestrial magnetic field. The magnetotail then becomes a region favouring charged-particle acceleration; trapped by the magnetic field lines, they precipitate down over the polar regions, in two symmetrical *cusps*. Excitation of atmospheric molecules by these particles creates the *polar aurorae*, that are observed at high northern and southern latitudes. The most spectacular aurorae are therefore observed at times of major solar eruptions. Thanks to measurements made on board satellites, zones of high concentrations of energetic particles (0.1 to 10 MeV) have been discovered. These are the *Van Allen belts*, which are symmetrical about the magnetic equator.

Just as the neutral atmosphere absorbs a large fraction of solar radiation, only allowing the visible spectrum and radiation in a few infrared "windows" to pass, the ionosphere is not transparent to electromagnetic radiation at very long wavelengths. This property is used, in particular, to transmit radio emissions (of kilometre wavelength – *long wave*), from one part of the Earth to another that is not in line of sight, by reflection from the ionosphere.

5.3.6 The Earth's Climate

The source of energy that is by far the dominant one in the Earth's surface and atmospheric thermal equilibrium is solar radiation. The major climatic result is therefore the presence of seasons, caused by the variation of the solar radiation received by a unit area of the surface over the course of the year as a result of the inclination of the Earth's axis to the perpendicular to the ecliptic.

Second, one characteristic of the troposphere's thermal equilibrium is that it is mainly heated from the base, i.e. from its contact with the ground. So the important factor is not the amount of energy that is incident on the top of the atmosphere, but the amount reaching the surface. With the high albedo of clouds, the degree of the Earth's coverage by clouds is therefore a very important factor. One feature of the Earth is that it is neither totally covered by clouds, nor totally free from them, but has an average amount of cloud cover, which is thought to be maintained (and has apparently been maintained for a long time) by a self-regulating mechanism, in an equilibrium state.

Finally, the greater heating of the low-latitude zones compared with that of the polar regions results in a *general circulation*, not only vertically but also horizontally. The permanent mixing of the troposphere by the winds reduces the differences in temperature with latitude.

Atmospheric motions are linked with ocean currents, which play a very important part in the horizontal circulation of the solar energy absorbed at the surface. The coupling between the oceans and the atmosphere is particularly shown by the exchange of energy in the form of latent heat as evaporation and later condensation, either in the form of clouds or ice.

These climatic changes with a yearly timescale are combined with variations caused by alterations in solar activity. A correlation has been established, for example, between a considerable reduction in solar activity during the second half of the 17th century (the *Maunder Minimum*) and a decrease in the Earth's surface temperature. The greatest variations in climate, however, come from changes in the Earth's orbital parameters. The most obvious effect is the recurrence of periods of global cooling and warming, which are accompanied by considerable variations in the extent of the zones covered in ice, and consequently with the mean sea level. Several cycles of major glaciations have been established, with timescales of a million years. Between two major glaciations, short glacial periods (a few thousand years long) alternate with interglacial periods of about fifty thousand years. The last glacial maximum was about 20 000 years ago. It was at this period, for example, that the African deserts were last covered in vegetation, and that the Behring Strait between Asia and North America was dry land. It is thought that this was how the American continent was populated from Asia, in other words at a extremely recent period when compared with the age of the oldest human remains that have yet been identified, in the region bordering Lake Turkana in East Africa, and which date back about three million years.

5.4 The Moon

From its movement in the sky the Moon obviously ought to be considered as a satellite of the Earth, which it accompanies in its annual revolution round the Sun. It seems, however, that none of the Moon's major characteristics are a result of its proximity to the Earth. It is convenient to treat the Moon as a planetary body like the other four inner planets because its large-scale features are of the same general type. The Earth-Moon system may be considered as a double system, as also with Pluto and its companion Charon. They are distinct from the other satellite systems, whether those of Mars, with Phobos and Deimos, or those of Jupiter, Saturn, Uranus or Neptune.

The Moon has been the subject of more, and probably the oldest, astronomical observations, because of the favourable conditions under which it can be observed. But detailed knowledge of this object only dates from the space age, and numerous questions still remain unresolved, beginning with that about its origin.

Because of its synchronous rotation, part of the surface is not available for observation from the Earth. We had to wait for the first interplanetary probes to obtain a photograph of the "hidden face" of the Moon. This proved to be different from the visible face, in that it does not have extensive dark maria that contrast, on the visible hemisphere, with the brighter surrounding areas. The whole of the lunar surface is pock-marked with meteoritic impact craters, which shows the absence of any geological activity and strong erosion for several thousand million years (Fig. 5.17).

The Moon has been the target of spectacular space explorations. In parallel to the Soviet automatic missions, which carried out analyses of the surface and returned some samples, the American Apollo programme led to six successful manned flights and the landing of astronauts, between 1969 July (Apollo 11) and 1972 December (Apollo 17). An accident forced the premature return of Apollo 13 without reaching the Moon. The scientific, cultural, economic and strategic impact of this programme was considerable. Hundreds of scientific teams around the world have worked, and hundreds are still working, on data and samples (more than 400 kg) brought back by these missions.

Despite the immense progress made in the last fifteen years, there remains much to be done in the study of our satellite. In particular, only the regions close to the equator have been overflown and mapped at high resolution with a view to establishing their mineralogy and chemical composition. Projects for a lunar satellite in polar orbit that are currently under study would allow this coverage to be extended

Fig. 5.17. The Moon as seen from an Apollo mission. (By courtesy of NASA)

to regions at high latitudes. Such systematic observation would moreover test the hypothesis that certain volatile components, such as water, may be trapped in certain polar craters that are always in shadow, and therefore remain below 200 K. Such a result would be essential for long-term projects for the exploitation of our satellite.

5.4.1 Telescopic Observation of the Moon

Even before the invention of the telescope, two types of surface region were distinguished: the *maria*, dark (with albedo $< 10\,\%$), which are surrounded by the brighter *highlands*. With the appearance of the first astronomical telescopes, Galileo and his successors discovered the circular formations that covered the surface: these are the lunar *craters*. The development of optical techniques progressively allowed resolution of the lunar surface to reach about one kilometre. The overall characteristics of the lunar features could therefore be studied for the visible face, that is 59 % of the surface, taking the Moon's *librations* (see Sect. 5.4.5) into account. More and more craters have been discovered, as well as other structures, much less numerous – sinuous *rilles* and escarpments (*rupes*). The maria appear much less rich in craters than the highlands, whence the idea that they are "younger", i.e. formed later. On the other hand, it was discovered that certain of the largest craters (Tycho, Copernicus) have *rays*, narrow, bright bands that may reach 1000 km in length. Yet scientific interpretation of these observations remains sketchy. In particular, it is not known if some craters have a volcanic origin, or if they were produced by giant impacts. Finally, it has been possible to analyze the polarization of sunlight diffused by the Moon, its variation with phase angle and also the changes in albedo with wavelength.

These observations have given three sorts of important results:

1. *lunar mineralogy*: the mass of the Moon, only 1/81 of that of the Earth, and its size, determined from Earth, allow the mean density to be estimated. It is 3.3, considerably less than that of the Earth (5.5). This excludes the possibility of there being a metallic core of any appreciable size, and indicates a low overall iron-content. On the other hand, the most plausible explanation for the origin of the maria through floods of basaltic lava, suggests that the Moon, like the Earth, has a feldspathic crust and a mantle. This has been confirmed by spectra that plot reflectivity with respect to wavelength, which show absorption bands in the near infrared that are linked to presence of different silicates, which are identifiable by reference to laboratory spectra. In particular, the spectra of the lunar maria show a strong band centred on 0.95 μm, which can be attributed to the presence of Fe^{++} in pyroxenes;
2. the existence of a *regolith*: the variation in the polarization with the phase angle, and the variation in albedo with wavelength indicate that the surface is covered with fine dust, with an average size of about ten microns. The layer of debris that covers the whole surface of the Moon is known as the regolith;
3. its *maturity*: the fact that the surface of the Moon is directly exposed to the interplanetary medium leads to progressive darkening of the soil, which explains, in particular, why only young craters show systems of *rays*, which fade as they age.

5.4.2 Spacecraft Observations

The Soviet programmes for exploring the Moon have been restricted to automatic missions. Luna 3, in 1959, showed the hidden face to be very different from the visible side: it has practically no maria. The Luna 16, Luna 20 and Luna 24 probes collected samples for return to Earth (310 g). Luna 24, in particular, carried out a deep core (2.4 m). The most revealing lunar studies, however, result from the American Apollo programme. They were arranged in three major classes: observations from orbit, in-situ experiments, and the analysis of samples brought back by the astronauts.

The observations from orbit were carried out from the command and service modules, after separation of the lunar landers. Photography from orbit and radar altimetry allowed precise determination of the Moon's mean radius (1738 km), its mass and volume, and thence its mean mass per unit volume (3.34 g/cm^3). The moment of inertia is equal to 0.99 times that of a homogeneous sphere of the same mass, whilst for the Earth it is only 0.8. This sets an upper limit of 400 km for the radius of any metallic core. The far side is 4 km higher than the near side with respect to the centre of mass. This asymmetry is not correlated with a positive gravity anomaly. It is therefore explained by the existence of a crust, in isostatic equilibrium, which is thicker (100 km), and therefore lighter, on the far side than on the near side (where it is only 50 km thick).

Optical imagery of the Moon has enabled complete cartography to be carried out to a resolution of at least one kilometre, and more than 50 % to a resolution of 100 m. Because of the orbits of the Apollo missions, the best-documented regions are those that lie close to the equator. These regions have also been studied for their chemical composition through X- and γ-ray spectrometry experiments.

The irradiation of the lunar surface by solar X-rays gives rise to X-ray fluorescence, detectable in orbit by using proportional counters. By taking the energy spectrum of solar X-rays and the detection threshold of the detectors into account, this technique enabled the presence of Al, Si and Mg to be determined, with a spatial resolution of about twenty kilometres. The depth probed by this technique is typically a few micrometres. Figure 5.18 shows the variations in the Al/Si and Mg/Si ratios over half of an orbit around the Moon. It is obvious that these ratios are excellent tracers of mare and highland materials: the highlands, rich in anorthositic feldspars show an Al/Si ratio greater than 1.2, contrasting with the maria, which, containing basalts rich in pyroxenes, are deficient in Al and enriched in Mg when compared with the highlands. These experiments were carried out on the Apollo 15 and 16 flights.

The Moon emits γ-radiation from two principal sources: the first is the natural disintegration of long-lived radioactive elements (U, Th, K). The second comes from the irradiation of the surface by high-energy cosmic rays, i.e. mainly galactic cosmic rays (see Sect. 2.2.4). γ-ray spectrometry can therefore determine the composition of the soil to a depth of the order of a few centimetres (the characteristic depth of diffusion for γ-rays), with respect to both its radioactive elements and its major elements. Figure 5.19 shows the results of the analysis of the surface of the Moon by Apollo 15. It is the integrated count for all energy between 0.54 and 2.7 MeV, i.e. a region mainly encompassing emission from natural disintegration. The count

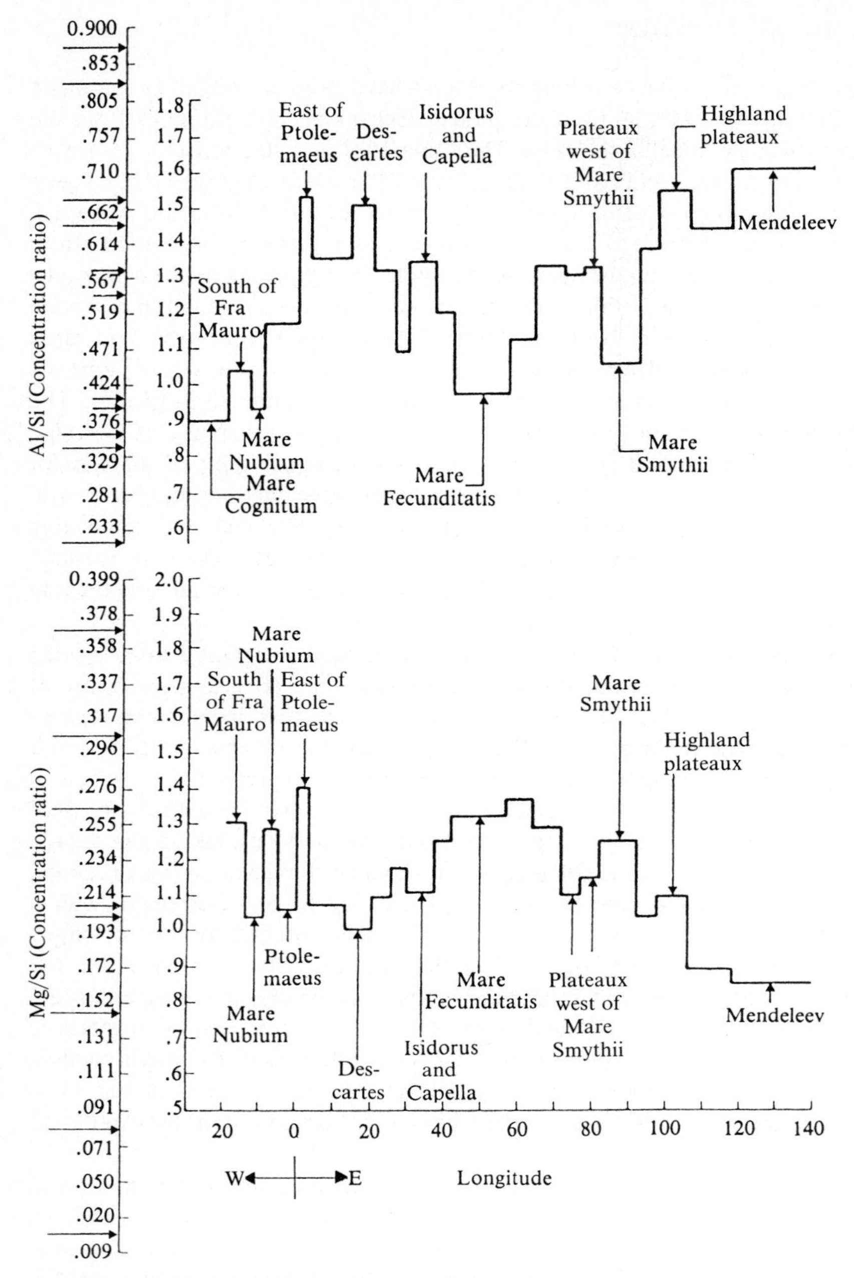

Fig. 5.18. Histograms showing variations in the Al/Si ratios (*top*) and Mg/Si (*bottom*) as a function of longitude, obtained from the Apollo 16 orbiter. (From *Apollo 16 Preliminary Science Report,* NASA SP-315, p.19–6)

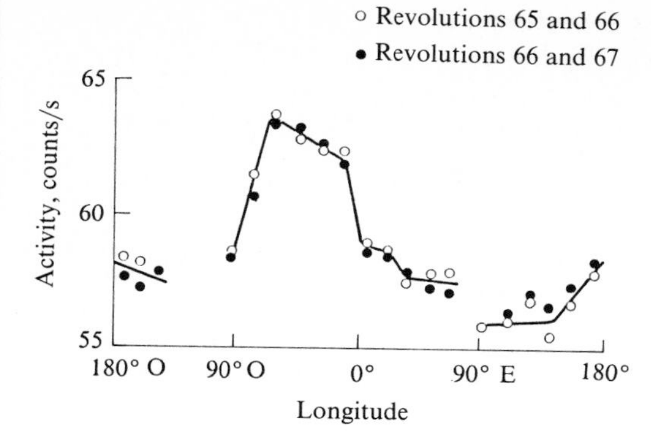

Fig. 5.19. γ-ray emissions between 0.54 and 2.7 MeV, detected by the Apollo 15 orbiter on successive orbits. Regions of high radioactivity (corresponding to Oceanus Procellarum and Mare Imbrium) can be distinguished from the surrounding highlands. (From *Apollo 15 Preliminary Science Report,* NASA SP-289, p.16–4)

rate therefore shows the variations in concentration with a spatial resolution of rather better than a hundred kilometres: the maximum emission corresponds to the maria in the west, Oceanus Procellarum and Mare Imbrium. Mare Tranquillitatis and Mare Serenitatis show weaker activity, and the highlands even less. If, on the other hand, one is interested in the signatures of individual elements, the signal must be integrated over several hours. No individual lunar formation is then resolved.

Other experiments from orbit allowed the magnetic field of the Moon (in particular) to be determined. The field measured from orbit shows important variations in longitude (Fig. 5.20), indicating that it mainly derives from remanent sources local-

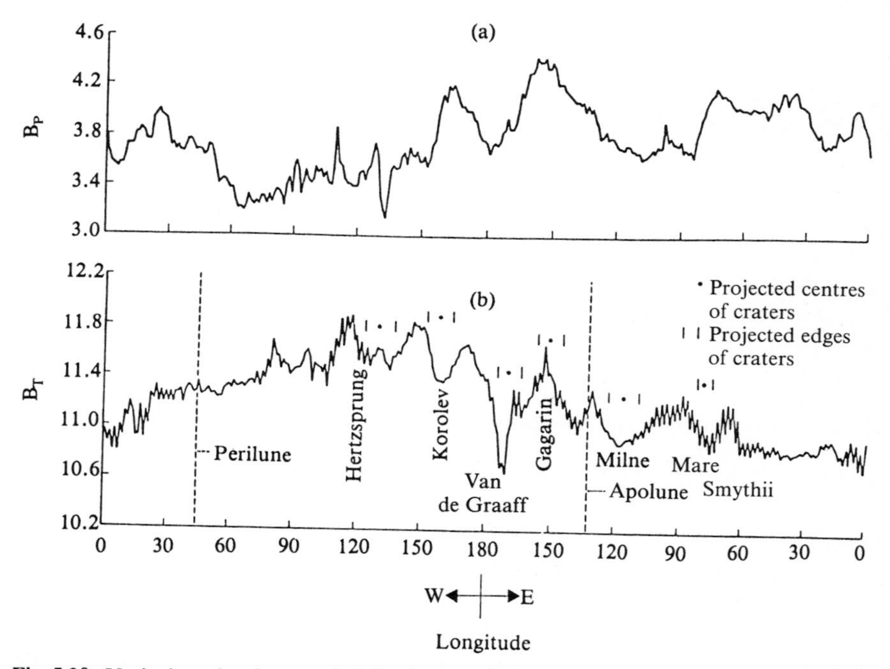

Fig. 5.20. Variations in the parallel (B_p) and transverse (B_t) magnetic fields detected from lunar orbit by the Apollo 15 mission. (From *Apollo 15 Preliminary Science Report,* NASA SP-289, p. 22–5)

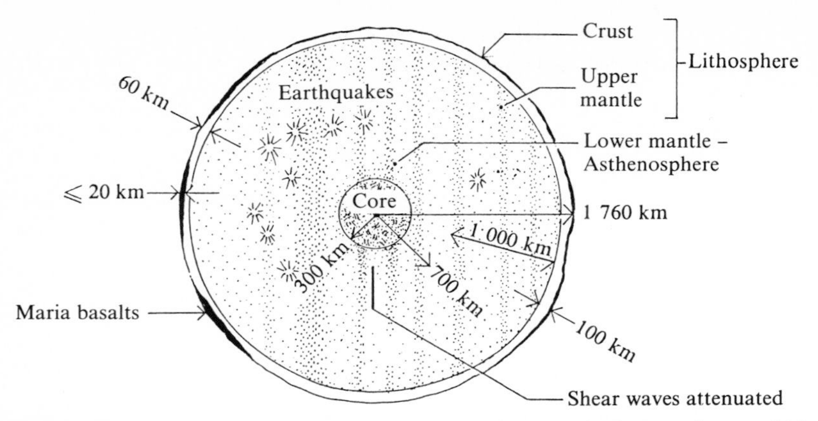

Fig. 5.21. Schematic structure of the Moon, according to seismic experiments. [After P. Thomas: "The Moon", in *Le grand Atlas Universalis de l'Astronomie* (Encyclopedia Universalis 1983). By courtesy of the publisher]

ized at the surface and extending over a few tens of kilometres. They are combined with a possible central magnetic dipole, an upper value for which has been set at 4×10^{19} G/cm^3. The corresponding field intensity at the surface of the Moon does not exceed 2 to 3 gammas.

Among the in-situ experiments, we may mention:

1. determination of the internal structure by active and passive seismic experiments. The Moon has a lithosphere that is both dynamically inactive and very thick (about 1000 km) when compared with that of the Earth. It includes a crust that is between 50 and 100 km thick. The elastic properties of the underlying layers, the asthenosphere and a possible central core, are very different. However, the existence of a high-density, liquid metal core is excluded (Fig. 5.21);
2. measurement of the surface magnetic field, dominated by local sources of remanent magnetism, which may exceed one hundred gammas;
3. determination of the composition of the solar wind, by the exposure of aluminium foils, and subsequent analysis in the laboratory of the rare gases trapped within it.

5.4.3 Laboratory Analysis of Samples

The quantity and diversity of samples returned continued to increase throughout the Apollo missions. In particular, beginning with Apollo 15, the astronauts were able to move as far as several kilometres from their base (the lunar module, LEM) by using a lunar rover. Samples were not only obtained from the surface, but also in depth, thanks to cores several metres long, which preserved the relative position of the samples obtained. The lunar soil is stratified by successive deposits of layers of ejecta from meteoritic bombardment, so deep samples were exposed at the surface at more remote periods than higher samples. Thanks to models describing the progressive burial of grains in the regolith, it is possible to draw up a time-scale corresponding to the depth of the samples in the cores. It is possible to go back more than two thousand million years by studying grains from depths of more than two metres in the soil.

The scientific fields covered by the analysis of lunar samples included in particular:

1. the *dating* of the principal lunar metamorphic events. The date of the Moon's formation is comparable with that of the whole Solar System, 4.55×10^9 BP (Before Present). The differentiation of the feldspathic crust dates from the very beginning of the Moon's history. For hundreds of millions of years after that, the Moon underwent intense meteoritic bombardment by the numerous protoplanetary bodies that then populated the region (the *accretion tail*). In particular, the large circular basins with diameters several hundreds of kilometres across appeared between 4.4 and 3.8×10^9 BP. On the visible hemisphere, these basins were subsequently filled with deep, basaltic lava flows between 3.7 and 3.1×10^9 BP. The meteoritic bombardment rate declined between 3.8 to 3.5×10^9 BP, reaching a value close to the current one. In the absence of any internal metamorphic activity, it was meteoritic bombardment that was responsible for the main modifications to the surface of the Moon over the course of the last three thousand million years;
2. the *comparative mineralogy* of the various lunar soils that have been analyzed. The dichotomy between maria and highlands dominates lunar mineralogy. The highlands primarily consist of anorthositic feldspars, rich in Al and Ca ($Si_2Al_2O_8Ca$). They are the uncovered part of the lunar crust and form 80 % of the whole surface. The maria, on the other hand, are basaltic, and are formed of lavas derived from the mantle. They are almost entirely concentrated onto the visible hemisphere, where they form 35 % of the surface. Their feldspars are calcium-poor pagioclases. The pyroxenes contain exsolution lamellae of pigeonite and augite that are a few hundreds of ångströms thick, indicating very rapid crystallization. It would appear that the mineralogical composition of the Moon as a whole is very close to that of the terrestrial mantle as regards the major elements Si, Mg and Fe. The volatile elements are, on the other hand, considerably impoverished, whilst refractory elements may be much more abundant on the Moon. This is particularly true of titanium, which appears in ilmenite, a double oxide of iron and titanium, $FeTiO_3$. Finally, an essential difference from the Earth is the absence of water in the composition of lunar minerals;
3. the *exposure history of grains* to meteoritic bombardment, to the solar wind, and to cosmic rays. The incessant impacts by meteorites of all sizes have pulverized the surface rocks, producing a layer of debris several metres thick (the regolith). The smallest meteorites, which are very numerous, perpetually reduce the size of grains in the regolith, which now consists of a very fine powder. The violence of some impacts melts some of the grains and sprays out fine droplets, giving rise to a wide range of partially or totally vitrified forms: "spherules", "vitreous aggregates", and "spatter" on lunar rocks. The regolith is frequently overturned by impacts, and this perpetual "gardening"

repeatedly brings different grains to the very surface, where they are exposed to space. The smaller the grain, the shorter the time it remains on the surface: a 1-μm grain stays there for a only few thousand years, whilst for a 100-μm grain the exposure age is hundreds of thousands of years. The stacking of successive layers of debris from neighbouring craters produces a stratified structure, which enables the evolution of the interplanetary medium to be followed, notably that of the charged particles of which it is composed: there is no atmospheric or magnetic shield to protect the Moon, which is always exposed to the solar wind and cosmic rays.

The solar wind is by far the most intense flux of particles: at present this flux is more than 10^8 protons $cm^{-2}s^{-1}$, and its velocity varies from 350 to 650 km/s, which corresponds to an energy of about a keV/nucleon. At this energy, the particles penetrate into only the first few hundred ångströms of the grains; as a result only grains at the very surface of the regolith are irradiated by the solar wind. Analysis of grains from different levels in lunar cores show that the isotopic composition of nitrogen, for example, has changed over time, reflecting evolution of the solar corona. The nitrogen has become progressively enriched with isotope 15 over the last two thousand million years. The average energy of solar-wind ions also seems to have increased over that period.

Solar cosmic rays, a million times less intense, have an energy that is a thousand times greater (1 MeV/nucleon). They penetrate a few tens of microns into the regolith. The "heavy ions" (the iron group) leave radiation *tracks* at the end of their paths, which can be observed under the microscope. The number of tracks allows the duration of the grains' exposure at the surface of the regolith to be measured. The tracks' characteristics enable the energy-spectrum and relative abundances of major groups of elements in the cosmic radiation to be determined. Solar cosmic radiation seems to have changed little as far as its chemical composition and energy-spectrum are concerned over the past two thousand million years.

Galactic cosmic rays, 100 times less abundant that the previous group, have sufficient energy (1 GeV/nucleon) to penetrate more than a metre into the regolith, producing nuclear reactions with atoms forming the grains. These effects enable the history of the successive layering observed in lunar cores, in particular, to be traced.

The combined effects of meteoritic bombardment and the charged-particle flux produce progressive alteration of the macroscopic properties of the surface, or *aging*, characterized by a *maturity* index. The incessant abrasion by the smallest meteorites progressively softens relief. A newly formed impact crater, with sharp edges and an initial size of 100 metres, is eroded away in a few hundred million years. Aging is also shown by changes in the characteristic optical properties.

The contrast between light and dark areas of the Moon has several causes. First, the maria have a different mineralogical composition to that of the highlands, which means that their albedo is on average only about half. Aging also produces progressive darkening of the surface in both maria and highlands. This phenomenon may be explained by the increase in the glass content, where iron is reduced, and by the appearance on a large fraction of the grains of a layer strongly altered by solar-wind ions. Initially, therefore, a ray from a large highland crater shows a strong contrast with a mare area that it crosses. Slowly the particles of which it is formed darken, meteoritic excavation mixes in surrounding material, and it disappears after a few hundred million years.

5.4.4 The Origin of the Moon

Three models have been proposed for the origin of the Moon:

1. *fission* of part of the mantle of a fluid Earth in rapid (2-hour) rotation. This model relies on the similar densities (3.3) and similar primary-element composition. In particular, it explains the low content of iron by this element being concentrated into the terrestrial core at the time of fission. Analysis that takes the distribution of siderophile and refractory elements into account, however, imposes serious restraints on this model. Even more important, there are objections to this model on dynamical grounds. In particular, it seems difficult to account for the 81:1 mass-ratio and the angle of 5° between the orbit of the Moon and the ecliptic;
2. *capture* by the Earth, with the Moon having been formed elsewhere in the Solar System. This model is suggested by study of the evolution of the orbit of the Moon through tidal effects, which indicates that the Moon was closer in the past. The primary objection is statistical: such an event is highly unlikely. Moreover, the initial orbit has to be very close to the Earth's, with a semi-major axis of between 0.95 and 1.05 AU. It is difficult to understand how an object formed at the same heliocentric distance as the Earth should have such a different overall composition, without a core of any considerable size;
3. *accretion* in orbit around the Earth. The principal merit of this model is that it does not require any cataclysmic event. On the other hand it does not resolve the objection raised for the capture model regarding overall composition: one would expect these two objects, the Earth and the Moon, formed in the same region of the Solar System, to have accreted materials that were similar in composition.

Each of these models therefore encounters problems, which means that more complex solutions have to be sought, perhaps with elements from more than one of them. For example, it has been suggested that a differentiated proto-Moon was broken up by a close encounter with the Earth. Fragments from the mantle could then have been captured and accreted in orbit. These later models cannot be tested in the absence of new data about the internal structure and overall mineralogy of the Moon, which may come from space missions in the 1990s.

5.4.5 The Earth-Moon System

At present the average Earth-Moon distance is 384 402 km. This distance is increasing with time as a result of tidal effects, by about 4 cm/yr at present (a similar effect is also causing Deimos to progressively recede from Mars, and Phobos to approach – see Sect. 5.6). Assuming that the amount of energy dissipated has remained constant over time, it is calculated that the Moon was closest to the Earth about two thousand million years ago. Considerable tidal effects ought to be found in lunar rocks of that epoch, because the youngest rocks solidified about three thousand million years ago. But this is not the case. In fact, there is no reason to suppose that the amount of energy dissipated has remained constant: we know that it is mainly exerted through the Earth's oceans, and everyone accepts that the distribution of the terrestrial continents has varied considerably throughout the Earth's history. There is therefore nothing to contradict the assumption that the recession of the Moon from the Earth dates back to the very beginning of the history of the Solar System.

Tidal effects are equally responsible for the fact that the Moon always shows the same face to the Earth: they have braked the Moon's rotation until the revolution and rotation periods are the same, as happens in every satellite system.

As the eccentricity of the Moon's orbit around the Earth is not zero (0.054), the orbital velocity is not constant, as would be the case for a circular trajectory. As a result, the rotation is periodically slightly in advance of or behind the orbital motion. This oscillation of the Moon relative to the Earth is known as *libration* in longitude, which is accompanied by a libration in latitude, because the axis of rotation lies at an angle of 5.8° to the normal to the ecliptic. The combined effects allow rather more than half (59 %) of the Moon's surface to be observed from Earth.

The Sun and the planets perturb the Moon's orbit. We may mention the advance in lunar perigee, which results in the line of apsides rotating with a period of 8.85 years. The line of nodes itself has a retrograde movement with an 18.6-year period.

Several different lunar months may be defined according to whether the Moon's position relative to the Earth or to the stars is used (Fig. 5.22). The sidereal month corresponds to the Moon returning to the same position with respect to a fixed reference, and is equal to 27.32 mean solar days. The period between two identical lunar phases is a synodic month (29.53 days), which is longer than the former because of the movement of the Earth in its orbit around the Sun.

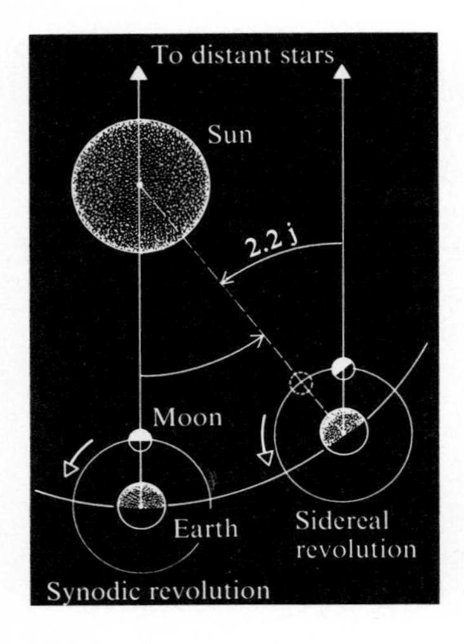

Fig. 5.22. Diagram illustrating synodic and sidereal rotation of the Moon. [After A. Cazenave: "La Lune", in *Le grand Atlas Universalis de l'Astronomie* (Encyclopedia Universalis 1983). By courtesy of the publisher]

5.5 Mars

The planet Mars has numerous points in common with the Earth. Although its diameter is only slightly more than half the Earth's, its structure is that of a differentiated planet: a central metallic core is surrounded by a mantle, which is itself covered by a crust. White cirrus clouds, mists and frosts occur in a very tenuous atmosphere, rich in carbon dioxide, whilst the considerable inclination of the axis of rotation to the plane of the orbit around the Sun causes marked seasons. A wide variety of geological formations and gigantic volcanoes are evidence of intense internal activity that has now disappeared. The reddish colour of the soil indicates hydration of the surface rocks, and is evidence of the role of water in the mineralogical evolution of Mars. For all that, we still do not know whether the water existed in liquid form, or – like that remaining today – was only in solid form (as frost and ice) or as vapour. Might the presence of vast dry gorges (the *canyons*) be evidence of past flows? In order for liquid water to have existed, the partial pressure of water vapour would have had to be greater than 6.1 mbar, which is considerably higher than the current value. Being more massive that the Moon or Mercury, Mars retained a high central temperature for longer, giving rise to considerable internal activity and leading to intense volcanism. The latter injected considerable quantities of water and carbon dioxide into the atmosphere. This, in turn, raised the surface temperature through a greenhouse effect. Overall, it is conceivable that the thermodynamic conditions (i.e. the pressure and the temperature) allowed water to exist in all three phases. One of the possible consequences is that Mars might have been the site of biochemical evolution.

Because it arose from radioactive decay, the internal energy decreased with time. Today, Mars is probably extinct, whilst the Earth, more massive, is still very active. From the point of view of comparative planetology, Mars is therefore unique in that it allows us to trace the complete evolutionary lifetime of an active planet. It is sufficiently close to be accessible for space exploration (Fig. 5.23), and is the

Fig. 5.23. The surface of Mars: desert and dunes seen from Viking 1. (By courtesy of NASA)

target of an international programme of automated missions, which will be launched by the Soviet Union, and which should culminate around the year 2000 with the return of samples of the martian soil, obtained by "intelligent" mobile robots.

5.5.1 The Observation of Mars

The first maps of Mars, drawn from astronomical observations, go back to the 17th century. It should, however, be noted that contrary to the case of the Moon, which is more than one hundred times closer, extremely little was known about the "red planet" even at the dawn of the space age. The impossibility of confirming by telescopic observation the existence of any geological formation at all – with the notable exception of the vast polar caps – did not even allow the planet's degree of present-day activity (internal, tectonic, chemical, or possibly even biological), to be determined. In particular, numerous authors identified optical changes as being evidence of life (indicating the presence of lichens, algae, etc.).

In the 1960s, the Americans and Soviets took turns, with alternating *Mars* and *Mariner* probes, in photographing the surface, analyzing the atmosphere, and measuring some of the overall physical properties. The first 21 photographs transmitted by Mariner 4 in 1965 revealed a body pock-marked with impact craters, just like the Moon: Mars appeared to be a geologically dead planet, and had perhaps been so since its birth. The Mariner 6 and 7 missions confirmed this view. It was the Mariner 9 probe, which orbited the planet in 1971–2, that transmitted the first complete set of photographs of the surface (some 7300 pictures), giving almost complete coverage, with a resolution better than 5 km. Because of its orbit, it was even possible to obtain a resolution of just a few hundreds of metres for a portion of the surface. These photographs, transmitted in digital form, were the basis of the first detailed geological maps of Mars. It then became obvious that the earlier missions had not photographed regions of Mars that were representative of the planet as a whole. Mariner 9, on the other hand, revealed a wide range of geological features, in particular gigantic volcanoes and giant "canyons". It should be noted that none of the canyons corresponds to the *canals* (Schiaparelli's *canali*) that earlier astronomers thought they saw on Mars.

The next stage in the exploration of Mars was reached with the spectacular, American *Viking* programme, which consisted of two spaceprobes, each with a landing probe (a *lander*), able to obtain samples and analyze them in situ, and an *orbiter*, designed to observe the planet from a low orbit and to transmit to Earth all the data obtained, including those from the lander. The first landing took place on 1976 July 20 in the region of Chryse Planitia. The second probe landed on the following September 3, in a region called Utopia Planitia, 6500 km from the first. To be more precise, the latitudes (northern) and longitudes of the two sites were respectively: 2.27° and 47.97° for Viking 1; 47.7° and 225.7° for Viking 2. Of the observations made from orbit, particular mention should be made of the more than 50 000 pictures of the surface that were obtained, with a resolution varying from 10 to 150 m according to the particular zones examined. As far as the in-situ experiments were concerned, which would, for example, determine the chemical composition, they had one primary object: to detect the presence of organic, or even living, material in the soil.

Since the cessation, in 1982 November, of the Viking programme, space exploration of Mars has had a respite. With the launch of the Soviet *Phobos* mission in 1988 and with the American MOM (*M*ars *O*rbiter *M*ission) planned for 1992, the exploratory stage will move on to the systematic study of Earth's sister planet.

5.5.2 Orbital Parameters, Overall Physical Properties and Internal Structure

The principal characteristics of Mars are summarized in Tables 1.1 and 1.3 in Chap. 1. It will be noted that the average density of Mars is distinctly less than that of the Earth. Although the different degrees of internal compression have to be taken into account, this probably indicates a lesser iron content in Mars: about 25 %, compared with 33 % for Earth. This smaller overall content is, however, accompanied by a higher concentration in the external layers (the crust and mantle). Should we therefore conclude that no central, metallic core exists?

Apparently not, if we go by the overall moment of inertia, which is estimated to be 0.365. This is the value for a body with a strong, central mass-concentration – we may recall that the moment of inertia of a hollow sphere is 1, that of a homogeneous sphere 0.4, and that it approaches 0 as the mass is concentrated towards the axis of rotation. The problem may be approached in another way, because the Soviet Mars 3 and Mars 5 probes measured a very weak magnetic field: their ferromagnetic-probe magnetometers found a field of about 60 gammas near the equator, at periapsis (at an altitude of 1500 km). This would correspond to a dipole moment of 2.4×10^{22} gauss/cm^3, about 1/4000 of the terrestrial value. We may therefore expect the liquid iron core of Mars, if it exists, either to be very small, or for its dynamical properties to prevent any significant dynamo effect. This does not exclude, however, the presence of a solid iron core.

We should emphasize, however, that very few of the measurements that would be required to unambiguously define the internal structure of Mars have, as yet, been made. Our evaluation of the mineralogical structure and thickness of the martian crust and mantle rest on just a few analyses of the composition of Mars, the overall properties available (density, moment of inertia) and geological maps of the surface. It is generally thought that the planet is covered by a crust that primarily consists of hydrated silicates. Its thickness, although not the same all over the planet, is around 50 km. An underlying mantle, a few hundred kilometres thick, doubtless contains olivine, with a general formula of $(Mg,Fe)_2SiO_4$, and iron oxide FeO. The thickness of this mantle would prevent any tectonic activity of the type found on Earth. Finally, the central core would consist of a mixture of iron and iron sulphide FeS.

5.5.3 The Composition of the Martian Soil

Part of the surface is covered by water and carbon dioxide in the form of ice. The polar caps spread as far as latitude 60° in the winter. In spring, they partially sublime, recondensing at the other pole. Under the pressure regime reigning on Mars, the condensation temperatures of water and carbon dioxide are respectively 190 K and 150 K. Because of this, there are permanent H_2O ice-caps – larger in the north than in the south – whilst the carbon dioxide (probably) evaporates completely, condensing alternately at one cap or the other. In addition, the water condenses in winter, probably at all latitudes, in the form of frost. This covers only part of the soil, because precipitation of the whole of the atmospheric water would correspond to a film of only about one hundredth of a mm. Finally, the subsurface temperature, at latitudes higher than 45°, should allow ice to remain throughout the year. At present we do not know if such layers of subsurface ice do exist, or their possible depth or thickness. They could form the equivalent of terrestrial *permafrost* layers.

The existence of such permafrost is, however, very likely. One argument has a theoretical basis: although thermodynamic conditions today prevent water from existing in its liquid form, it is probable that things were different in the past, when internal activity injected large quantities of water vapour into the atmosphere. The presence of gorges seems to indicate that there has been flowing water; it could have subsequently soaked into the surface and formed thick masses of subterranean ice. Another argument is observational: some impact craters are surrounded, not by the pattern of rays that are found when impacts occur on rock or dust, but by "fluid" ejecta. These are interpreted as having arisen from a frozen, or even liquid, substrate: beneath the permafrost porous rocks could contain large quantities of liquid water. So some areas of the surface of Mars may form a "hydrolithosphere", several kilometres in thickness.

As with water, so the existence of traps for carbon dioxide is an important question that the future exploration of Mars will seek to answer. Despite it being the major atmospheric component, the overall abundance of CO_2 is very low when it is compared, for example, with that of Venus. We can envisage a cycle, like that occurring on Earth, where a significant fraction of carbon-dioxide gas becomes trapped in surface rocks as carbonates, after having gone through a liquid phase. The detection, and mapping of such carbonates would give very important information about the existence of such traps – which were probably liquid – during the times of intense activity.

The composition of the "sand" grains on the surface was obtained for the first time by the Mariner 9 mission, thanks to a Michelson interferometer ("Iris"), which analyzed the martian atmosphere. It obtained the infrared spectrum between 5 and 50 μm, with a spectral resolution of 2 cm^{-1}. This was how evidence for an intense absorption, centred at about 1000 cm^{-1} was discovered, and which was attributed to the presence of particles lifted into the atmosphere during the 1971–2 dust-storm. Attempts were made to reproduce the structure with terrestrial minerals. It was found that the best fit came from a mixture of basalts and montmorillonite, a clay formed by the aqueous alteration of silicates.

Although related to just individual points on the surface, the experiments carried out on board each of the Viking Landers led to important results. The instruments used were X-ray spectrometers. These instruments analyze a sample by determining its X-ray emission, which is characteristic of the chemical composition. This emission is not spontaneous. On the surface of planetary bodies devoid of atmospheres, solar X-rays reaching the surface induce fluorescence in the surface particles. But the atmosphere of Mars is sufficiently dense to protect the surface from solar X-rays. So fluorescence of the grains had to be induced by local irradiation. It was decided to irradiate the soil with two radioactive sources, ^{55}Fe and ^{109}Cd. The first gives, by electron capture, a 5.9-keV emission, and the second two emissions, one by electron capture, of 22.2 keV, and the other by γ-ray emission, of 87.7 keV. As a result, an extremely wide range of atomic masses could be excited and analyzed, but the detectors themselves imposed a lower limit about mass 20 (H, C, N, O, and Na were not detectable). Table 5.4 indicates the elementary compositions of four martian samples analyzed, the first three having been obtained close to the Viking 1 Lander, and the last near Viking 2. These data have been obtained after counting times of the order of a thousand hours. Turned into oxide concentrations these measurements give the values tabulated in Table 5.5.

Table 5.4. Elementary composition, determined by X-ray spectroscopy of four martian samples at the Viking 1 (1, 2 and 3) and Viking 2 (4) sites

	1	2	3	4
Mg	5.0 ± 2.5		5.2	
Al	3.0 ± 0.9		2.9	
Si	20.9 ± 2.5	20.8	20.5	20.0
S	3.1 ± 0.5	3.8	3.8	2.6
Cl	0.7 ± 0.3	0.8	0.9	0.6
K	< 0.25	< 0.25	< 0.25	< 0.25
Ca	4.0 ± 0.8	3.8	4.0	3.6
Ti	0.51 ± 0.2	0.51	0.51	0.61
Fe	12.7 ± 2.0	12.6	13.1	14.2
L	50.1 ± 4.3			
X	8.4 ± 7.8			
Rb	$\leqslant 30$ ppm			$\leqslant 30$ ppm
Sr	60 ± 30 ppm			100 ± 40 ppm
Y	70 ± 30 ppm			50 ± 30 ppm
Zr	$\leqslant 30$ ppm			30 ± 20 ppm

Table 5.5. Mineralogical composition of martian soil samples, calculated from the elementary composition

	1	2	3	4	Estimate of the absolute error
SiO_2, m %	44.7	44.5	43.9	42.8	5.3
Al_2O_3, m %	5.7		5.5		1.7
Fe_2O_3, m %	18.2	18.0	18.7	20.3	2.9
MgO, m %	8.3		8.6		4.1
CaO, m %	5.6	5.3	5.6	5.0	1.1
K_2O, m %	< 0.3	< 0.3	< 0.3	< 0.3	...
TiO_2, m %	0.9	0.9	0.9	1.0	0.3
SO_3, m %	7,7	9.5	9.5	6.5	1.2
Cl, m %	0.7	0.8	0.9	0.6	0.3
Total	91.8		93.6		...
Rb, ppm	$\leqslant 30$			$\leqslant 30$	
Sr, ppm	60 ± 30			100 ± 40	
Y, ppm	70 ± 30			50 ± 30	
Zr, ppm	$\leqslant 30$			30 ± 20	

The composition of the soil is very similar at both the Viking 1 and 2 sites, although these are very distant from one another. We should not conclude from this that there is considerable uniformity in composition of the martian soil, however, because the two sites were chosen precisely because of their similarities.

The major finding is the importance of SiO_2 (45 ± 5 %) and of Fe_2O_3 (19 ± 3 %). Also notable is the high concentration of sulphur, which was unexpected, and which is about one hundred times the sulphur concentration in the terrestrial crust. On the other hand, potassium is rare: an upper limit of 0.25 % has been established, about five times less than the concentration in the Earth's crust. Finally we may note that the density of the dust is 1.1 ± 0.15 g/cm^3, which indicates that its porosity is about 60 %.

From all this data it is possible to try to construct a mineralogical model of the dust. Its composition has been compared with about a thousand terrestrial minerals and, using a computer, with many plausible mixtures. It appears that the best fits are those obtained by mixing different types of clay, with montmorillonite $Al_2Si_4O_{10}(OH)_2{\cdot}nH_2O$ as a basis, mixed with nontronite (where part of the Al is replaced with Fe^{+++}) and saponite (where Al is replaced with Mg).

The presence of clays on the surface of Mars, which seems to be borne out by these measurements, could be explained in two possible ways: first, one could envisage silicates being altered by water, particularly in the form of vapour, i.e. by the martian atmosphere, the reaction being accelerated by solar ultraviolet radiation. Equally, one could envisage that we are dealing with the results of volcanic activity that took place beneath a thick layer of ice. We know that a volcanic eruption that occurs beneath a glacier, producing a violent reaction between magma rich in iron and subterranean ice, can produce these types of minerals, as shown by the palagonites in Iceland.

5.5.4 The Atmosphere of Mars

Before the space age, estimates of martian atmospheric pressure, based on the diffusion of sunlight, gave values close to 100 millibars. Carbon dioxide CO_2, which was detected spectroscopically from Earth, appeared to be a minor atmospheric component. It took the Mariner 6, 7 and 9 missions to determine the total pressure, by occultation of radio waves when the probes passed behind the planet, and to discover that the atmosphere consists of practically pure carbon dioxide. We now know that the latter forms more than 95 % of the atmosphere, the total pressure of which varies between 7 and 10 millibars. In terms of the number of molecules, the following abundances apply: 95.3 % of CO_2, 2.7 % of N_2, 1.6 % of A, 0.13 % of O_2, 0.07 % of CO and some 0.03 % of H_2O, the last abundance varying greatly with season and latitude. The abundance of water may seem very low. This is a result of the very low temperature found at the surface. In the phase diagram of water, the triple point lies at $T = 273$ K and $P = 4.6$ mm Hg (6.1 millibars). Under martian conditions, the consequences are that liquid water cannot be encountered if the partial pressure of H_2O is less than 4.6 mm Hg. Currently, maximum water partial pressures do not reach 0.5 mm Hg. Water can therefore only exist as vapour or ice, according to the temperature. For example, at the Viking 2 site, the night-time

Fig. 5.24. In winter, frost covers the desert at the Viking Lander 2 site. (By courtesy of NASA)

temperature in winter drops to 160 K. The saturation vapour pressure for ice is then lower than 10^{-6} mm Hg: water therefore precipitates in the form of frost (Fig. 5.24). The atmosphere is extraordinarily dry. As warming occurs, the frost disappears. At a temperature of 180 K, the saturation pressure of water climbs to 5×10^{-5} mm Hg, which corresponds to an abundance close to 10^{-3} %. At a temperature of 200 K, it has increased to 0.02 %.

It seems that with the low O_2 content, very little ozone O_3 is present: 0.3 ppm. The atmosphere of Mars is therefore highly transparent to solar ultraviolet radiation. As a result there is a significant atmospheric photochemistry. The most abundant components, CO_2 and H_2O, will be partially dissociated into $CO + O$ and $OH + H$ respectively. Molecular oxygen will mainly arise from the recombination of individual atoms of oxygen. The fact that the abundances of CO and O_2 currently observed are low – although future space missions may possibly determine that the levels are somewhat higher – seems to indicate that there is a very effective mechanism creating CO_2 from CO under martian conditions. It has been suggested that the OH radical plays a part in this reaction. This assumes the existence of a vertical transport mechanism capable of ensuring rapid mixing, because the distribution profiles for CO_2 and H_2O are probably different. In fact, this is what we lack for understanding the aeronomy of the martian atmosphere: vertical distribution profiles for the principal molecular, radical and atomic components. In addition, the presence and the distribution of aerosols must be taken into account, as these probably play a major part in allowing mixed-phase chemical reactions to take place.

As we have seen earlier, although carbon dioxide is the principal atmospheric component, apart from existing as a gas it may be present in the form of carbonates in the surface rocks. In the same way, we do not know in what form the majority of martian nitrogen is to be found. The existence of nitrates is quite plausible, even though none have yet been detected. As regards the H_2O molecule, we have discussed earlier the possible existence of thick permafrost layers. However, it is by no means impossible that some water vapour has escaped from the planet into space, as has happened with Venus. An indicator of this process is the D/H ratio (see Sect. 5.2.4). Recent detection, by a team working with the CFH telescope on Hawaii, of HDO lines in the infrared spectrum of the martian atmosphere seem to indicate an enrichment of the D/H ratio by a factor of 6 when compared with the terrestrial value. This may indicate that some of the water vapour has escaped.

As far as the rare gases are concerned, helium is not retained gravitationally. Apart from argon, the others are only present in trace amounts: Ne (2.5 ppm), Kr (2.5 ppm) and Xe (0.08 ppm). Their elementary and isotopic abundances, which should be established by future space missions, are important tracers of the evolution of the atmosphere of Mars, and of the planet as a whole.

The greenhouse effect is weak, and the surface temperature variations are considerable, given the low mass of the atmosphere and the absence of the heat "reservoirs" that are provided by the oceans on Earth: the temperature may vary by more than 50°C between day and night at the Viking 1 and 2 sites. Taken overall, the climate on Mars is cold. The temperature may drop below $-120°$ at the southern pole in winter, and only rise about 0° near the equator in summer.

The martian atmosphere undergoes great climatic changes throughout the year, which are strongly dependent on latitude. In fact, the dynamical properties of Mars are notable for a large obliquity of the rotational axis (25.1°) and a high eccentricity (0.0934). As a result, the martian year (which lasts 669 martian days, or 687 terrestrial days) is divided into very marked seasons, and these seasons have unequal lengths, because of Kepler's Third Law. In the northern hemisphere, spring, summer, autumn, and winter last 194, 178, 143, and 154 martian days respectively. The highest temperature is therefore reached during the southern-hemisphere summer, because the distance from the Sun is then the least. That summer season is, on the other hand, shorter than the northern hemisphere's, where it lasts 24 martian days longer.

These dynamical properties of Mars are correlated with significant variations in temperature and pressure, causing the precipitation of ices (water and carbon-dioxide snow) or their vaporization. Direct effects are that clouds, primarily cirrus, as well as mists, form at different times according to the season. Polar "hoods" of freezing fog cover the polar caps in winter, whilst in the summer cloud formation is largely confined to relief features (such as the tops of the volcanoes). The atmospheric pressure may vary by 30 % over the year. Violent winds sweep the surface at speeds measured in tens of m/s and cyclones and anticyclones (high and low pressure systems) develop. The storms raise large quantities of dust into the atmosphere, lifting it as high as 50 km. This dust in turn influences the martian meteorology by increasing atmospheric opacity. Precipitation of water and carbon dioxide therefore accompany the deposit of dust, most notably in the polar caps. It is these movements of material that cause periodic variations in the planet's optical properties, which were long interpreted as being signs of seasonal changes in vegetation.

5.5.5 The Search for Organic Material

Contrary to the observations made by earlier space missions, Mariner 9 proved the existence of surface formations that can be interpreted as having resulted from a flow of water. Nothing else was needed to give a new lease of life to the theories that organic evolution could have produced biological systems. The Viking programme was therefore given one primary objective: to find evidence for chemical evolution, or organic activity, even if fossil, on the surface of Mars. At first, the detection of nitrogen in the atmosphere was taken to indicate that conditions were suitable for biological evolution. Each of the two landers carried a sample arm, capable of collecting samples and of placing them into the experimental chambers, where they would be analyzed and submitted to specific tests. The experiments were of two types: chemical and biological.

The chemical experiment consisted of the analysis of the soil by gas chromatography followed by mass spectrometry. The aim was to detect complex organic compounds, the decision having been taken to only consider carbon-based chemistry. The result was more than just disappointing: no fragment of a nucleic acid or of proteins was discovered, not even a hydrocarbon. The total amount of organic matter detected was indeed far less than on the Moon – where it only results, however, from the direct implantation of solar-wind carbon. The detection threshold was an abundance of 10^{-9}: with terrestrial samples, the same equipment is easily capable of detecting organic compounds in the most extreme desert soils, where the abundances are of the order of 10^{-6} or even 10^{-7}. This seems to indicate that organic compounds, such as those originating from the fall of a carbonaceous chondrite, for example, are rapidly destroyed by solar radiation.

The biological experiment would not just detect organic material, but would determine whether it was living by means of its metabolic activity. Three instruments were carried. The first, the Pyrolytic Release experiment, tested the ability of the martian soil to fix carbon. In the presence of the martian atmosphere, it was allowed to react with carbon dioxide and carbon monoxide that had been labelled with radioactive ^{14}C, before it was heated and analyzed for the radioactivity of any possible organic compounds that might have been synthesized. This experiment was repeated nine times, with and without the presence of sunlight, simulated by a xenon lamp. Seven of the experiments, including that carried out in the dark, gave a positive result: part of the carbon had been fixed by the samples, without the help of photosynthesis. The quantities fixed were infinitesimal, which might explain why the chemical experiments did not detect anything.

The other two experiments, known as the Gas Exchange and the Labelled Release experiments, involved reacting the martian soil with a water-based nutrient medium. They therefore required the samples to be introduced into pressurized and heated chambers, where water would remain liquid. The Gas Exchange experiment analyzed, by chromatography, the gases released after the sample reacted with the nutrient solution. The analyses were repeated regularly throughout the incubation period, which lasted seven months. Immediately after the solution was introduced, large quantities of carbon dioxide and oxygen were detected, but these declined progressively with time. The explanation was soon evident: it was not a biological reaction but a chemical one that was freeing the carbon dioxide trapped in the soil. As for the oxygen, that was being formed by decomposition of peroxides by water vapour in the presence of iron compounds.

The Labelled Release experiment reacted samples with simple organic compounds, notably formic acid, where the carbon was radioactive (^{14}C). Any metabolic activity would have been accompanied by the synthesis of gaseous radioactive compounds, detectable by their radioactivity. As in the behaviour found in the Gas Exchange experiment, carbon dioxide was released at the beginning of the experiment. It was $^{14}CO_2$, which indicated that it was a true reaction and not just simple degassing of atmospheric CO_2 trapped in the soil. However, this production of $^{14}CO_2$ can be explained by the theory that the martian soil contains peroxides: these would react with formic acid, releasing carbon dioxide and water.

In the light of these last two experiments, how should the first one, which appeared to show some form of metabolic activity, be interpreted? One result seems to indicate chemical, rather than biological, behaviour: one experiment had been carried out after the soil sample had been heated to 90°C for two hours. Under these conditions, one would have expected any possible organisms to have been destroyed. Yet the result was identical with that obtained without any prior heating. The fact remains that it is not easy to interpret the positive results obtained as purely chemical reactions. An element of mystery still surrounds this Pyrolytic Release experiment.

This element of doubt does not alter the more-or-less unanimous opinion that life does not exist at either of the two sites explored. Were they correctly chosen? Might not the constraints imposed by having to secure a safe landing, in particular having to avoid any major relief feature, any canyon, in favour of a flat, desert area, have been responsible for this failure to detect any extraterrestrial life? It is possible to maintain this point of view, and future missions will help to answer it. But it is still true that the biological experiments showed, unambiguously, nothing more than a high oxidizing capacity, and an astonishing lack of even the simplest organic compounds.

5.5.6 Geological Formations

Mars is not a perfect sphere: its flattening is greater than that of the Earth. The polar radius is 18 km less than the equatorial radius, which corresponds to a ratio of 1:192, compared with 1:298 for the Earth. In addition, the differences in altitude are much more pronounced, and show a strong north-south asymmetry.

In the southern hemisphere, two large, deep basins, Argyre Planitia and Hellas Planitia, lie below the mean surface level of the planet. The bottom of Hellas, at −6 km, is the lowest point on Mars. These basins were probably produced, like the large lunar basins, by the impact of very large meteorites some 4 thousand million years ago. In general terms, the southern hemisphere contains ancient terrains, which are covered in craters, and its average altitude is low.

In the north, on the other hand, there are areas that are very high, including the Tharsis Ridge, where the elevation of the crust (by about 6 km) is at its greatest. It is on this plateau that three of the gigantic volcanoes of Mars are found, in a line and all of similar heights. Not far away rises Olympus Mons, more than 25 km high, and 700 km across at the base. It is probably the largest volcano in the Solar System. According to crater-counting methods, the uplift of the Tharsis plateau dates from 4.1 to 3.3 thousand million years ago. The volcanic complex did not appear until later, about 3 to 2.5 thousand million years ago. The immense size of the volcanoes indicates the presence of a lithosphere that is sufficiently thick to support them. This may be explained as follows: in the absence of plate tectonics – and this appears to be unchallenged – the volcanoes stayed above the zones of magma and as a result lava flows accumulated as long as the local magma sources remained in existence. It seems that this volcanism lasted until recently, a few hundreds of millions of years ago. The activity of smaller volcanoes (all lying in the northern hemisphere), on the other hand, would have ceased much earlier in the planet's history. Perhaps we should not completely exclude the possibility of some residual – albeit limited – volcanism today.

Mars has a gigantic network of canyons, organized around Valles Marineris, a canyon that cuts across a whole hemisphere of Mars, with a length of 5000 km, depth that reaches 7 km, and an average width of 200 km. It originates in the western part of the Tharsis Ridge and stretches towards the east, staying approximately on the equator. Photographs clearly show these canyons as having resulted from a fluvial type of erosion, which implies that water must have been able to flow on the surface. One model that is based on geological observation of the surface, has been proposed, by Masson in particular. This suggests that the initial accretion of the planet was followed by intense degassing, producing a dense, warm atmosphere, particularly rich in water vapour. As the surface cooled, this water condensed, causing rain to deluge the surface, with resulting intense erosion. This water percolated into the soil; once freezing point had been reached, there was progressive growth of a thick permafrost. The eventual melting of these masses of ice, probably triggered by volcanic activity, produced violent floods, which excavated the network of canyons that we see (Fig. 5.25).

Apart from the large-scale geological formations – impact craters, volcanoes and canyons – a whole series of morphologically complex structures can be seen on Mars: faults, landslides and valleys. So tectonic activity has taken place on Mars, but this has been different from the plate tectonics found on Earth. None of the associated geological formations (mountainous regions, in particular), are observed on Mars, nor are there two types of crust, oceanic and continental.

Large regions of sand dunes are visible, especially near the north polar cap. They are a direct result of the violent winds that sweep the surface. The latter are

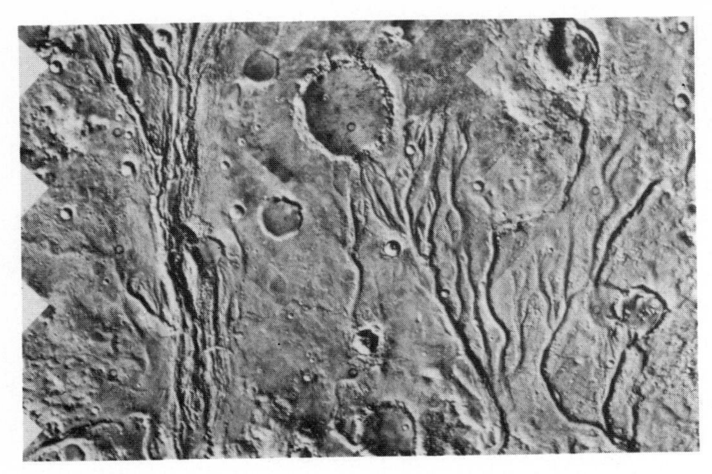

Fig. 5.25. This photograph, taken from Mars orbit, clearly shows the channels bearing witness to past floods. (By courtesy of NASA)

also responsible for the constant aeolian erosion that takes place. This, in particular, is what destroys impact craters with diameters below about a hundred metres. The redistribution of particles also produces the stratified deposits ("laminated terrain"), particularly in the polar caps. These deposits act as tracers for the accumulation of the ice-sheets, and may therefore be used as chronometers for martian palaeoclimatology.

5.6 Phobos and Deimos

Mars has two satellites: *Phobos* and *Deimos*. They are both small objects, neither being spherical in shape. They are approximately ellipsoidal, being 27 × 21 × 19 km, and 15 × 12 × 11 km for Phobos and Deimos, respectively (Figs. 5.26 and 5.27). Discovered in August 1877 by Asaph Hall, their exploration primarily dates from the American Mariner and Viking space missions. Thanks to the latter we have very high-resolution images of them, reaching a few metres for Phobos.

The orbits of Phobos and Deimos are circular, lie in the equatorial plane of Mars, and are direct. Their radii are 9354 and 23490 km, corresponding to orbital periods of 7 h 39 m and 30 h 18 m, for Phobos and Deimos, respectively.

The tidal effects acting between Mars and its satellites have two main consequences:

1. the satellites have stabilized in *synchronous rotation*, i.e. their rotational periods are equal to the orbital periods around Mars: Phobos and Deimos always present the same face to Mars, pointing their major axes towards its centre;
2. the equatorial "bulges" of Mars have a different axis of symmetry to the Mars-satellite axis. This is partly because of the high viscosity of the interior of Mars, and partly because Mars' rotation period 24 h 37 m, is different from

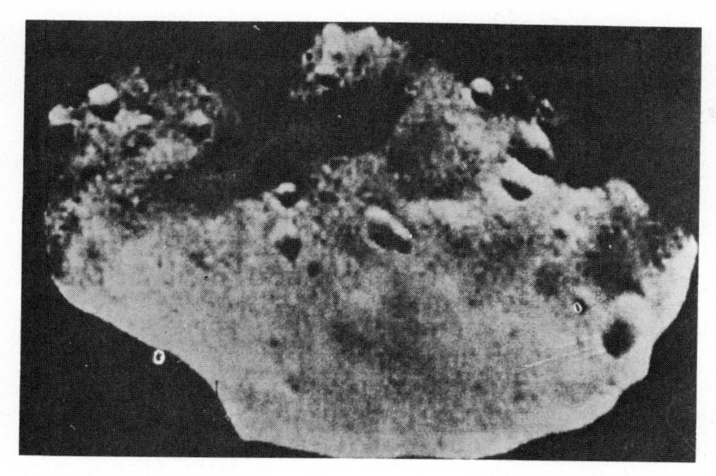

Fig. 5.26. Phobos, as seen by Viking 1 in 1977. (By courtesy of NASA)

Fig. 5.27. Deimos, as seen by Viking 1 in 1977. (By courtesy of NASA)

that of the satellites. As a result the gravitational force exerted on the latter is not strictly radial. In the case of Phobos, which orbits faster than Mars rotates (it rises in the west and sets in the east), the tidal component is equivalent to a braking force, which in turn inexorably causes the orbit of Phobos to shrink. In a few tens of millions of years, Phobos will crash into Mars. Deimos, on the other hand, makes less than one revolution in one martian rotation, which produces a slow, but progressive increase in the radius of its orbit, similar to the effect occurring in the Earth-Moon system.

The average density of the satellites is about 2. It is therefore decidedly different from that of Mars, and this probably reflects a difference in overall composition. This is also apparent in the surface composition: Phobos and Deimos have a very low albedo (around 5 %), less even than that of the darkest lunar soils. They appear to be dominated by presence of opaque, carbon-rich, hydrated minerals, analogous to the matrix found in carbonaceous chondrites, the most primitive meteorites (see Chap. 11).

There is general agreement that Phobos and Deimos must have had a different origin than Mars. They could not have been formed by accretion in orbit of material similar to that that gave rise to Mars. They could be objects from the family of asteroids – more precisely, type C asteroids, which are numerous in the outer part of the main belt and among the Apollo-Amor objects – that have been finally captured by Mars. The dynamical problems of such a capture have not yet been settled.

The surfaces of both Phobos and Deimos are covered by a *regolith* and are pitted with meteoritic impact craters. The density of these allows the surfaces to be dated at more than three thousand million years. But there are various differences between these two satellites. The thickness of the regolith is distinctly greater (more than 200 m) on Deimos than it is on Deimos. Phobos has two large craters, *Stickney*, 10 km in diameter, and *Hall*, 6 km in diameter. The diameter of Stickney is more than a third of Phobos as a whole. The effects of the impact responsible for this crater can be seen in a series of parallel *grooves*, several hundreds of metres long and twenty-odd metres deep, that cover a large part of the surface. As for Deimos, it shows small-scale albedo contrasts and blocks of rock that are not seen on Phobos. Among the scientific questions to be solved are the origin and the evolution of these satellites, their degree of internal differentiation, and the presence of carbon compounds as well as that of water.

6. The Asteroids (or "Minor Planets")

Observation of planetary orbits, the radii of which increase in a geometric progression (the Titius-Bode law, see Sect. 1.1.2), shows a "gap" between Mars and Jupiter. On 1st January 1801, the Sicilian, Giuseppe Piazzi (1746–1826) discovered *Ceres*, a small planet about a thousand kilometres in diameter, orbiting at about 2.8 AU. The orbit of Ceres corresponded exactly to that of the "missing planet" predicted by the Titius-Bode law. But the mass of Ceres was far smaller than that of Mercury, the smallest of the planets. It was soon discovered that Ceres was not the only body orbiting between Mars and Jupiter: three other asteroids (or "*minor planets*") were discovered at the beginning of the last century: *Pallas, Juno* and *Vesta*. Further discoveries were made as observational techniques improved, and several thousand asteroids have now been catalogued (Tables 1.4 and 6.1), the total mass of which hardly amounts to twice that of Ceres. We may expect observations by the *Hipparchos* satellite to lead to a considerable increase in the number of asteroids studied.

Table 6.1. Principal characteristics of the ten largest asteroids

Asteroid	Type	Apparent magnitude	Diameter (km)	Semi-major axis (AU)
1 Ceres	C	7.5	1032	2.767
2 Pallas	U	8	588	2.772
4 Vesta	V	6.5	576	2.361
10 Hygeia	C	10	430	3.134
704 Interamnia	U	11	338	3.061
511 Davida	C	11	324	3.175
65 Cybele	C	12	308	3.433
52 Europa	C	11	292	3.103
451 Patienti	C	11.5	280	3.063
15 Eunomia	S	9.5	260	2.643

6.1 Orbital and Physical Characteristics

The orbital characteristics of the asteroids show a wide range of values for their semi-major axes, eccentricities, and inclinations to the ecliptic: the last may reach as much as 68°. Most asteroids have semi-major axes lying between 2 and 3.5 AU, forming the *main belt* (Fig. 6.1). Beyond 3.5 AU, all the bodies fall into just two families, the *Hildas* (at 4 AU) and the *Trojans* (at 5 AU). Closer to the Sun, marked

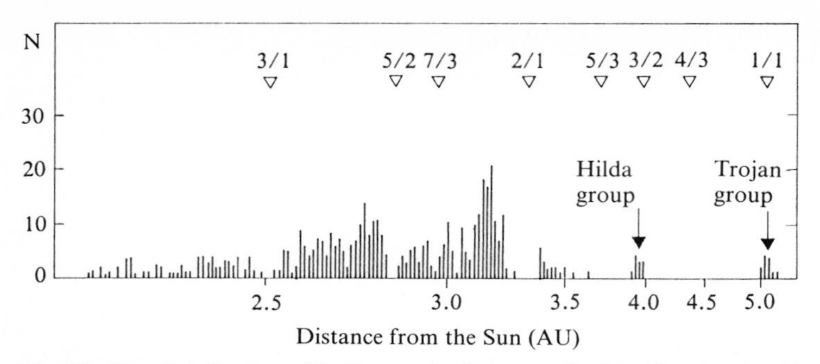

Fig. 6.1. The distribution of heliocentric distance of asteroids, showing the ***main belt***, the ***Hilda*** and ***Trojan*** groups, as well as the location of the ***Kirkwood gaps***. [After J.-P. Bibring and Y. Langevin: "Les astéroides", ***Le grand Atlas Universalis de l'Astronomie*** (Encyclopedia Universalis 1983). By courtesy of the publisher]

discontinuities occur between regions rich in asteroids: these are the *Kirkwood gaps*, lying at heliocentric distances (2.5 AU, 2.83 AU, 3.0 AU, 3.3 AU), which correspond to orbital periods that each have a simple relationship (1/3, 2/5, 3/7, 1/2) with that of Jupiter. In addition, roughly a hundred asteroids of very small size (a few kilometres) have orbits that intersect, or closely approach, that of the Earth. These form the *Apollo-Amor* family, which mainly consists of asteroids derived from the main belt, having been ejected from the Kirkwood gaps by Jupiter's gravitational perturbations. Some others are probably the nuclei of extinct, short-period comets.

Figure 6.2 is a representation, in the plane of the ecliptic, of typical asteroid orbits. The main belt contains practically all the known bodies. Most of these asteroids have low eccentricities, with the notable exception of a few such as *Hidalgo*, where aphelion is close to Saturn's orbit. Only a few asteroids have orbits that enter the inner Solar System, the *Amor* objects, which cross the orbit of Mars, and the *Apollo* objects, which cross that of the Earth.

Over the course of time, the magnitude of an asteroid varies, sometimes significantly, for two primary reasons. First, it depends not only on the distance of the Earth, but also on the phase angle (between the Earth-asteroid, and Sun-asteroid lines). The slow variation is accompanied by oscillations with a much shorter period, caused the object's rotation. It has been possible to establish the rotational period for many asteroids, and these typically range between two and twenty-four hours. Some of the light-curves, such as those of *Camilla* and *Eunomia* suggest that the asteroids are quite rectangular in shape; others may even consist of two bodies in contact (*Hector*). Apart from this, some recent results from stellar occultations appear to suggest that some asteroids may have satellites.

The size-distribution of asteroids may be determined by two different, optical methods. The most direct one, recently developed, consists of determining separately the amounts of energy that are reflected and absorbed by the object, from the flux received in the visible and infrared regions respectively. The sum of these two quantities is equal to the solar flux received by the asteroid, which is proportional to it projected surface area, and therefore to the square of its size. A more indirect method uses an empirical relationship that links the variation in polarization of the

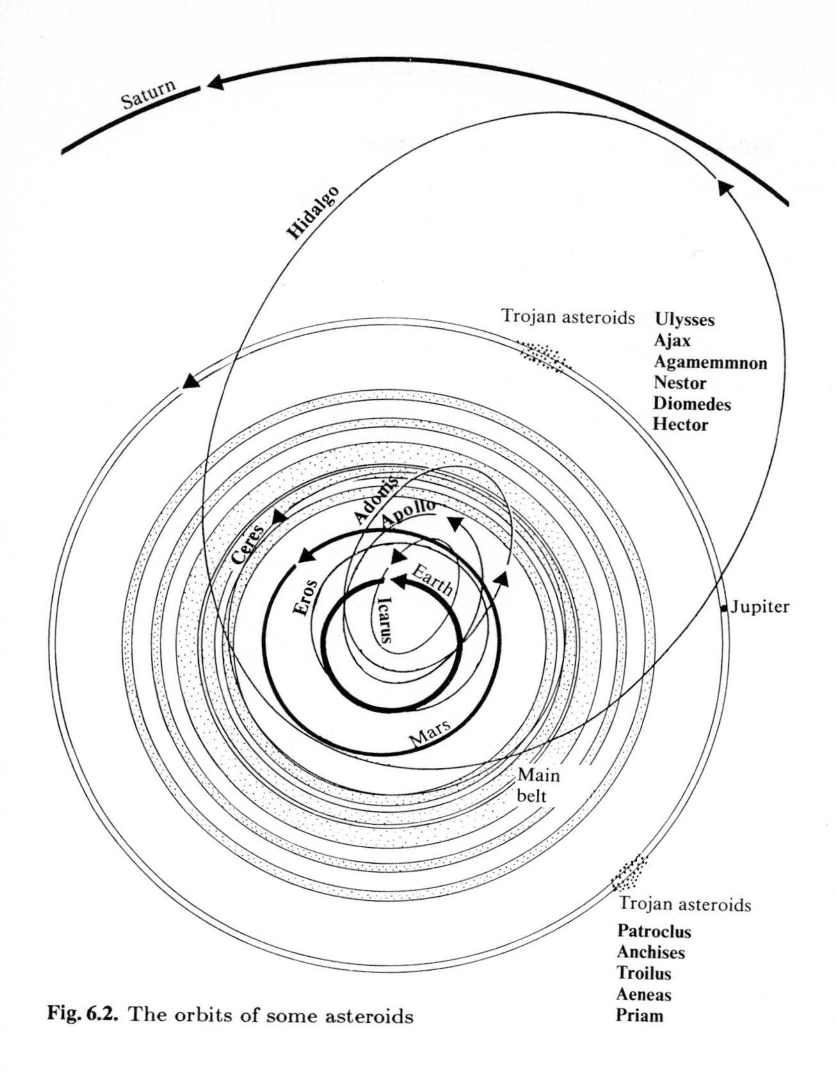

Fig. 6.2. The orbits of some asteroids

light emitted as a function of phase angle on the one hand, with the percentage of solar radiation that is reflected (the albedo) on the other. Measurement of visual magnitude then allows the size of the object to be determined. These measurements show two very significant characteristics: first, the number of asteroids increases very rapidly as their size decreases: if the latter is divided by ten, then the former has to be multiplied by one hundred. Second, asteroids can be divided into two major classes according to their albedo: bright, with albedo above 0.1, and dark, with albedo below 0.05. The latter are therefore among the darkest objects in the Solar System (the lunar maria, which are very dark, have an albedo of 0.07).

6.2 Mineralogical Characteristics and Classification

Over the last few year, progress in infrared astronomy has enabled information about the surface composition of asteroids to be obtained. Numerous minerals show characteristic absorption bands in the near infrared between 0.6 and 4 μm (Fig. 6.3). In particular, it is possible to determine the relative abundance of major groups of silicates: pyroxenes (*a*), olivines (*b*), and feldspars (*c*) each show a band in the near infrared, the position and shape of which are easily identifiable. Curve (*d*) corresponds to Fe-Ni alloys. It is also possible to determine the presence of hydrated minerals (*e*), thanks to the strong water band at 3 μm. Figure 6.4 shows the near-infrared spectra of typical asteroids. It can be seen that asteroids differ as to the mineralogical composition of their surfaces. In particular, the characteristic silicate bands are clearly apparent in the spectrum of Vesta, whilst that of Ceres is dominated by opaque minerals and hydrates.

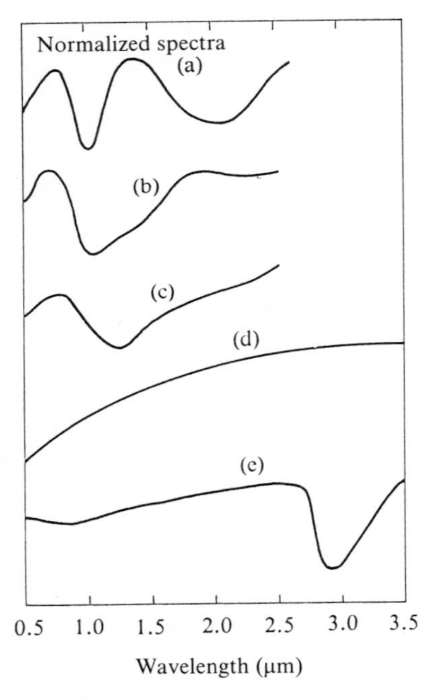

Fig. 6.3. Characteristic near-infrared spectra of the main minerals forming asteroids: pyroxene (*a*) shows a narrow band at 1 μm and a wide one at 2 μm, whilst olivine (*b*) produces an asymmetrical band between 1 and 1.5 μm. Feldspar (*c*) shows a band at 1.25 μm. An iron-nickel alloy shows a steady increase in albedo as the wavelength increases. Finally hydrated minerals (*e*) show a characteristic absorption band of water at 3 μm

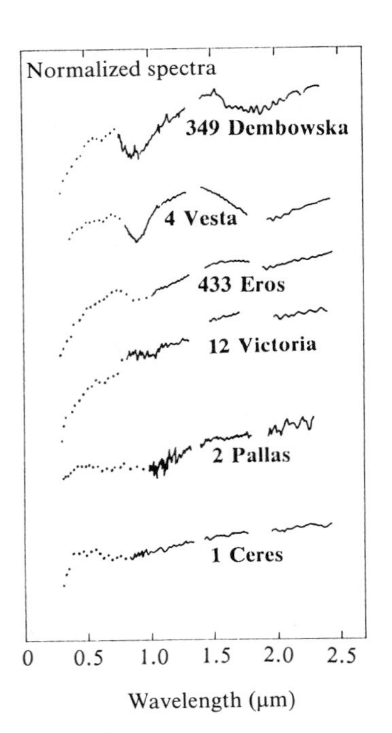

Fig. 6.4. The near-infrared spectra of typical asteroids show profound differences in mineralogical composition. Bands characteristic of silicates are clearly seen in the spectrum of Vesta, for example, while the spectrum of Ceres is dominated by opaque and hydrated minerals. [After M.J. Gaffey, T.B. McCord, H.P. Larson and G.J. Veeder: *Asteroids*, ed. by T. Gehrels (© University of Arizona Press, Tucson 1979)]

Spectral characteristics therefore allow us to group asteroids into different classes:

1. *C-type asteroids* (*C*arbonaceous) correspond to the darkest asteroids, rich in hydrated silicates and in carbon. These are the most numerous (about 60 %), in particular in the outer portion of the main belt;
2. *S-type asteroids* (*S*iliceous) show the spectral characteristics of rocky bodies that principally consist of pyroxenes and olivines, as well as a metallic phase (iron and nickel). They are abundant among the Apollo-Amor objects and in the inner part of the main belt, and represent about 30 % of catalogued asteroids;
3. *M-type asteroids* (*M*etallic) seem to be entirely metallic (iron and nickel);
4. *U-type asteroids* (*U*nclassified) do not fit this classification. For example, Vesta has a particularly high albedo (40 %), and a spectrum dominated by pyroxene and feldspar absorption bands.

6.3 Asteroids and Meteorites

Similarities have been sought between the absorption spectra of asteroids and those of meteorites, of which they are probably the parent bodies (Fig. 6.5). This showed that the carbonaceous chondrites may have been derived from C-type asteroids, and some stony meteorites from S-type asteroids. Iron meteorites may derive from M-type asteroids, and stony-iron meteorites from either M- or S-type asteroids. But ordinary chondrites, which are the most numerous in collections, do not seem to have any analogue among observed asteroids. This could arise because these meteorites

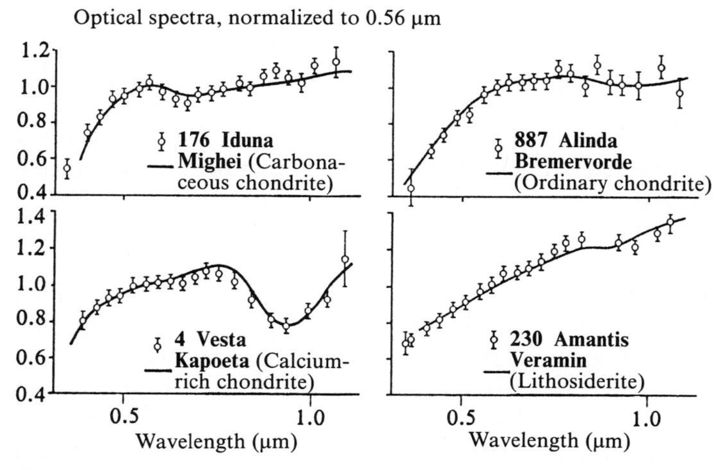

Fig. 6.5. Asteroids and meteorites. The principal spectral characteristics of meteorites (continuous lines) greatly resemble those of certain asteroids (points with error bars). These similarities indicate that the asteroids are the most likely parent bodies for meteorites. [After C.R. Chapman, M.J. Gaffey: *Asteroids*, ed. by T. Gehrels (© University of Arizona Press, Tucson 1979)]

are not representative of the surfaces of these asteroids, but of their interiors. The achondrites, also relatively abundant, can apparently be associated only with a very small group of asteroids, such as *Vesta*.

6.4 The Origin and Evolution of Asteroids

Two principal theories have been put forward to explain the origin of the asteroids in the main belt. The oldest assumes the existence of a parent planet, orbiting between Mars and Jupiter, which was broken up. Nowadays it is considered that it more likely that we are dealing with a population of objects the accretion of which into a planet was blocked at some intermediate stage by the primordial Jupiter's gravitational perturbations. Under this hypothesis, the asteroids would be the sole survivors of the myriad small bodies from which the inner planets were formed.

After their accretion, bodies larger than a few kilometres in size could have reached high temperatures through the disintegration of short-lived radioactive elements; in some of these bodies a metallic core would have formed, surrounded by a silicate crust. These processes took place over just a few million years, right at the beginning of the Solar System's existence. After this brief period of internal activity, only collisions between objects in the main belt would be able to modify their characteristics and cause them to fragment and, together with planetary perturbations, change their orbits. The relative velocity of these collisions would have been a few kilometres per second.

Bombardment by smaller objects pulverizes surface rocks, leading to the accumulation of a surface layer of debris, the *regolith*. Because of the intensity of the bombardment and the low gravitational field, the thickness of this layer could attain about a kilometre. Certain meteoritic breccias may have been formed within this bed of dust.

Collisions between bodies of similar size lead to the complete fragmentation of both objects. Because of this process, the lifetime of asteroids that do not exceed about a hundred kilometres in size is low in comparison with that of the Solar System. Most asteroids are therefore fragments of large, primordial bodies, metallic asteroids coming, for example, from the core of a differentiated body.

Among the fragments produced by a collision, some may find themselves in an unstable zone, like those of the Kirkwood gaps. They become the source of the Apollo-Amor objects. But these remain few in number, because they soon collide with the inner planets (after a few hundred million years). In the case of the Earth, objects smaller than about a metre generally end up in meteorite collections. The very rare collisions with asteroids several kilometres in size have produced vast impact craters and raised large quantities of dust or water vapour – from oceanic impacts – into the atmosphere. Such catastrophic events may have had considerable effects on the evolution of life on the Earth.

Because of the small size of asteroids, our knowledge of them can hardly increase significantly by using just conventional astronomical methods, either from the ground or even with the future Space Telescope, with which a large minor planet of 300 km in diameter would still remain more or less a point source. Only a spaceprobe mission will allow this family of objects to be studied in detail. We may expect them to show points of similarity with the small rocky bodies already observed at close quarters: *Phobos* and *Deimos* (see Chap. 5), and *Amalthea* (see Chap. 8), the innermost of Jupiter's satellites. Such exploration could have economic repercussions, because it is possible to envisage exploitation of small minor planets in the medium term as sources of minerals (in particular of metals), which could be placed in Earth orbit.

7. The Giant Planets

We have already seen (Sect. 1.2.2) how observation of the basic physical properties of planets in the Solar System led to their being classified in two major categories: the inner planets (see Chap. 5), and the giant planets. The latter, numbering four – Jupiter, Saturn, Uranus, and Neptune – have large diameters (whence their name), a relatively low density of the order of 0.7 to 1.5 (see Table 1.3), and a large number of satellites. A system of rings has been discovered around each giant planet.

We have also discussed (Sect. 1.2.2) how the chemical composition of planetary atmosphere could, to a first approximation, be explained: the giant planets formed at temperatures that were sufficiently low for them to retain ices in their cores. They therefore acquired so much mass that not even the lightest elements were able to escape from their atmospheres, which were mainly gravitationally accreted about their cores. This is why their atmospheres are described as being primary, in contrast to those of the inner planets. Their composition should reflect, at least partially, the composition of the primordial solar nebula from which the Solar System was formed (see Chap. 3).

The atmospheres therefore consist mainly of hydrogen (of the order of 90 %) and helium (about 10 %); other components only being present in trace amounts. The most abundant minor components are CH_4 (about 0.1 %) and NH_3 (0.02 %).

It should be noted, however, that the concept of primary and secondary atmospheres is somewhat theoretical, useful for describing the large-scale features of the evolution of planetary atmospheres, but that these ideas must be used with some caution. What we know of the formation of the giant planets does not, at present, allow us to conclude that the composition of their atmospheres accurately reflects that of the primordial solar nebula that gave birth to them; on the contrary, evidence of differentiation has been discovered.

The crucial question is: how, by studying the atmospheres of the giant planets, can we work backwards to the physical and chemical conditions prevailing in the solar-nebula material, from which they were formed? Answering this question would imply knowledge of all the changes that have occurred in the meantime, during both the formation phase of the giant planets, and their subsequent evolution. Nevertheless, the giant planets do appear to bear witness to the early stages in the Solar System's evolution; we merely have to find ways of understanding the information that they provide.

The giant planets have no surfaces in the sense that we have described for the inner planets and satellites. From density and gravity measurements, as well as from theoretical models, a giant planet should have a central core, doubtless consisting of rocky and icy material, surrounded by hydrogen and helium under

Table 7.1. Cosmic abundances of the most common elements

Element	Abundance	Element	Abundance
H	3.18×10^{10}	B	3.50×10^{2}
He	2.21×10^{9}	V	2.62×10^{2}
O	2.17×10^{7}	Ge	1.15×10^{2}
C	1.18×10^{7}	Se	6.72×10^{1}
N	3.74×10^{6}	Li	4.95×10^{1}
Ne	3.44×10^{6}	Ga	4.8×10^{1}
Mg	1.06×10^{6}	Kr	4.68×10^{1}
Si	1.00×10^{6}	Sc	3.5×10^{1}
Fe	8.3×10^{5}	Zr	2.8×10^{1}
S	5.0×10^{5}	Sr	2.69×10^{1}
Ar	1.17×10^{5}	Br	1.35×10^{1}
Al	8.5×10^{4}	As	6.6
Ca	7.21×10^{4}	Te	6.42
Na	6.0×10^{4}	Rb	5.88
Ni	4.80×10^{4}	Xe	5.38
Cr	1.27×10^{4}	Y	4.8
P	9.60×10^{3}	Ba	4.8
Mn	9.30×10^{3}	Mo	4.0
Cl	5.70×10^{3}	Pb	4.0
K	4.20×10^{3}	Sn	3.6
Ti	2.77×10^{3}	Ru	1.9
F	2.45×10^{3}	Cd	1.48
Co	2.21×10^{3}	Nb	1.4
Zn	1.24×10^{3}	Pt	1.4
Cu	5.40×10^{2}	Pd	1.3

(After R. Prinn and T. Owen, *Jupiter*, T. Gehrels, ed., Tucson, University of Arizona Press, © 1976.)

high pressure, probably in the metallic state. This globe is itself surrounded by a very thick atmosphere, mainly consisting of hydrogen and helium (Table 1.3), the elemental composition of which should, to a first approximation, reflect the cosmic abundances of elements (Table 7.1).

Only the outermost atmospheric layers of the giant planets, down to a level corresponding to a pressure about 10 bars, are accessible for observation. Almost all the results that we have for giant planets have been gained from these regions; the planets' internal structure remains essentially a problem in theoretical modelling.

7.1 The Neutral Atmosphere of the Giant Planets

This region, extending from a few millibars to about ten bars, is the best-known layer in the atmospheres of the giant planets. It has been probed by spectroscopy from Earth, from the near ultraviolet to radio waves. This is where atmospheric components have been detected and their abundances measured, where the thermal structure has been analyzed, where the radiation balance has been measured, and where the cloud layers have been studied (Figs. 7.1 and 7.2).

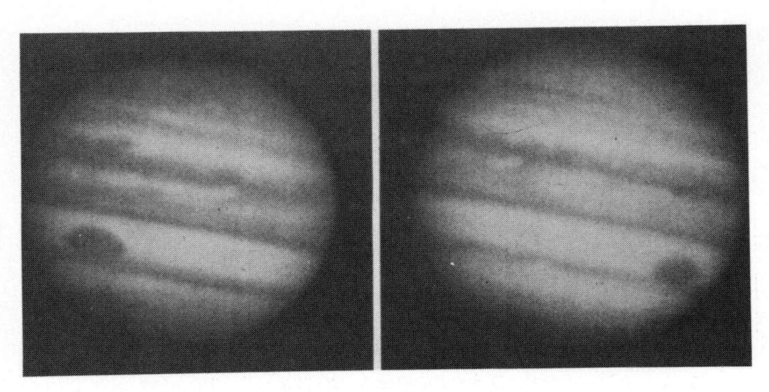

Fig. 7.1. Jupiter as observed by telescope from the Earth

Fig. 7.2. Saturn as observed by telescope from the Earth

The techniques used to probe planetary atmospheres have been described in detail earlier (see Sect. 2.2). It should be noted that spectroscopy, carried out at all wavelengths, is the most powerful tool for studying the physical and chemical properties of the atmospheres of the giant planets (the thermal profiles, and the abundances of the components). Imagery, in the visible or the infrared, allows variations between the centre and the limb to be studied, and is an indispensable means of following the spatial and temporal variations in a planetary atmosphere. We should also mention the observation of stars occulted by planetary atmospheres; this allows the thermal structure of upper atmospheres to be deduced. Finally, the Galileo mission, launched in 1989, will open the era of direct, in-situ observation of the giant planets, in particular the determination by mass-spectrometry of the atmospheric composition of Jupiter.

From the measurements of the abundances of minor components detected in the neutral atmospheres, it is possible to calculate, for each giant planet, elemental and isotopic abundance ratios. These ratios are better known the closer the planet, and therefore the brighter and easier to observe. In the case of Jupiter, measurements, or estimates, have been obtained for H/He, C/H, N/H, P/H, Ge/H, O/H, D/H, $^{12}C/^{13}C$ and $^{14}N/^{15}N$; in the case of Neptune, only C/H and D/H ratios have been established, with large degrees of uncertainty.

The determination of abundance ratios from the observation of two atmospheric components is relatively simple when the two components are uniformly mixed: the ratio that one measures is then constant with height throughout the atmospheric region probed. This is the case, for example, with non-condensable species such as CH_4 and CH_3D in the atmospheres of Jupiter and Saturn, whose ratios with respect to hydrogen remain constant with altitude. In many other cases, on the other hand, the ratio of an atmospheric component to hydrogen varies as the result of one or more physical and chemical processes: condensation (as happens with H_2O and NH_3), photodissociation (NH_3 and PH_3), chemical reactions (PH_3, GeH_4), etc. It is then much more difficult to measure the mixing ratio of the species concerned (relative to the most abundant gas, hydrogen), and even more difficult to interpret it correctly. It is essential to specify, for any measurement, the atmospheric region where the observation was made, and to understand the mechanism responsible for the variation in the mixing ratio. Such research requires a thorough knowledge of the planet's thermal structure, as well as certain information about the structure of its clouds.

7.1.1 Thermal Structure

Radiative and Convective Transfer

We have discussed earlier (Sect. 2.2.1) the different physical laws that link the various quantities involved in atmospheric structure (pressure, density, temperature, gravity): these are the perfect gas law, the law of hydrostatic equilibrium and, for the calculation of the flux emitted by the whole atmosphere, the radiative transfer equation. Radiation is only responsible for energy transfer in the upper part of an atmosphere, however, at pressure levels less than approximately one bar; at higher pressures, energy transfer is carried out by convection.

It can be shown that when the convective flux predominates, the gradient dT/dz may be likened to the adiabatic gradient in deep layers of the giant planets:

$$\frac{dT}{dz} = -\frac{g}{C_{\mathrm{p}}} \tag{7.1}$$

g being the gravitational acceleration and C_{p} the specific heat. A feature of the four giant planets is therefore a convective zone within which the adiabatic gradient, which is of the order of $-2°$/km, varies as the quantity C_{p}, a function of the H_2/He ratio. Above an atmospheric layer where the pressure is slightly less than one bar, the atmosphere ceases to be optically thick, and energy transfer takes place by radiation.

Models of Radiative Transfer

In the atmospheric region where transfer is radiative, it is possible to calculate the temperature at each level, by writing that flux is conserved: in other words, the flux emitted by any atmospheric layer is equal to the sum of the flux received from all the other higher or lower atmospheric layers. In the theory of local thermodynamic equilibrium (LTE, see Sect. 2.2.1), the flux emitted by an atmospheric layer is none other than σT_4, where σ is Planck's constant and T is the temperature of the layer. We therefore have the following equation, for layer i:

$$\sigma T_i^4 = \sum_{j>i, j<i} \int_0^\infty B_\nu(T_j) \mathrm{e}^{-\tau_{ij}/\mu} d\tau \, d\nu \tag{7.2}$$

where the quantity τ_{ij}, the optical thickness between layers i and j, is defined as follows:

$$\tau_{ij} = \int_i^j K_\nu \varrho \, dz \tag{7.3}$$

(K is the coefficient of absorption and ϱ is the density).

In order to solve this problem numerically, one starts by taking an initial thermal gradient, generally isothermal, and proceeds by successive iterations until the difference between the two terms in (7.2) is reduced to zero.

There is a simple analytical solution to (7.2) when the coefficient of absorption K is independent of frequency ν: this is the grey-body case.

Making certain simplifications, (7.2) becomes:

$$T^4(\tau) = T_\mathrm{e}^4 \frac{1 + 3\tau/2}{2} \tag{7.4}$$

where T_e is the effective temperature, defined in Chap. 1.

For the gas giants, numerical calculation has been carried out by several workers, taking the absorption coefficients of various atmospheric components into account. Hydrogen and helium, the principal components, do not have direct spectral signatures, but have a continuous absorption spectrum, owing to H_2–H_2 and H_2–He collisions; this continuum reaches into the infrared, with a maximum between 15 and 30 μm. If there were no other contribution to the absorption, the thermal profile would decrease with height, and would then become isothermal, with $T = T_\mathrm{e}$. But there is another contribution, which comes from the absorption, in the upper layers, of infrared solar radiation by atmospheric methane and also by the aerosols present in the atmosphere. As a result there is an increase in the temperature of the outer layers, which has been effectively confirmed by observation, above an atmospheric level corresponding to a pressure of about 100 mb.

Inversion of the Brightness Integral

We have seen that the far infrared spectrum of the giant planets is dominated by the continuum component caused by H_2–H_2 and H_2–He collisions. As the continuum spectrum is not the same for H_2–H_2 collisions as for H_2–He collisions, this continuum component more or less depends on the H_2/He ratio in the atmosphere; it varies, on the other hand, very slowly as a function of the temperature T.

From this it will be seen that it is possible to relate a specific set of frequencies ν_i in the infrared (for example between 10 and 100 μm) to a set of atmospheric levels P_i; the value P_i (see Sect. 2.2.1) corresponds to the atmospheric layer being probed by radiation of frequency ν_i, when the whole disk is observed (as is the case with observations from the Earth). This applies because hydrogen is the major atmospheric component, the quantity of He present being only about 10 %. This relationship is, to a first approximation, independent of the profile $T(P)$.

In addition, we saw earlier (Sect. 2.2.1) that the brightness temperature T_{Bi} measured at a given frequency is, to a first approximation, equal to the temperature of the atmospheric level at which $\tau = 0.66$. It will therefore be seen that by linking levels P_i with brightness temperatures T_{Bi} measured at frequencies ν_i, we obtain a set of samples giving, to a first approximation, the profile $T(P)$. It can be shown that by using a double iteration it is possible to obtain simultaneously the $T(P)$ profile and the H_2/He ratio. Taking a series of values for this ratio (varying the proportion of hydrogen between, for example, 0.8 and 1), the iterative method is applied to obtain $T(P)$ for each value of H_2/He; it can be shown that the "true" H_2/He value and the "true" $T(P)$ profile are those that give the fastest convergence.

This method has been successfully used by several authors (Gautier et al.). The region of the atmosphere that is probed extends from about 100 mb to 1 bar: it is therefore the region where the inversion minimum occurs. For Jupiter, Saturn and Uranus, where precise infrared measurements have been obtained, in particular thanks to the Voyager mission, it has been possible to obtain simultaneously $T(P)$ and H_2/He to a high degree of precision.

Determination of $T(z)$ by the Stellar Occultation Method (see Sect. 2.2.3)

When a planet passes in front of a star, the starlight, before being occulted by the planet, is refracted by the latter's atmosphere. By measuring the star's light-curve at immersion and emersion, it is possible to deduce the particle density as a function of altitude, and, by assuming a mean molecular density, the $T(z)$ profile. The region probed by this method is that of the stratosphere between about 10^{-4} and 10^{-2} bar.

For observation to be possible from Earth, a bright star has to be involved, and the events are correspondingly rare. The first observation that allowed a determination of $T(z)$ was the occultation of the star β Sco by Jupiter, in 1971. The same technique has been used to determine $T(z)$ for Uranus and Neptune, from several observations made since 1977. The observations were made in the visible or in the near infrared, with filters corresponding to methane absorption bands, in order to lessen the flux from the planet itself.

The occultation method has also been used to obtain thermal profiles of the outer atmospheres of Jupiter, Saturn, Uranus and Neptune, by measuring radio signals from the Pioneer and Voyager probes.

Results

Figure 7.3 shows the four $T(P)$ profiles for Jupiter, Saturn, Uranus, and Neptune, obtained from radiative models by using the temperature T_e measured for each planet. The similarity of the profiles is striking: in the deep layers there is a convective zone with an adiabatic gradient, which differs little from planet to planet; around

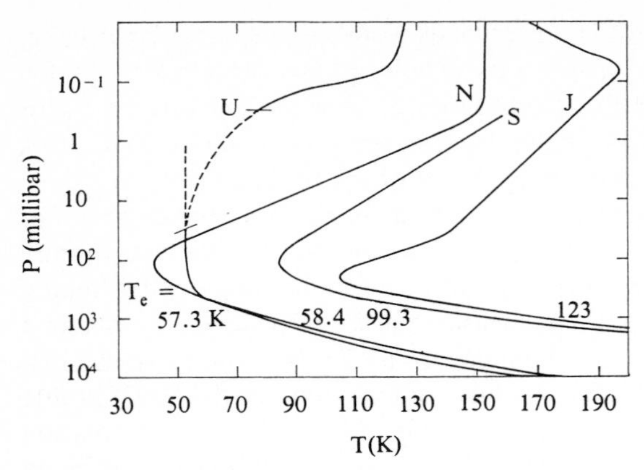

Fig. 7.3. Temperature profiles for the giant planets. [After D. Gautier, R. Courtin: Icarus **39**, 28 (Academic Press, 1979)]

the 100-mb pressure level (the tropopause), there is a temperature inversion; above the tropopause there is an increase in temperature caused by the absorption of solar radiation by the neutral atmosphere (gas or particles). This similarity is explained by the fact that the chemical composition of the atmosphere is more or less the same for the four planets, and also because the coefficient of absorption produced by pressure depends strongly on P (because it is proportional to the square of the density), and only weakly on the temperature.

The profiles shown in Fig. 7.3 are average profiles calculated for the planet as a whole. It is now possible to determine $T(P)$ profiles for different parts of the disk, thanks to data from space missions, either by inversion of the brightness integral, or by the occultation method. With Jupiter, in particular, considerable variations have been found in the $T(P)$ profile (see Fig. 7.4) according to which parts of the disk are observed (the equatorial zone, the tropics, or the Great Red Spot). Possible variations (both spatial and with time) in the thermal profile are a factor that must be borne in mind in interpreting spectroscopic data, in particular those obtained with a degree of spatial resolution.

7.1.2 Cloud Structure

There is a striking contrast between our considerable knowledge about the thermal profiles of the giant planets and the uncertainty that prevails with regard to the structure of their clouds. The main reason for this is undoubtedly the difficulty in interpreting the spectral signatures from particles, in contrast to those from gaseous molecules.

Thermochemical Equilibrium Models

Cloud structure exists in the atmospheres of the giant planets. The first indication of this is given by models of thermochemical equilibrium: if we know that a component is present in a gaseous state, by using its saturation curve, its abundance, and the thermal profile, we can calculate the atmospheric level at which it will condense into

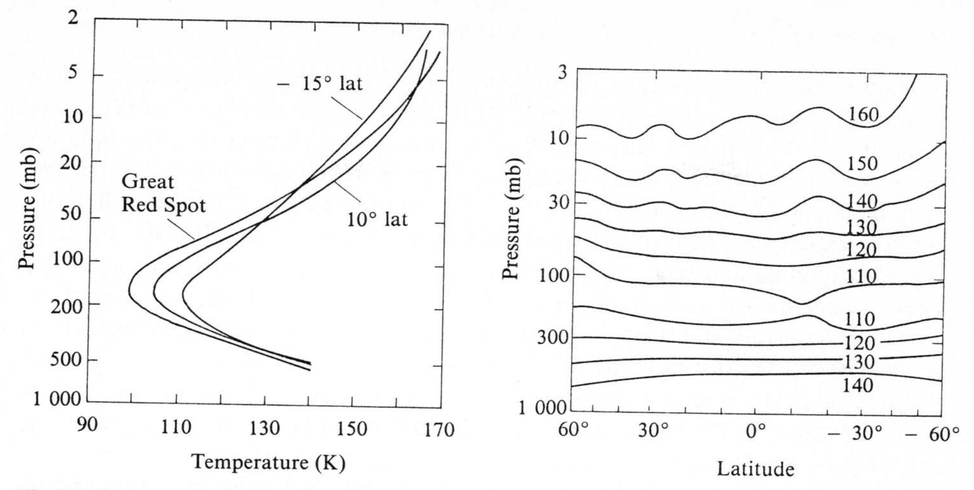

Fig. 7.4. Temperature profiles for Jupiter obtained by Voyager 1. ***Left***, typical profiles at latitudes +10° and −15°, and at the Great Red Spot. ***Right***, pressure levels corresponding to different temperatures, as a function of latitude. [After R. Hanel et al.: Science **204**, 972 (© AAAS 1979)]

cloud: this occurs with NH_3 on the four giant planets. Calculations of thermochemical equilibrium have been made for Jupiter by Lewis, who predicts three layers of clouds (Fig. 7.5): NH_3 at 0.5 bar, NH_4SH and NH_4OH around 2 bars and H_2O around 5 bars. It should be noted, however, that this model assumes that the various components have relative abundances like those found in the Sun, which does not seem to be absolutely correct (see Sect. 7.1.4): Fig. 7.5 should therefore be taken as figurative only. According to similar calculations CH_4 and C_2H_2 may possibly condense on Uranus and Neptune.

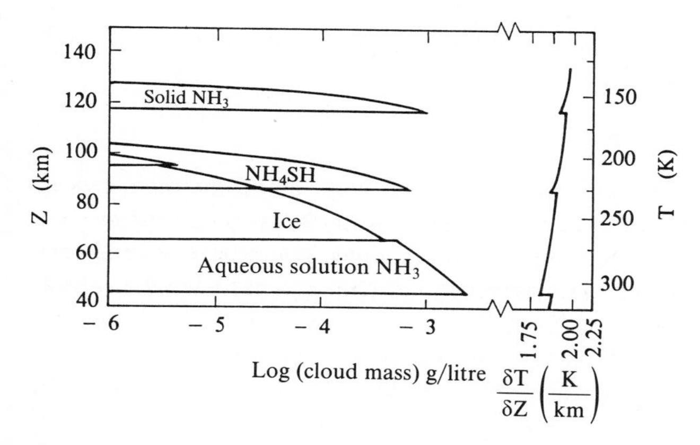

Fig. 7.5. Cloud structure on Jupiter, and the saturated adiabatic curve. [After R. Prinn, T. Owen: ***Jupiter***, ed. by T. Gehrels (© University of Arizona Press, Tucson 1976)]

Observational Evidence Obtained by Spectroscopy

In reality, the situation is more complex than implied by this simple scheme, which reduces the cloud structure of the giant planets to a certain number of well-defined cloud layers. This model does not, indeed, allow us to interpret spectroscopic observations of Jupiter and Saturn made in the visible in any quantitative manner. We have mentioned earlier (Sect. 1.2.2) the "reflecting layer model" (RLM), where the cloud is assumed to behave like a reflecting surface, and where diffusion by particles is negligible above the cloud. If this theory were valid, we would obtain the same abundance for a given atmospheric component whatever the intensity of the transition observed. But this is not the case: on the four giant planets, the calculated abundance of CH_4, for example, is less, the stronger the transition being considered. The interpretation is simple: weak lines, which have a low coefficient of absorption, are more strongly affected by scattering. This has the effect of increasing the mean free path of the photons. There is therefore a scattering medium above the cloud layers, or between them. From measurements of limb darkening at visible and ultraviolet wavelengths, and using different types of scattering models, some authors (West, Tomasko) have concluded that a thin veil exists in the upper atmosphere of Jupiter, at a pressure level of about 100 mb. Other authors have tried to interpret measurements of Jupiter in the visible and infrared on the basis of a model with several cloud layers, scattering taking place within the cloud layers.

Other information is given by measurements in the infrared. A particularly interesting spectral domain is the region around 5 μm, where there is no absorption by CH_4 or NH_3; the radiation originates in the deeper layers of the atmosphere (where the pressure is of the order of a few bars), so minor atmospheric components may be detected. For Jupiter, the spectrum at 5 μm has been studied in detail by observation from the ground and from space by the Voyager probes. Interpretation of the spectrum at various points on the disk shows that a cloud layer strongly absorbing infrared radiation must exist at a pressure of about two bars; this cloud layer may be caused by the condensation of NH_4SH and/or NH_4OH, as suggested by Lewis. At the same time, in the region around 10 μm the spectrum of Jupiter is continuous, probably caused by the presence of NH_3.

Observation of Cloud Structure in the Visible

Large-scale Structure: Zones and Belts. For Jupiter and Saturn, the diameters of which are sufficiently great for them to be studied from Earth, even without modern instrumentation, we have a vast collection of observations, going back several centuries. Jupiter, in particular, which has a very rich morphology, has been the subject of systematic study since the time of Cassini.

A remarkable characteristic of Jupiter is the stability of its structure of *belts* and *zones*, as well as that of the *Great Red Spot* (see Table 7.2, Fig. 7.6 and Fig. 7.7). The stability of the belts and zones has its origin in the high rotational velocity of the planet, and in the asymmetric character of the global circulation. Measurements from spaceprobes have shown that the bright zones are colder than the dark belts in their upper layers and are also higher. The interpretation that is generally accepted is that the zones are anticyclonic regions, the site of ascending currents; at the top

Table 7.2. Nomenclature of bands and zones on Jupiter

Designation	Definition	Approximate latitude (degrees)
NPR	North polar region	47–90
NNTB	North-north temperature band	43
NTZ	North temperature zone	35
NTBn	North temperature band (northern component)	30
NTBs	North temperature band (southern component)	23
NTRZ	North tropical zone	15–20
NEBn	North equatorial band (northern component)	14
NEBs	North equatorial band (southern component)	10
EZn	Equatorial zone (northern component)	+ 3
EB	Equatorial band	0
EZs	Equatorial zone (southern component)	− 3
SEBn	South equatorial band (northern component)	−10
SEBs	South equatorial band (southern component)	−19
GRS	Great Red Spot	−22
STRZ	South tropical zone	−25
STB	South temperate band	−29
STZ	South temperature zone	−37
WOS	White oval south	−35, −37
SSTB	South-south temperature band	−41
SPR	South polar region	45–90

of these rising columns ammonia clouds form; the gases, freed from their condensable vapours, redescend along the belts, which are therefore cyclonic regions, driven by descending currents and free of clouds (see Fig. 7.8). This interpretation is confirmed by measurements of these regions made at 5 μm, which probe the deeper layers: in the belts, the radiation comes from deeper layers than in the zones, where there are no clouds, the brightness-temperature at 5 μm being higher there. Super-

Fig. 7.6. The planet Jupiter observed by the Pioneer 11 probe in 1974. (By courtesy of NASA)

Fig. 7.7. The planet Jupiter photographed by the Voyager 1 probe in 1979. (By courtesy of NASA)

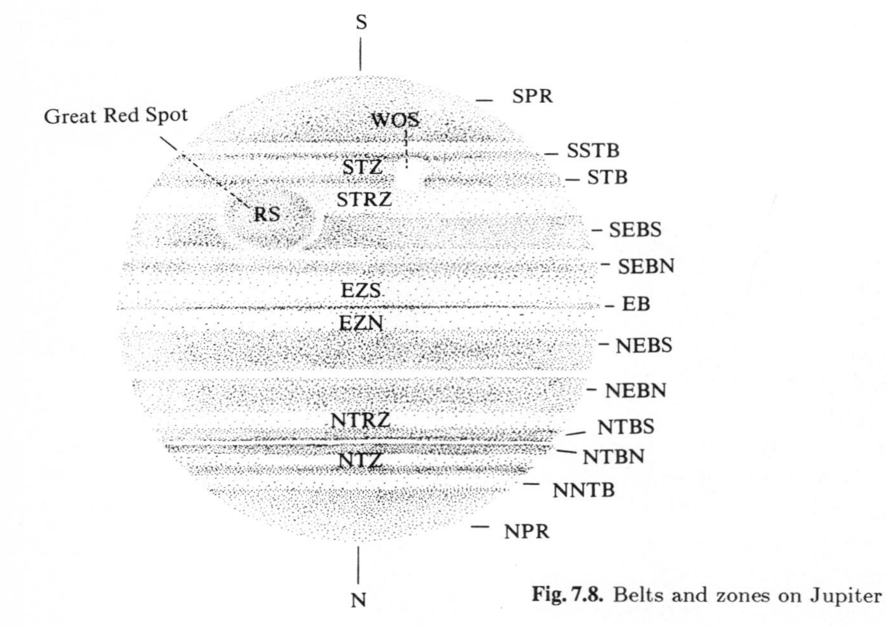

Fig. 7.8. Belts and zones on Jupiter

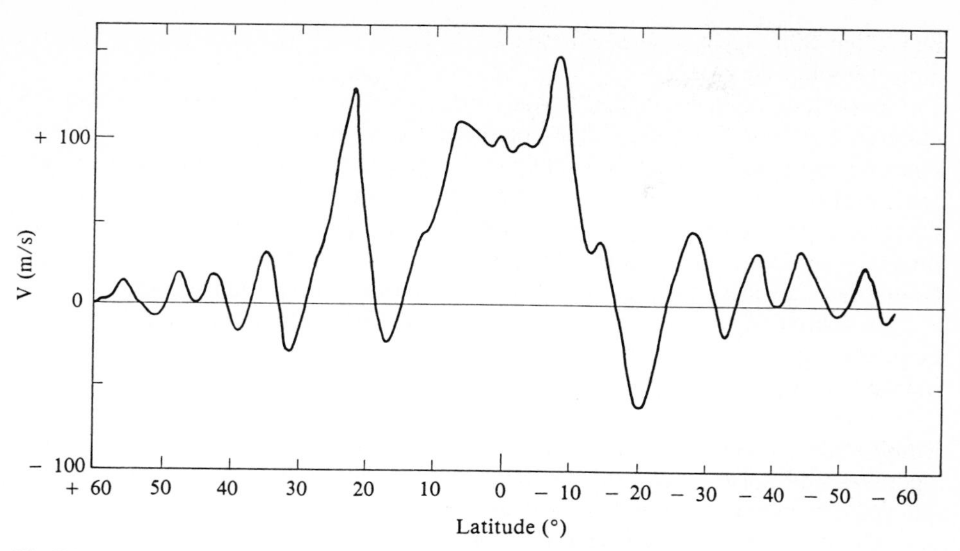

Fig. 7.9. Wind velocities on Jupiter as a function of latitude

imposed on this vertical circulation is a horizontal circulation, which, because of the strength of the Coriolis term, is of a very specific nature. The conversion of the north-south flow into east-west flow, under the action of the Coriolis forces, gives rises to zonal winds in opposite directions (E-W and W-E) in the northern and southern portions of each zone. In addition, the width of the zones is inversely proportional to the Coriolis term, and decreases with latitude.

Although it can be described in these mainly qualitative terms, the circulation on Jupiter and on Saturn is far from being understood. The various theories advanced do not allow us to explain in a satisfactory manner all the observed phenomena, at both large and small scales, and in particular the zonal winds observed on Jupiter and Saturn (see Fig. 7.9). Two classes of models have been considered by various authors. In the first, it is assumed that the circulation on Jupiter and Saturn, like that on the inner planets, is mainly governed by variation in the incident solar flux as a function of latitude. These models, however, appear incapable of explaining the equatorial jet, which is about four times as intense on Saturn as on Jupiter, whilst the incoming energy is four times weaker. This is why other models have been developed, in which it is assumed that the circulation is convective in nature and caused by an internal heat source. The existence of an internal source has been established for Jupiter and Saturn: this result is discussed in detail later (Sect. 7.2). Unfortunately, the amount of internal energy that would be required to explain the equatorial jet appears to be significantly greater than the measured flux. Both these models are probably too simple: understanding of the circulation of planetary atmospheres requires a model that includes both convective processes and baroclinic mechanisms. In the case of Jupiter, we may hope that research in this field will develop significantly with the availability of data from the Galileo mission.

For Uranus, the images obtained by the Voyager 2 probe in 1986 January indicate that the banded structure is poorly visible. Only the south pole, which is facing the Sun, appears somewhat darker. Interpretation of these images requires more elaborate treatment.

In the case of Neptune, the Voyager images have shown evidence for a complex circulation system. The deep blue colour of the planet is attributed to clouds of methane and possibly hydrogen sulphide. In the southern hemisphere, a large dark spot, probably anticyclonic like the Great Red Spot on Jupiter, stands above this cloud deck and is driven by upward motions. This spot is topped by white cirrus, probably composed of methane. Smaller spots are also observed in the southern hemisphere; all these features have significantly different rotation speeds.

In summary, one is both struck and puzzled by the extreme diversity of the circulation systems observed on the four giant planets. Understanding them will require more sophisticated theoretical modelling and further long-term observations.

Small-scale Structure: Spots, Barges, Ovals and Plumes. Jupiter's Great Red Spot was well-known, but images obtained by the Pioneer and Voyager spaceprobes showed a enormous variety of other meteorological phenomena on Jupiter (Fig. 7.10), as well as the presence of several spots, not nearly so easily seen, on Saturn (Fig. 7.11).

Discovered in 1664, the Great Red Spot lies in the South Tropical Zone, and has a longitudinal extent of about one sixth of Jupiter's diameter (Fig. 7.12). It drifts slightly in longitude when compared with the general circulation at the same latitude. Its top lies at a great height in the atmosphere, about eight kilometres above the cloud deck. It is the centre of intense activity, as confirmed by the Voyager photographs. It has a very powerful, internal circulation, which causes it to interact with its surroundings, either incorporating or shedding material. What is the nature of the Great Red Spot? At first it was thought to be a solid island suspended in the atmosphere, then a Taylor column, that is, a stationary fluid column. In the light

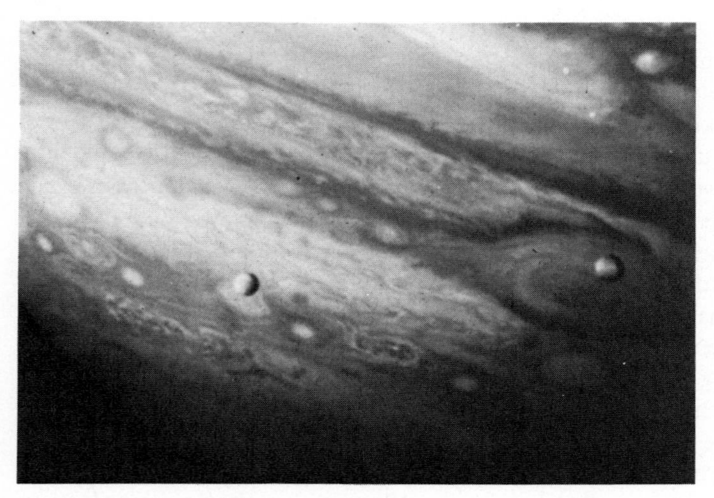

Fig. 7.10. The satellites Io and Europa in front of the Great Red Spot on Jupiter; Voyager 1 photo. (By courtesy of NASA)

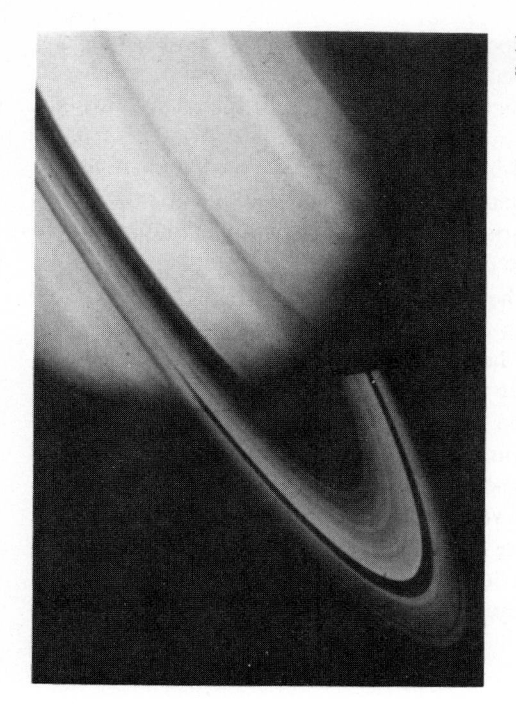

Fig. 7.11. Atmosphere and rings of Saturn as seen from Voyager 2. (By courtesy of NASA)

Fig. 7.12. Detail of Jupiter's atmospheric structure near the Great Red Spot; Voyager 1 photograph. (By courtesy of NASA)

of the Voyager data, however, it now appears that we are dealing with an immense vortex, born at the boundary between the oppositely-directed horizontal currents in the north and the south of the South Tropical Zone. But we are still unable to explain its stability.

Other spots, comparable to the GRS but smaller in size, have been identified on Jupiter, as well as other types of phenomena: the *plumes*, the *white ovals*, the *brown barges* and blue-grey regions. By comparison of measurements made in the infrared and visible regions it has been possible to establish a scale of altitudes for these objects: according to Owen and Terrile, the plumes are high-altitude ammonia cirrus, brought up by vertical currents and then spread out by horizontal ones; the white ovals also correspond to NH_3 clouds; the brown barges may be NH_4SH clouds, and so be at a level where the pressure is higher; finally the blue-grey regions, which are most frequently present in the North Equatorial Belt, can be correlated with the highest infrared emission at 5 μm: they therefore correspond to the deepest atmospheric layers that can be probed when clouds are absent; the blue-grey colour is thus caused by Rayleigh scattering of visible photons by atmospheric hydrogen.

On Saturn, the small-scale structure is far simpler than on Jupiter. The few spots that have been identified, with far less contrast than those on Jupiter, appear to be vortices, similar in nature to the Great Red Spot.

7.1.3 Molecular Abundances

As we have mentioned earlier, the atmospheric composition of the giant planets reflects, to a first approximation, the composition expected from cosmic abundances. Hydrogen is therefore the main component, followed by helium (about 10 % of the total by volume – we shall see later that this quantity varies from one giant planet to another), then by the most abundant of the heavier elements (C, N, O...) which are present in their reduced forms (CH_4, NH_3, H_2O). Other components have been detected in still lesser quantities: PH_3, GeH_4, isotopic components ($^{13}CH_4$, $^{15}NH_3$, CH_3D...) and dissociation products (C_2H_2, C_2H_4, C_2H_6, HCN...). In order to obtain the mixing ratios of these components it is essential to define the atmospheric level at which the measurement is taken. In most cases the mixing ratios are not constant with height, for different reasons: some elements condense at a specific atmospheric level (e.g. NH_3); others are dissociated at different heights (NH_3, PH_3, CH_4....). In both cases the distribution density decreases very rapidly above a specific atmospheric level. Other components that are dissociation products, however, have maximum abundances in the upper atmosphere: this is the case with hydrocarbons, which are photodissociation products of CH_4. Finally some components may have their distribution affected by the atmospheric circulation (this appears to the be case for PH_3, and perhaps also for NH_3 and H_2O).

Some atmospheric components have mixing ratios that are constant with altitude (the ratios being measured relative to hydrogen, the main component): this applies to He, CH_4 (photodissociated at very high altitudes), $^{13}CH_4$, and CH_3D. The H/He, D/H, $^{12}C/^{13}C$ (and equally $^{14}N/^{15}N$) ratios may be determined independently of altitude. In the other cases values are referred, for preference, to the deepest layers (corresponding to a pressure of several bars), in order to avoid, as far as possible,

Table 7.3. Molecular abundances in the giant planets

Molecule	Primordial abundance	Jupiter	Saturn	Uranus
(1) H_2	0.83	0.89	0.94	0.85
He	0.17	0.11	0.06	0.15
HD	5×10^{-5}	$5\text{–}10 \times 10^{-5}$	10^{-4}	$6\text{–}10 \times 10^{-5}$
CH_4	10^{-3}	$1\text{–}3 \times 10^{-3}$	$1\text{–}4 \times 10^{-3}$	$3\text{–}10 \times 10^{-3}$
NH_3	2×10^{-4}	$2\text{–}5 \times 10^{-4}$	$2\text{–}5 \times 10^{-4}$	–
H_2O	1.3×10^{-3}	$10^{-6} - 10^{-5}$	–	–
PH_3	7×10^{-7}	8×10^{-7}	$\sim 2 \times 10^{-6}$	–
CH_3D	$\sim 2 \times 10^{-7}$	$\sim 2 \times 10^{-7}$	3×10^{-7}	$\sim 2 \times 10^{-6}$
GeH_4	1.6×10^{-9}	$7\text{–}10 \times 10^{-10}$	4×10^{-10}	–
CO	–	2×10^{-9}	$\sim 2 \times 10^{-9}$	–
$^{13}CH_4$	1.1×10^{-5}	$\sim 1.1 \times 10^{-5}$	$\sim 1.1 \times 10^{-5}$	–
$^{15}NH_3$	7×10^{-7}	$\sim 7 \times 10^{-7}$	–	–
(2) C_2H_2	–	$\sim 7 \times 10^{-8}$	$\sim 2 \times 10^{-8}$	$\sim 5 \times 10^{-8}$
C_2H_4	–	–	–	–
C_2H_6	–	$\sim 3 \times 10^{-5}$	$\sim 5 \times 10^{-6}$	–
HCN	–	10^{-9}	–	–

(1) The mixing ratio measured in the deep layers of the atmosphere
(2) The mixing ratio measured in the upper atmosphere and corresponds to the peak distribution

the effects of condensation and photodissociation. This excludes the dissociation products, whose measured mixing ratios naturally refer to the upper atmosphere. Finally, evidence is beginning to be found, notably in the case of Jupiter, of spatial variation: even in the deep layers, the mixing ratios for certain components (NH_3, H_2O...), appear to be different between the zones and belts, this variation being caused by the planet's general circulation.

By way of example, Table 7.3 gives the relative abundances of atmospheric components detected on the giant planets. In some cases these abundances have been estimated by various methods and at various wavelengths, so there is a certain amount of scatter in the results. The values given in Table 7.3 should therefore be considered as indicating orders of magnitude. Each result is discussed in greater detail later (Sect. 7.1.4).

Figures 7.15 and 7.20 give examples of vertical distributions that are not homogeneous, the former showing components affected by condensation, and the latter others affected by dissociation.

7.1.4 Elemental and Isotopic Abundance Ratios

History

It was in the 1970s that is was realised that measurements of the abundance ratios in the giant planets had considerable significance for both cosmogony and cosmology. We should remember that in 1970 only three components had been identified in the atmosphere of Jupiter, CH_4, NH_3, and H_2; the presence of helium was suspected,

Table 7.4. Detection of molecules in the atmosphere of Jupiter

Molecule	Spectral region in which detected (μm)	Year of first discovery
CH_4	0.8	1932
NH_3	0.8	1932
H_2	0.8	1960
$^{13}CH_4$	1.1	1972
CH_3D	4.5	1972
HD	0.7	1973
C_2H_2	13.7	1974
C_2H_6	12.1	1974
H_2O	5.3	1975
CO	4.7	1975
PH_3	10.0	1976
$^{15}NH_3$	11.0	1978
GeH_4	4.7	1978
HCN	13.8	1981
$^{12}C^{13}CH_2$	13.2	1985
C_2H_4	10.5	1985
C_3H_4	15.8	1985
C_6H_6	14.8	1985
$^{12}C^{13}CH_6$	14.0	1987

Detection of molecules in the atmosphere of Saturn

Molecule	Spectral region in which detected (μm)	Year of first discovery
CH_4	0.8	1932
H_2	0.8	1960
NH_3	1.25×10^4	1970
$^{13}CH_4$	1.1	1975
C_2H_6	12.1	1975
PH_3	10.5	1975
CH_3D	4.5	1977
HD	0.6	1978
HCN	14.0	1981
C_3H_4	30.8	1981
C_3H_8	13.4	1981
CO	4.7	1986
GeH_4	4.7	1987

Detection of molecules in the atmospheres of Uranus and Neptune

Planet	Molecule	Spectral region in which first detected (μm)	Year of first discovery
Uranus	CH_4	0.8	1952
Uranus	H_2	0.8	1966
Uranus	HD	0.6	1978
Uranus	CH_3D	1.6	1981
Uranus	C_2H_2	0.17	1984
Neptune	CH_4	0.8	1952
Neptune	H_2	0.8	1974
Neptune	C_2H_6	12.1	1977

but not confirmed. Comparison with the list of molecules detected in Jupiter's atmosphere up to 1987 (Table 7.4) shows the significant and rapid progress that has taken place since then. This success is largely owing to the development of infrared astronomy: with the exception of the HD molecule, all the atmospheric components discovered since 1970 have been observed in the near or far infrared.

As we have mentioned earlier (Sect. 2.2) the abundances of atmospheric components, from which abundance ratios are deduced, are measured by spectroscopy, using either the reflected component of the solar flux ($\lambda < 3\,\mu$m), or the thermal flux ($\lambda > 3\,\mu$m). In the first case the principal problem arises from scattering by atmospheric particles, which cannot be neglected (see Sect. 7.1.2), but which it is impossible to model properly: as even the nature of the scattering medium is often unknown, it is impossible to determine its distribution, and still less its scattering factors. Scattering models that incorporate a large number of poorly-known factors often fail to give unequivocal determination of abundance ratios. They have, nevertheless, been used by numerous authors, particularly for the determination of the C/H ratio on Jupiter and Saturn. Another method applied to the reflected flux consists of defining criteria for comparing the lines that seek to minimize the effects of scattering (Owen, Combes and Encrenaz); this method has been employed to determine several ratios (C/H, D/H, and $^{12}C/^{13}C$) on Jupiter and Saturn.

The thermal emission spectrum of the giant planets has also been used to determine abundance ratios. In this case modelling is simpler, as scattering by the clouds may, to a first approximation, be neglected; the flux is calculated by the radiative-transfer equation (see Sect. 2.2.1). Here, good knowledge of the $T(P)$ is indispensable; the only exception is the H_2/He ratio, which may be obtained at the same time as $T(P)$ by a double iteration method, as we have already seen (Sect. 7.1.1). The spectral regions observable from the Earth are the window at 5 μm, already described (Sect. 7.1.2) and the window at 10 μm; ground-based observations have been supplemented by those by the IRIS experiment on the Voyager probes. The D/H, Ge/H, O/H, P/H, and $^{14}N/^{15}N$ ratios have been measured in this way. In addition, the Voyager IRIS data allowed spectral regions unobservable from the ground to be studied: the 6–7 μm region (measurements of C/H and $^{12}C/^{13}C$), and the 15–40 μm region (measurement of H_2/He).

Measurement of the H_2/He Ratio in the Giant Planets

We have described earlier (Sect. 7.1.1) the method of inverting the brightness integral that allows the $T(P)$ profile and the H_2/He ratio to be determined simultaneously. The most suitable spectral range lies between 15 μm and 100 μm, depending on the planet; for Jupiter it is limited to the 15–40 μm region, because beyond that wavelength the continuum caused by the translational component of the spectrum produced by H_2–H_2 and H_2–He collisions is masked by the strong absorption band of ammonia NH_3; with the other planets NH_3 condenses more or less completely, because of their low temperatures, so absorption in this band becomes negligible. The 15–40 μm band used for Jupiter corresponds to the maximum intensity of the rotational component of the pressure-induced spectrum; the strongest lines are at 17 μm and 28 μm. First estimates of the H_2/He were made using data obtained from an aircraft, and then from data returned by Pioneer 11. The IRIS experiment on the Voyager probes provided a more accurate measurement of the H_2/He ratio for Jupiter and also one for Saturn. The results are as follows, the parameter q being the abundance relative to hydrogen ($q = (H_2)/(H_2 + He)$):

1. $q = 0.89 \pm 0.03$ for Jupiter (Gautier et al.) ; (7.5)

2. $q = 0.93 \pm 0.03$ for Saturn (Conrath et al.) ; (7.6)

3. $q = 0.85 \pm 0.03$ for Uranus (Conrath et al.) . (7.7)

For Neptune, present measurements are not sufficiently precise to give a measure of the H_2/He ratio.

Measurement of the D/H Ratio in the Giant Planets

Deuterium was detected on the giant planets in 1973 in the form of two molecules CH_3D and HD; CH_3D in the infrared and HD in the visible. In both cases, determination of the D/H ratio is difficult; at visible wavelengths the HD lines are weak, and have to be compared with the quadripolar lines of H_2, which are much stronger and created under very different scattering conditions; in the infrared, the CH_3D/H_2 ratio, obtained from the data, has to be converted to the D/H ratio, requiring knowledge of CH_4/H_2 (that is, the C/H ratio), as well as of the fractionation coefficient f, which gives the relation between CH_3D/CH_4 and D/H:

$$\mathrm{D/H} = \frac{1}{4f}\mathrm{CH_3D/CH_4} = \frac{1}{8f}\frac{\mathrm{CH_3D/H_2}}{\mathrm{C/H}} \quad . \tag{7.8}$$

In the literature there are a large number of determinations of the D/H ratio for Jupiter, and a wide scatter in the results. With HD, the D/H ratio has been deduced by the direct use of RLM (see 7.1.2), by scattering models, or by comparison with weak methane lines to obtain a measurement of the D/C ratio (D/H then being derived from the C/H ratio). With CH_3D, the D/H ratio has been determined from data at 5 μm and at 10 μm, obtained either from the ground or from the Voyager IRIS experiment. Overall, recent results appear to be converging on a value of:

$$(\mathrm{D/H})_{\mathrm{Jupiter}} = 2\text{-}4 \times 10^{-5} \quad . \tag{7.9}$$

For the other giant planets, the uncertainty is even greater. The values measured for Saturn are comparable with those for Jupiter. With Uranus, the D/H ratio determined from CH_3D appears to be higher, which might correspond to an enrichment in deuterium in the CH_4 ice in the core. D/H ratio measurements in the giant planets are summarized in Table 7.5.

Astrophysical Implications of Measured Abundances of Helium and Deuterium

It is interesting to examine how the values of the H/He ratio for Jupiter, Saturn and Uranus compare with other determinations of H/He obtained for the Sun and other sources, and which might represent the primordial helium abundance. According to the most recent results, it appears that the helium abundance on Uranus is, to a first approximation, comparable with the solar and "primitive" helium abundances, but

Table 7.5. Determination of D/H ratio in the giant planets

Method used	Molecule used	Spectral region	Jupiter	Saturn	Uranus
Reflecting-layer method	HD	0.7 μm	5×10^{-5}	5×10^{-5}	$3\text{–}5\times10^{-5}$
Comparison method	HD	Visible and near IR	2×10^{-5}		
Diffusion model	CH_3D	1.6 μm			$\sim10^{-4}$
Thermal emission (observed from the ground)	CH_3D	4.5 μm	5×10^{-5}		
Thermal emission (Voyager-IRIS)	CH_3D	4.5 μm 8.6 μm	$2\text{–}4\times10^{-5}$	3×10^{-5}	

that the helium abundance on Saturn is appreciably lower. The value obtained for Jupiter lies between the other two.

A plausible explanation for this lack of helium on Saturn is offered by the following process: at the high pressures reigning within Saturn, and at a sufficiently low temperature, helium would no longer mixed uniformly with metallic hydrogen; differentiation would therefore take place, and the helium, as it condensed, would be concentrated into the core. This process would result in an impoverishment in helium in the overlying atmosphere, which is what is observed on Saturn. It also results in the liberation of energy. This might be responsible for the fact that Saturn radiates twice as much energy as it receives from the Sun (see 7.2.1). The measurements of internal energy flux and H_2/He abundances for Jupiter, Saturn and Uranus are in agreement with evolutionary models for the giant planets that imply that helium differentiation exists within Saturn, but that this does not occur (or is far less significant) in Jupiter, and does not take place in Uranus (see below, Sect. 7.1.5). On this basis, the helium abundance in the atmosphere of Uranus may represent the helium abundance in the Primitive Solar Nebula and therefore set an upper limit for primordial helium. It should be recalled that in Big-Bang theories, the helium now observed mainly originated in the Big Bang itself, with a smaller contribution arising from the burning of hydrogen to helium within stars that has been ejected into the interstellar medium by novae and supernovae.

The measurement of D/H obtained for Jupiter also has cosmological implications. Like the H/He ratio on Uranus, it seems to represent the D/H ratio in the Primitive Solar Nebula, because current theories of the formation and evolution of the giant planets do not envisage any process that could modify the deuterium abundance. This result seems to be confirmed by the good agreement that exists between Jupiter's D/H ratio and that determined by Geiss and Reeves for the Primitive Solar Nebula from measurements of the $^3He/^4He$ ratio in the solar wind.

When Jupiter's D/H value, which is therefore assumed to represent that of the Primitive Solar Nebula some 4.6 thousand million years ago, is compared with the D/H value measured in the interstellar medium now, we find (see Fig. 7.13) that the abundance of deuterium has decreased by a factor of between 5 and 8. This result is not surprising for Big-Bang theory, because it implies that deuterium is destroyed in stars and can therefore only decrease in abundance with time. It is

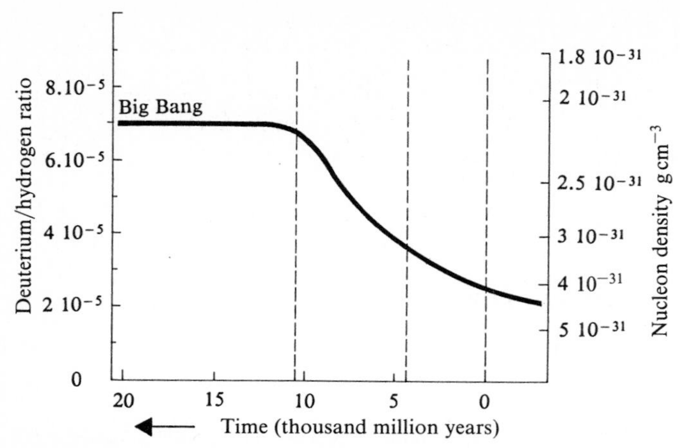

Fig. 7.13. The evolution of deuterium abundance in the universe with time, according to one model of the evolution of galaxies. [After D. Gautier: "Jupiter", *Le grand Atlas Universalis de l'Astronomie* (Encyclopedia Universalis 1983). By courtesy of the publisher]

now possible to use models of galactic evolution (Fig. 7.13) to obtain an estimate of the amount of primordial deuterium at the time of the Big Bang. According to the model by Audouze and Tinsley, the primordial deuterium abundance was of the order of twice that found in the Primitive Solar Nebula, and therefore about ten to sixteen times the current deuterium abundance. With this value for the primordial deuterium abundance, and using Wagoner's cosmological models, it is possible to derive the present density of baryons in the universe. The value thus obtained, 2×10^{-31} g cm^{-2} proves to be in agreement with a completely independent determination obtained from measurements of the lithium abundance in the oldest stars in the Galaxy.

These results and comments illustrate the interest that these measurements of helium and deuterium in the giant planets have, and the importance of their implications for both cosmogony and cosmology.

The C/H Ratio

Like deuterium, the determination of the carbon abundance in the giant planets has been the subject of numerous controversies and, apart from the case of Jupiter, the situation is still uncertain.

In the reducing atmosphere of the giant planets, most carbon is in the form of methane. CH_4 is present in the visible and infrared spectra, where it is the predominant feature. The spectra of Jupiter and Saturn show very similar spectra as far as methane is concerned; differences mainly arise from NH_3 absorptions, which are present on Jupiter and absent from the other giant planets. The spectra of Uranus and Neptune, on the other hand, show absorptions caused by CH_4 that are very much stronger (see Fig. 7.14).

As for deuterium, the abundance of CH_4 has been estimated from the reflected component of planetary spectra by using the reflecting layer model (RLM), scattering models, and the comparison model that minimizes scattering effects (see 7.1.3).

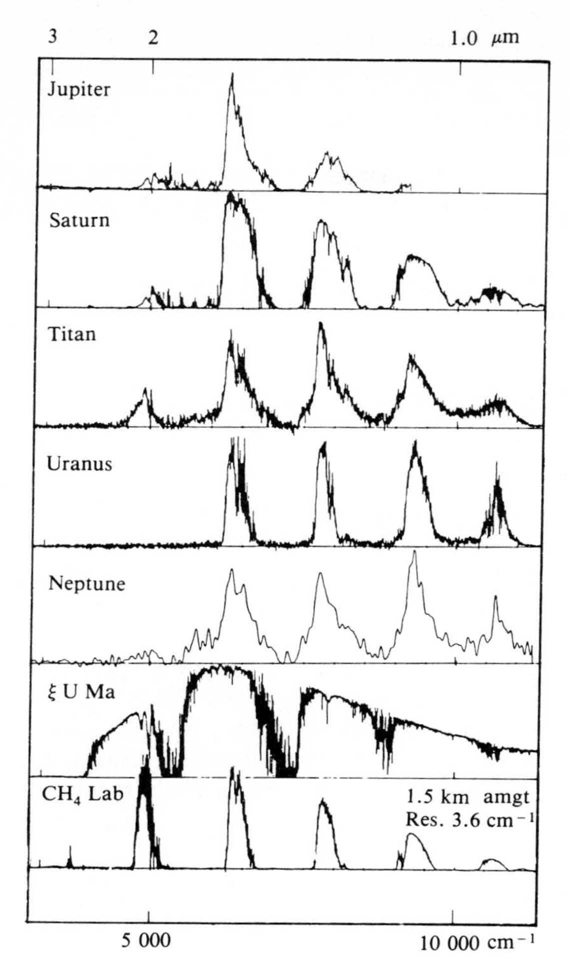

Fig. 7.14. Spectra in the near infrared. It will be seen that the spectra are marked by methane absorptions. [After H. Larson, by courtesy of *Annual Review of Astronomy and Astrophysics*, Vol. **18** (© Annual Reviews Inc. 1980)]

More recently, thanks to the data from the Voyager IRIS experiment, the ν_4 band of methane at 7.7 μm has been used to obtain an estimate of C/H on Jupiter and Saturn. For Jupiter, there is now a reasonable agreement between the values determined by the various methods; the results converge on the value:

$$(\mathrm{C/H})_{\mathrm{Jupiter}} \sim 10^{-3} \tag{7.10}$$

which corresponds to a carbon enrichment by a factor of two over the solar value. But for Saturn, on the other hand, there is a certain scatter in the results. At present it is not possible to obtain accurate estimates for Uranus and Neptune. The thermal component is too weak to be used, and as far as the reflected component is concerned, scattering effects are so significant that the methods used for Jupiter and Saturn are not applicable. The only thing that appears certain is that the abundance of carbon on Uranus and Neptune is considerably higher than that of Jupiter and Saturn, perhaps by a factor of as much as 10. A summary of the results for C/H for the giant planets is given in Table 7.6.

Table 7.6. Determinations of the C/H ratio in the giant planets (Value of the C/H ratio for the Sun: 5×10^{-4})

Method used	Molecule used	Spectral region	Jupiter	Saturn	Uranus	Neptune
Reflecting-layer model	CH_4	1.1 µm	5×10^{-4}			
Reflecting-layer model	CH_4	Visible	10^{-3}	10^{-3}	10^{-2}	10^{-2}
Comparison method	CH_4	Visible + near IR	10^{-3}	10^{-3}	$>3 \times 10^{-3}$	
Diffusion models	CH_4	Visible + near IR	$1\text{–}3 \times 10^{-3}$	$1\text{–}4 \times 10^{-3}$	$3\text{–}10 \times 10^{-3}$	
Thermal emission (Voyager)	CH_4	8 µm	10^{-3}	2×10^{-3}		

The N/H Ratio

The three cases that we have so far considered concern elements that are uniformly mixed with hydrogen in the atmosphere, where the mixing fraction remains constant throughout the atmosphere that is being probed. With N, P, O and Ge, we come to elements that have a vertical distribution that varies with altitude, and the study of which thus becomes far more complicated.

Ammonia, NH_3, is present throughout Jupiter's spectrum, from the ultraviolet to radio waves. Ultraviolet measurements, carried out around 2000–2200 Å by the IUE satellite, and infrared measurements that correspond to the strongest bands, probe the top of the atmosphere, above the level of the temperature inversion, and tell us about the condensation and photodissociation of NH_3. In the far infrared and in the radio region (in the wings of the inversion band centred on $\lambda = 1.25$ cm), it is the deep layers, below the NH_3 clouds, that are being examined. The NH_3 inversion band is also observed on Saturn. As for the NH_3 bands in the visible and near infrared, they appear to be formed at different atmospheric layers, as a function of the strength of the transitions.

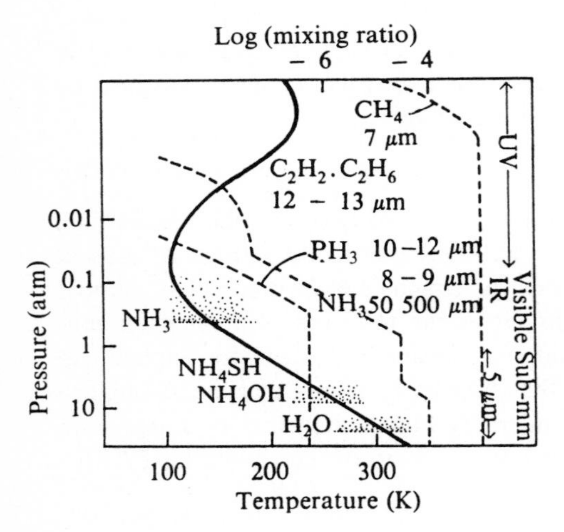

Fig. 7.15. NH_3 and PH_3 distribution profiles for Jupiter. The continuous line indicates the temperature profile, and the broken lines the distributions of the different atmospheric components. A schematic indication of the various cloud layers is also shown

The most recent results appear to indicate that, on both Jupiter and Saturn, the N/H ratio below the NH_3 clouds is less than the solar value by a factor of about two. The NH_3 profile above the ammonia clouds on Jupiter is shown in Fig. 7.15. With Uranus and Neptune, radio measurements also seem to indicate that N is underabundant when compared with the solar value. There is a simple explanation for this behaviour: nitrogen is probably trapped, with other elements, in the deepest cloud layers. From calculations of thermochemical equilibrium on Jupiter, Lewis predicted the formation of NH_4SH and NH_4OH, which ought to condense into clouds; the same mechanism can be envisaged for the other giant planets.

On Jupiter, at very high pressure levels (above 5 bars), the N/H ratio appears, however, to be superabundant by a factor of two, and variations are observed between the belts and zones. These results were obtained by radio measurements at 1.3, 2 and 6 cm, and also appear to be confirmed by IRIS data at 5 μm.

The P/H Ratio

In the atmospheres of Jupiter and Saturn, phosphine PH_3 is a particularly interesting molecule because it undoubtedly acts as a tracer for significant dynamic activity. In fact, according to thermochemical calculations, phosphine ought not to be observable on the giant planets: at temperatures below 2000 K, it ought to react with H_2O to form P_2O_5. Yet PH_3 has been detected on both Jupiter and Saturn, at 2, 3, 5, and 10 μm. As with NH_3, there is a vertical distribution where the abundance decreases with height. This gives rise to the following interpretation: PH_3 could be transported by rapidly ascending vertical currents, and would be progressively destroyed as it ascended. Above the temperature minimum, the remaining PH_3 is destroyed very rapidly by photodissociation (see Fig. 7.15). At a pressure of several bars on Jupiter, the abundance of phosphorous is close to the solar value; on Saturn it is enriched by a factor of about three. In addition, it appears that the PH_3 abundance on Jupiter varies, both from place to place and with time. There can be little doubt that systematic study of PH_3 would give a lot of information about the atmospheric circulations on Jupiter and Saturn; the Galileo mission should enable this research to be carried out for Jupiter in 1995–96.

The O/H and Ge/H Ratios

Jupiter is the only planet for which these measurements have been possible, using the spectral window at 5 μm. H_2O and GeH_4 have been identified in spectra obtained from the ground, and also from the Voyager IRIS data (see Fig. 7.16). The abundances measured are decidedly less than cosmic abundances. For GeH_4, this result is probably linked, as with NH_3, with the presence of a low-lying cloud layer. The condensation of H_2O is predicted by thermochemical models, and in the presence of this H_2O cloud, GeH_4 may be expected to be destroyed in favour of pure germanium Ge, or the compound H_2GeO_3.

For H_2O, an underabundance by a factor of the order of thirty has been measured for certain regions of Jupiter's disk; in addition, spatial variations have also been detected. The underabundance of H_2O, if it were to be confirmed as applying

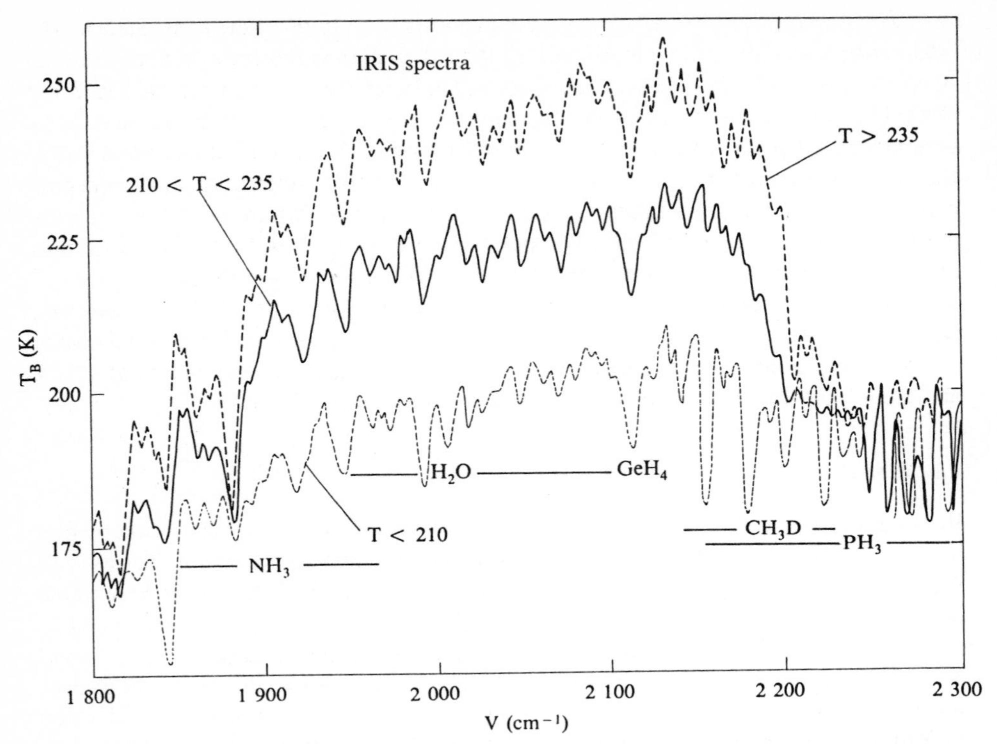

Fig. 7.16. Spectra of Jupiter at $5\,\mu$m recorded by the Voyager IRIS experiment. [After P. Drossart et al., Icarus **49**, 416 (1982)]

to the whole planet, would pose a problem difficult to solve by thermochemical modelling.

Because of their variation with location, and possibly also with time, the GeH_4 and H_2O molecules are probably, like NH_3 and PH_3, useful tracers of the circulation on Jupiter.

Astrophysical Implications of Measurements of Elemental Abundances

Models of the formation of the giant planets that have been developed over the last few years may be grouped into two principal categories.

In the first class of models, it is believed that the giant planets were created by homogeneous accretion of gaseous material from the Primitive Solar Nebula. When the volume of accreted gas became sufficient (several thousands of times the current volume of the planets), a hydrodynamic collapse took place, followed by a slow contraction, which took the planets to the state in which we see them now. In this view, the overall composition of the giant planets remains the same as that of their initial material, the Primitive Solar Nebula.

In the second theory, known as the "nucleation theory", it is assumed that the formation of grains took place, starting with ices (H_2O, NH_3, CH_4, etc.), but also including silicates and metals, and that then these grains accreted into a core. This core would also have contained significant quantities of *clathrates*, that is hydrated compounds that can form at low temperatures incorporating certain molecules and atoms such as CH_4, N_2, CO, A, etc. When the core became sufficiently massive, its gravity was able to retain gases that did not condense, beginning with H_2 and He. During this phase of the accretion, the temperature of the core could have risen sufficiently to cause gases to sublime, or to liberate the gases trapped in the clathrates, which would have had the effect of enriching the atmosphere in elements such as C, N, O, etc., in comparison with hydrogen and helium. Obviously, in this theory the elemental abundance ratios obtained for the atmospheres of the giant planets should be enriched in C, N, O, etc. in comparison with the abundances in the primordial gaseous material.

From what we have said previously, carbon appears to be the only reliable test that can be applied at present, as measurements of the other elements are still too imprecise. On the four giant planets, carbon appears to be enriched by a factor of two or more, in comparison with the current solar value. This result appears to favour nucleation models rather than those of homogeneous gravitational collapse. We shall see later that this conclusion is confirmed by theoretical work concerning the internal structure of the giant planets (Sect. 7.2). Nevertheless, if the underabundance of oxygen on Jupiter is confirmed, this conclusion could be open to question.

$^{12}C/^{13}C$ *and* $^{14}N/^{15}N$ *Ratios and Their Astrophysical Implications*

The study of isotopic ratios, such as $^{12}C/^{13}C$, $^{14}N/^{15}N$, and $^{18}O/^{16}O$, is particularly interesting for research into nucleosynthesis and into the chemical evolution of the Galaxy, because these ratios ought to be less affected by chemical fractionation than elemental ratios. The element ^{13}C is mainly produced in the "cold" CNO cycle, which occurs in the inner regions of massive, main-sequence stars. This is why the $^{12}C/^{13}C$ ratio measured in the galactic centre, which is rich in massive stars, is lower than in the local interstellar medium, the latter value itself being slightly less than in the Solar System. In the Sun, the inner planets, meteorites, and the Moon, the value found is remarkably constant, and equal to 89. In contrast to ^{13}C, ^{15}N is formed in the "hot" CNO process, which occurs in nova and supernova explosions. In the Solar System, the $^{14}N/^{15}N$ ratio has been measured on the inner planets. but there is uncertainty about the solar $^{14}N/^{15}N$ ratio measured from solar wind ions trapped on the lunar surface. A determination of $^{12}C/^{13}C$ and $^{14}N/^{15}N$ in the most primitive regions of the Solar System would, by comparison with current values in the interstellar medium, allow us to determine the chemical evolution of our galactic neighbourhood in the past 4.5 thousand million years.

$^{12}C/^{13}C$ has been measured on Jupiter and Saturn in three different ways: from the $^{13}CH_4$ lines in the $3\nu_3$-band at 1 μm; from the Q branch of the ν_4 at 8 μm, and from the lines of $^{12}C/^{13}CH_2$ at 13 μm. The first determination, on Jupiter and Saturn, led to a solar value; the second, on Jupiter, produced a $^{12}C/^{13}C$ ratio significantly higher, by a factor of about two, than the solar value. Finally, the third method has given, for Jupiter, a $^{12}C/^{13}C$ ratio that is significantly less than the solar value. At present it is not possible to decide between these results, nor to explain their differences. On the other hand, it is not easy to explain how the value of $^{12}C/^{13}C$

for Jupiter could be significantly enriched with respect to the value found for the Solar System as a whole.

We have one estimate of $^{14}N/^{15}N$ for Jupiter, from the $^{15}NH_3$ lines in the ν_2 at $10\,\mu m$. Unfortunately, this estimate is strongly dependent on the atmospheric model, in particular on the cloud structure. Nevertheless, it would seem reasonable to assume that there is a certain enrichment in ^{15}N with respect to the terrestrial value, by a factor of about two. The solar abundance of ^{15}N may itself be enriched over the terrestrial value, but in a smaller proportion. Here again, there is no simple explanation of the variation in ^{15}N abundance between the Sun and the giant planets. However, we should mention that the $^{14}N/^{15}N$ for Jupiter is still very uncertain. As for $^{12}C/^{13}C$, we must undoubtedly await mass spectrometry by the Galileo probe to give an exact, and definitive answer. There have been no $^{14}N/^{15}N$ determinations for the other giant planets.

7.1.5 The Upper Atmospheres of the Giant Planets

The atmospheric regions probed by study of radiation in visible, infrared and radio wavelengths, lie, in most cases, at pressure levels ranging from 10 mb to about ten bars. Above these regions there are the *upper troposphere* and then the *thermosphere* and the *exosphere*, which reach altitudes where the pressure is of the order of a nanobar. A characteristic feature of these atmospheric regions is that they are subject to solar ultraviolet radiation, which penetrates to a greater or lesser extent, and induces intense photochemical reactions. Ultraviolet spectroscopy of the giant planets is therefore a powerful tool in analyzing their upper atmospheres.

Determination of Temperature and Density Profiles

We have mentioned earlier (see Sect. 7.1.1) how the stellar occultation method allows the thermal structure between 0.1 and 10 mb to be determined. The same method, in the ultraviolet, probes the temperature distribution at very high altitudes (about 1000 km above the temperature minimum, and therefore at a pressure of around one nanobar). This technique has been used on Jupiter and Saturn by the Voyager 1 and 2 probes, with the Sun or bright stars as targets. On both Jupiter and Saturn, ultraviolet radiation is absorbed by H_2 between 600 and 800 Å, by CH_4 between 900 and 1150 Å, and by C_2H_2 and C_2H_6 between 1425 and 1675 Å. The light-curves recorded at these various wavelengths enable the vertical distribution of the different components to be established (Fig. 7.17). The temperature of the thermosphere is high: it is of the order of 1000 K or more on Jupiter, and between 400 and 1000 K on Saturn.

The Eddy Diffusion Coefficient

If photochemical processes were absent, the atmospheres of the giant planets would be homogeneously mixed, up to the height of the homopause, at a pressure of about one nanobar. Photodissociation occurs at much lower altitudes, however, which leads to density profiles where the mixing fraction decreases as the total pressure. For Jupiter in particular, photodissociation of NH_3 takes place as low as the temperature minimum, at a pressure of about 100 mb.

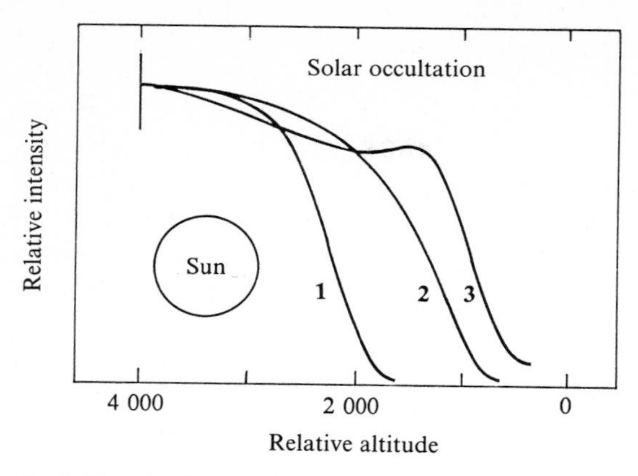

Fig. 7.17. Relative intensity of the UV flux as a function of altitude for three wavelength regions: *(1)* 600–800 Å, *(2)* 900–1150 Å, *(3)* 1425–1675 Å. In the first case the absorption is caused by H_2, in the second by CH_4, and in the third by C_2H_2 and C_2H_6. [After Broadfoot et al.: "Extreme Ultraviolet Observations from Voyager 1 Encounter with Jupiter", Science **204**, pp. 979–82, 1st June 1979 (© AAS 1979)]

In dealing with problems of aeronomy, it is essential to know the significance of vertical transport, as this determines the distribution of minor components with altitude. This value is expressed by the *eddy diffusion coefficient* K. At the homopause, the coefficient K is equal to the molecular-scattering coefficient; above that level species are separated according to their molecular weight. Several indirect methods may be used to determine the coefficient K: observation of the Ly-α flux (which gives the abundance of atomic hydrogen above the penetration level, defined by CH_4), measurement of helium at 584 Å (which gives the helium abundance in the overlying atmosphere), the vertical distribution of CH_4. A discussion of these methods has been given for Saturn by Atreya. It should be noted that a helium emission at 584 Å was observed on Saturn by Voyager; on Jupiter, on the other hand, helium emission observed by Pioneer 10 was not confirmed by Voyager; in this second case, strong variation with time of the jovian thermosphere cannot be excluded.

The most probable value of K for Jupiter is of the order of $10^6\,cm^2\,s^{-1}$. The value for Saturn appears to be at least ten times higher: between 10^7 and $10^8\,cm^2\,s^{-1}$. An indication of this difference can be seen in the ratio, of the order of 10, between the abundances of C_2H_2 measured on Jupiter and Saturn, both in the ultraviolet (Fig. 7.18) and in infrared emission at 13 μm. For Uranus, the value of K seems to be about $10^6\,cm^2\,s^{-1}$. In addition, variation of the coefficient with latitude, as occurs in the Earth's atmosphere, cannot be ruled out. As yet there is no determination of K for Neptune.

Photochemistry of the Giant Planets: Dissociation and Ionization

Photochemistry on the giant planets is mainly caused by the absorption of solar ultraviolet photons by molecules. The penetration depth, which is set for each wavelength by Rayleigh scattering, fixes, for any specific component, the altitude at which the

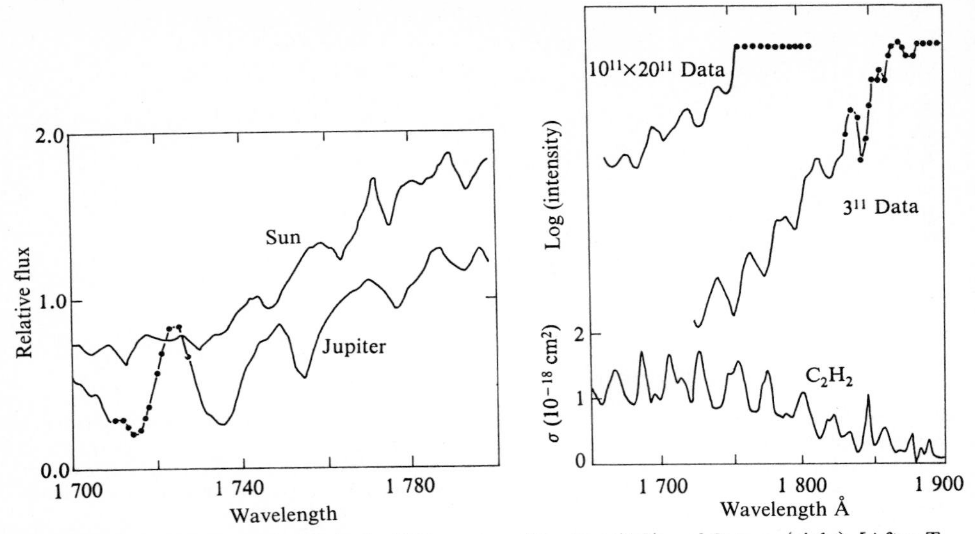

Fig. 7.18. Identification of C_2H_2 is the UV spectra of Jupiter (*left*) and Saturn (*right*). [After T. Owen et al.: Astrophysical Journal **236**, L39 (1980), and W. Moos, J. Clarke: Astrophysical Journal **229**, L107 (1979). By permission of the authors and of the *Astrophysical Journal*, published by University of Chicago Press (© The American Astronomical Society 1979,1980)]

component is dissociated. For example, the altitude at which NH_3 is dissociated by energetic photons of 1800–2000 Å is between 100 and 10 mb. For CH_4, dissociated at 1400–1600 Å, the destruction level is higher in the atmosphere; for H_2 it occurs at even greater heights (Fig. 7.19).

The photochemistry of CH_4 has been studied in detail by Strobel and Atreya. The principal dissociation products are C_2H_2 and C_2H_6 (already identified) and C_2H_4. Strobel has also considered the photochemistry of NH_3, which has an important role on Jupiter; in particular, coupling of the two processes leads to the formation of HCN, observed on Jupiter. For Saturn (and perhaps also for Uranus

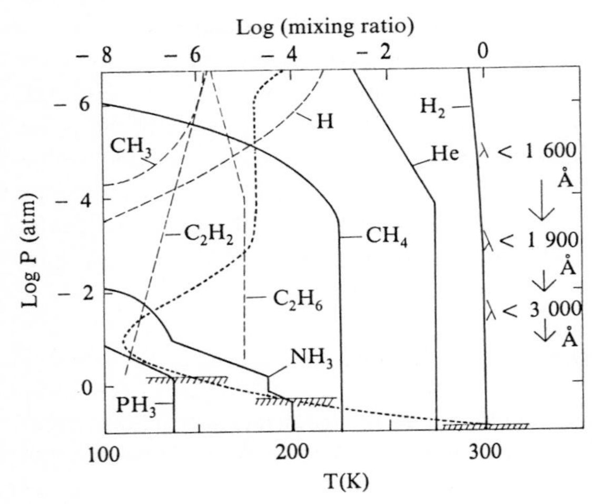

Fig. 7.19. Model of the upper atmosphere of Jupiter, showing the different levels to which solar UV radiation penetrates

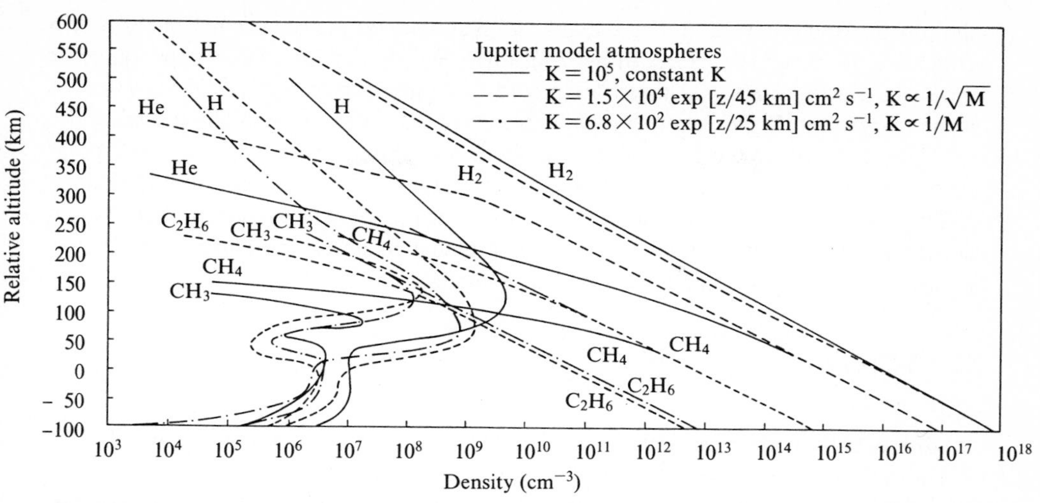

Fig. 7.20. Model of photodissociation on Jupiter. [After D. Atreya, T. Donahue: *Jupiter*, ed. by T. Gehrels (© University of Arizona Press, Tucson 1976)]

and Neptune), the photochemistry of PH_3 may play an significant role, leading to the formation of components that have yet to be detected: such as HCP and CH_3PH_2; Fig. 7.20 shows an example of a model of the dissociation of methane and hydrogen on Jupiter.

7.2 The Internal Structure of the Giant Planets

Observations of the giant planets generally involve only the exterior of these objects; their internal structure is not accessible to direct observation. In addition, because of the very considerable pressures and temperatures that are involved, it cannot be the subject of laboratory investigations. This is why study of the interior of the giant planets takes the form of theoretical modelling of the states of matter at high pressure. We do, however, have some observational data that may serve to provide limits for these models.

7.2.1 The Experimental Data

Mass, Radius, Rotation and Gravitational Field

We have precise information about the masses and radii of the giant planets (see Table 1.3). Determination of their rotation periods is more ambiguous. For Jupiter, the apparent rotation period of the external meteorological phenomena is slightly different from that of the inner layers, which has been determined from Jupiter's decametric radiation. It is the latter, linked to the magnetic field, which is taken as a reference, and which gives "System III" longitudes l_{III}. It has also been possible to measure the rotation of the internal layers of Saturn, by studying the periodicity in the planet's radio emission. For Uranus and Neptune, on the other hand, there is still considerable uncertainty (see Table 1.3); only changes in the visible appearance of the disks can be used, and these often lead to contradictory results. Observation of these two planets by Voyager should enable these values to be established accurately.

The rotation of the giant planets leads to a departure from sphericity and polar flattening, and from measurements of the gravitational field it is possible to deduce information about the distribution of mass within the planets. Assuming hydrostatic equilibrium, the gravitational potential of a planet in uniform rotation may be expressed with second-order Legendre polynomials:

$$\Phi(r,\theta) = -\frac{GM}{r}\left[1 - \sum_{1}^{\infty}\left(\frac{Re}{r}\right)^{2n} J_{2n}P_{3n}(\cos\theta)\right] \quad . \tag{7.11}$$

In this equation, θ is the colatitude (i.e. the complement of the latitude), Re the equatorial radius, M the planet's mass, G the gravitational constant and $P_{2n}(\cos\theta)$ the Legendre polynomial of order $2n$. The quantities J_{2n} are known as the gravitational moments. J_2 and J_4 have been measured for Jupiter, Saturn and Uranus; for Neptune only J_2 is known at present.

Internal Energy in the Giant Planets

The existence of an internal energy source within Jupiter was discovered in 1969, when the first infrared flux measurements indicated that the planet radiated twice as much energy as it received from the Sun (see (1.8) and (1.9)). Later measurements from Earth and from space, confirm this result and established the presence of internal energy sources inside Saturn and Neptune, but not within Uranus (Table 7.7). This difference in behaviour in two apparently similar planets, Uranus and Neptune, is accompanied by an appreciable difference between their atmospheres. The temperature on Neptune appears to be higher in the upper atmosphere above 100 mb. One possible cause for these differences could be that, at present, the south pole of Uranus is pointing continuously towards the Sun.

The existence of an internal energy source in Jupiter, Saturn and Neptune, which in all three cases is similar in scale to the solar energy absorbed, is certainly a basic factor that models of internal structure must take into account.

Table 7.7. Internal energy in the giant planets

Planet	Effective temperature (K)	Ratio of energy emitted to absorbed solar energy
Jupiter	124	1.67
Saturn	95	1.78
Uranus	58	1.1
Neptune	58	2.6

Thermal Structure and Chemical Composition of the Amospheres

Without wishing to repeat the discussion of the observations that has already been given (see Sect. 7.2), it is useful to recall the data that bear on the internal structure of the giant planets. There are two important points. First, for all the four planets the measurements seem to indicate that at pressures above one bar, the thermal profile is adiabatic, which suggests that the atmospheres are convective; the $T_0(P_0)$ value at a pressure of one bar provides a limit.

Second, measurements of the abundances of helium and carbon have, as we have seen, profound implications for models of internal structure: helium seems to be present on Jupiter with an abundance similar to the cosmic one, whilst it is noticeably less on Saturn. As for carbon, it seems to be overabundant on the four planets by comparison with the cosmic abundance.

7.2.2 Modelling the Internal-energy State

An idea of the pressure prevailing within any body of known mass and radius may be gained by calculating the quantity

$$p = \varrho q R \tag{7.12}$$

ϱ being the density, R the radius, and q the quantity

$$q = \frac{GM}{R^2} \quad . \tag{7.13}$$

We then find the following orders of magnitude: 2.6×10^7 bar for Jupiter, 4×10^6 bar for Saturn, and 3×10^6 bar for Uranus and Neptune. The immediate conclusion is that, at such pressures, hydrogen, the major constituent, has to be in metallic form. It is, however, extremely difficult to determine accurately the phase diagram of hydrogen at high pressures and temperatures, because laboratory experiments are not possible; this is why the approach is purely theoretical. It is the same for the equations of state for helium and for H_2O, CH_4 and NH_3 ices which should be present in amounts reflecting their cosmic abundances.

Two types of internal model can be considered: static models, in which the thermal profile is chosen a priori – which therefore fixes the thermal flux emitted by the planet – and evolutionary models, which try to explain the origin of this thermal flux.

Static Models

These are constructed from the following equations:

1. the equation for hydrostatic equilibrium in the presence of rotation:

$$\frac{dp}{dr} = -GM(r)\frac{\varrho(r)}{r^2} + \frac{2}{3}\Omega^2 r \varrho(r) + O(\Omega^4) \tag{7.14}$$

   where Ω is the rotational velocity and $M(r)$ is the mass within radius r;
2. the equations of state $p(\varrho, T, (\chi_i))$, $\chi_i(r)$ being the relative abundance of component i;

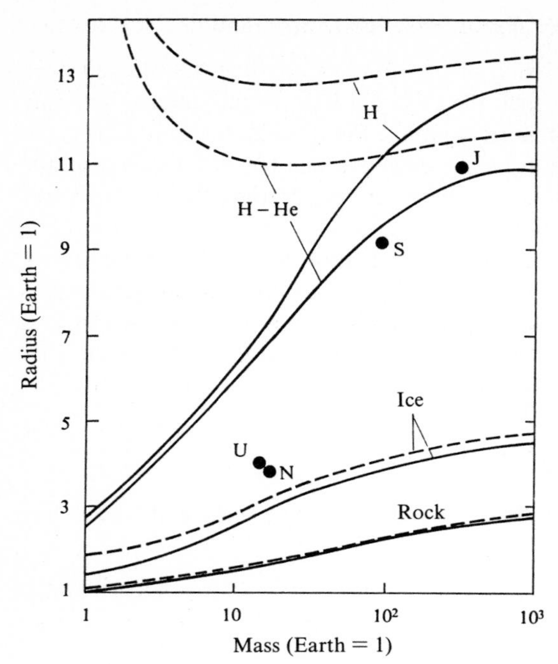

Fig. 7.21. The mass-radius ratio, for a self-gravitating body. Different compositions are considered: (1) pure hydrogen, (2) H–He (25 % of He), (3) ices and (4) rocks. [After R. Stevenson, by permission of *Annual Review of Earth and Planetary Science*, Vol. **10** (© Annual Reviews Inc. 1982)]

3. the profile $T(\varrho)$, assumed to be adiabatic in most of the models;
4. the relation between the mass-distribution and the gravitational moments J_{2n}.

A first, simple example of the calculations is given in Fig. 7.21, which corresponds to $T = 0$. It is immediately apparent that Jupiter and Saturn should consist mainly of hydrogen, with a small contribution from helium. Uranus and Neptune, in contrast, should have a considerable amount of ices.

The classical method of constructing static models is to assume successive layers of rock, ice, and gaseous envelope with cosmic abundances. The boundaries of the various layers are defined by adjusting the factors in accordance with the measured values of ϱ, J_2 and J_4.

The static models thus constructed do not necessarily give realistic planetary models; luckily, other constraints are met by evolutionary models.

Evolutionary Models

The most plausible explanation given for the excess of internal energy in Jupiter is that it derives from gravitation: the cause is either the gravitational collapse that occurred when the planet was formed, or internal differentiation of light and heavy elements. In the first case, an estimate of the time required for the planet to cool may be obtained, knowing the current energy excess L. Assuming that the body has cooled in a homogeneous manner since its formation:

$$L = 4\pi R^2 \sigma (T_{\mathrm{e}}^4 - T_0^4) = -\frac{d}{dt}[M\overline{C}vT_{\mathrm{i}}] \quad . \tag{7.15}$$

T_e is the effective temperature, T_0 the effective temperature in the absence of an internal source (and therefore related to solar flux by (1.8), T_i is the average internal temperature. $\overline{C}v$ is the mean specific heat per gramme. We may assume that R is approximately constant throughout contraction and at T_i is of the same order as T_e. We then obtain:

$$\tau = \frac{M\overline{C}v}{4\pi R^2 \sigma} \int_{T_{e_0}}^{T_{e_i}} \frac{d\tau}{T^4 - T_0^4} \quad . \tag{7.16}$$

τ is the time taken by the planet to cool from its initial temperature T_{e_i} to its present temperature T_{e_0}. As $T_{e_i} \gg T_{e_0}$, precise choice of T_{e_i} is not important in calculating the integral, which may be replaced by:

$$\tau = \frac{M\overline{C}v}{4\pi R^2 \sigma} \int_{T_{e_0}}^{\infty} \frac{dT}{T^4 - T_0^4} \quad . \tag{7.17}$$

We then obtain the simple result:

$$\tau = \frac{MCv}{4L} \quad . \tag{7.18}$$

For this model to be plausible, the value τ found by calculation must be of the same order as the age of the Solar System, i.e. 5×10^9 years.

It is not possible to calculate the energy liberated by a differentiation process easily: one can, however, show that it may be larger by several orders of magnitude than the energy liberated by homogeneous contraction.

There are two types of approach in evolutionary models that have currently been developed. In the first class, it is suggested that the giant planets formed by homogeneous collapse, starting with a gravitational instability (Cameron 1978). This is the hypothesis that we have assumed in calculating the planet's cooling time (7.15) to (7.18). In the second class of models, the giant planets are thought to have formed, like the inner planets, by successive accretion; first there was the formation of a core of ice and rock. When this core became sufficiently massive, it caused the collapse of the gaseous envelope that now surrounds it, and which is of primordial composition.

This second sequence of events, which has been developed by Mizuno in particular, has several advantages: the initial formation process would be the same for both the inner and the giant planets but, above all, it predicts the existence within the giant planets of cores that have masses of the order of ten Earth masses, which seems to agree with static models. Finally this hypothesis is able to explain the excess of carbon measured in the atmospheres of the giant planets: this carbon would derive from the evaporation, during the heating phase, of CH_4 ice and CH_4–$(H_2O)_7$ clathrates. The carbon thus freed from the core would have gone to enrich the atmosphere, which itself was of primordial composition (see Sect. 7.1.4). For all these reasons, the model of the formation of the giant planets by nucleation appear to be more likely than the homogeneous contraction model. But this scheme still leaves some observational facts unexplained (for example, the disparity in the masses of the four giant planets), and doubtless the true situation was more complicated than this simple model would suggest.

The Results

Models of internal structure are largely the work of Hubbard, McFarlane, Salpeter, Stevenson, Zharkov and Trubitsyn. The characteristics of each planet will now be described briefly.

Jupiter. The models that best account for all the observations are those of Hubbard. These have a core, equal to ten to twenty Earth masses, formed of a mixture of ice and rock; this core is surrounded by a layer of metallic hydrogen, which occupies about two-thirds of the jovian radius, and then by a convective atmosphere consisting of hydrogen and helium (see Fig. 7.22).

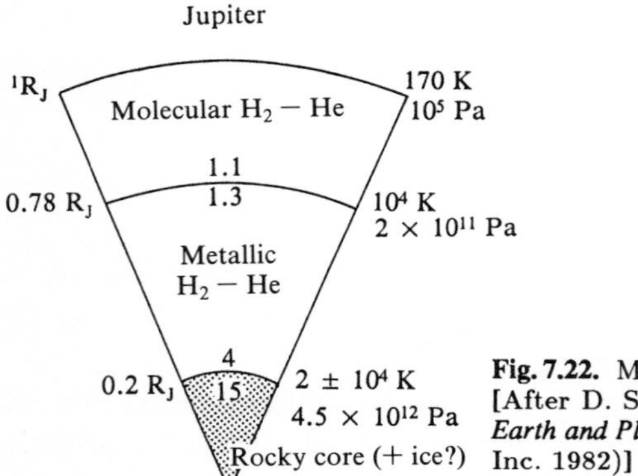

Fig. 7.22. Model of the internal structure of Jupiter. [After D. Stevenson, by permission of *Annual Review of Earth and Planetary Science*, Vol. **10** (© Annual Reviews Inc. 1982)]

If the planet's cooling time is calculated by (7.15) to (7.18), a value of the order of 5×10^9 years is obtained, compatible with the age of the Solar System. This result, coupled with the helium abundance, which appears close to the helium abundance in the primordial Solar Nebula (see Sect. 7.1.4), suggests that there has been little gravitational differentiation in Jupiter's interior.

Saturn. The case of Saturn is more complex: no current model is capable of accounting for all the observational data. Two factors are particularly important: the large energy excess, and the underabundance of helium in the upper atmosphere. The measured energy excess is too great to be ascribed to homogeneous cooling of the planet: the cooling time, calculated by (7.24), is considerably less than the age of the Solar System. It is tempting to consider the theory of helium differentiation by condensation within Saturn's metallic region. This process would enable one to explain, at least in qualitative terms, the energy excess and the lack of helium in the atmosphere.

Figure 7.23 shows the temperature-pressure diagram for conditions appropriate to the interiors of the giant planets, as well as the saturation curve for a "cosmic" mixture of hydrogen and helium. At pressures less than about 2 to 3 Mbars, the hydrogen is in molecular form, whilst it becomes metallic at higher pressures. As the planet cooled, the straight line $\log T(\log P)$ that describes adiabatic transfer, shifts downwards. It will be seen that the straight line corresponding to $T = 2 \times 10^9$ years

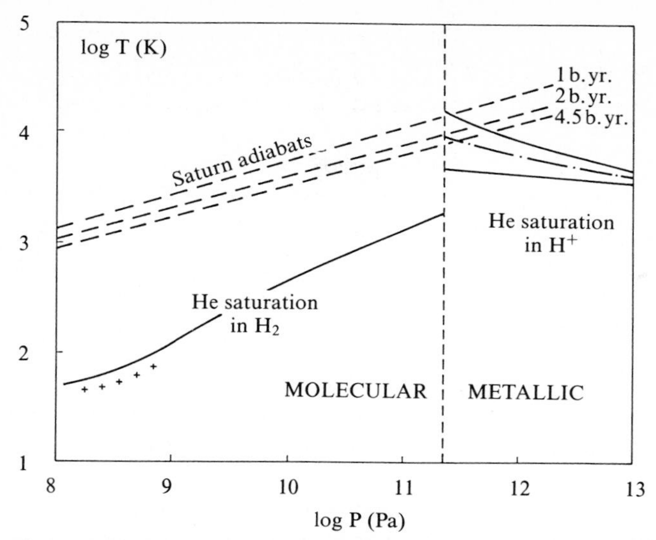

Fig. 7.23. The (T, P) helium saturation law for a cosmic-abundance, hydrogen-helium mixture. The pressure is in pascals (1 bar $= 10^5$ Pa). The crosses show experimental results. The broken vertical line indicates hydrogen's transition between the molecular and metallic forms. [After D. Stevenson, by permission of *Annual Review of Earth and Planetary Science,* Vol. **10** (© Annual Reviews Inc. 1982)]

intersects the curve for helium saturation in hydrogen. From this we may conclude that differentiation of helium takes place within Saturn at pressures above several Mbars. For Jupiter, the adiabatic lines lie higher in the $\log P - \log T$ diagram; differentiation, if it exists, ought to begin later and be less important that in Saturn. For Uranus and Neptune, which have significant cores, the pressure probably does not exceed 1 Mbar, so the hydrogen should all be in molecular form. The corresponding adiabatic lines, lying below those for Saturn, but above the curve for helium saturation in hydrogen, would not intersect it, so no differentiation is to be expected (see Sect. 12.1).

For Saturn we are therefore led to a structure like that shown in Fig. 7.24: a core, comparable to that in Jupiter, surrounded by a layer of metallic hydrogen, enriched

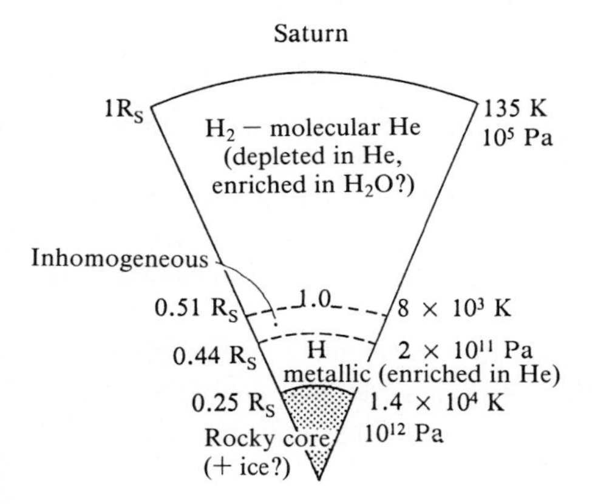

Fig. 7.24. Model of the internal structure of Saturn. [After D. Stevenson, by permission of *Annual Review of Earth and Planetary Science,* Vol. **10** (© Annual Reviews Inc. 1982)]

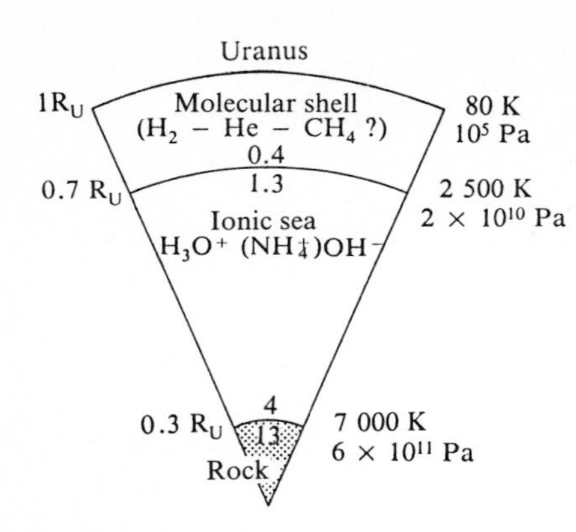

Fig. 7.25. Model of the internal structure of Uranus. [After D. Stevenson, by permission of *Annual Review of Earth and Planetary Science*, Vol. **10** (© Annual Reviews Inc. 1982)]

in helium, which is itself surrounded by a transition zone where differentiation occurs; above this zone there is the atmosphere of neutral hydrogen, depleted in helium.

Uranus and Neptune. As we have few observational data, the models for these two planets are still more uncertain. Figure 7.25 shows a possible model for Uranus, according to Hubbard: a core of rock with a radius of about one third of the planet's; then a convective layer of ionized "ocean", surrounded by a molecular envelope enriched in ices, particularly CH_4. The same could apply on Neptune.

7.2.3 Conclusion

Before drawing any conclusions from the examination that we have just carried out, it is important to remember that considerable uncertainty remains, because of our inadequate knowledge of high-pressure physics. This uncertainty concerns the nature of the metallic-hydrogen/molecular-hydrogen transition, the thermodynamics of molecular hydrogen, and the behaviour of ices at high pressures. Bearing this qualification in mind, it is nevertheless possible to summarize some of the significant results that seem to be established:

1. the four giant planets each have a core that is about ten Earth masses: this core is probably the source of the magnetic fields on Jupiter, Saturn and Uranus, and the one that Neptune quite probably possesses;
2. the four giant planets probably formed by a nucleation process, beginning with the accretion of a core of ices and rock, followed by the gravitational collapse of a gaseous envelope, of primordial composition, around the core thus formed. Degassing of the ices trapped in the core may have enriched the atmosphere in C, N, and O;
3. the internal energy-excess of Jupiter may be explained by the loss of thermal energy acquired during the gravitational collapse. In this case, the helium abundance on Jupiter ought to be comparable with that of the Primitive Solar Nebula;
4. for Saturn, the internal energy-excess has to have another source: differentiation of helium by condensation, followed by gravitational precipitation, can explain this energy excess, as well as the surrounding atmosphere's helium depletion.

7.3 The Magnetic Field of the Giant Planets

A magnetic field has been positively identified in the cases of Jupiter, Saturn and Uranus, for which a detailed study has been possible by virtue of the Pioneer and Voyager spacecraft. For Jupiter much important information has also been derived from observations of radio emissions in the decimetric and decametric regions.

7.3.1 The Magnetic Field of Jupiter

Description

Thanks to measurements from various instruments mounted on board the Pioneer 10 and 11 spacecraft, and the Voyager 1 and 2 probes, Jupiter's magnetic field is now very well known.

It may be described by a magnetic dipole with an axis inclined at about 11° to the planet's rotational axis; in addition its centre is displaced from the centre of the planet by about one tenth of the jovian radius in a direction perpendicular to the rotational axis. This asymmetry explains the different measurements of the strength of the magnetic fields at the poles: 14.8 gauss at the north pole, and 11.8 gauss at the south pole. At the equator the field is 4.2 gauss, ten times that found on the Earth.

The origin of the jovian magnetic field is thought to be, as with the Earth, a *dynamo* effect arising in the fluid, conducting, central regions of the planet. We have seen earlier (Sect. 7.2) that all four giant planets probably have cores; Jupiter's should be surrounded by a layer of fluid, metallic hydrogen, in convective motion, where electrical currents might be produced. The possibility of the dynamo effect arising in higher, non-metallic layers that still possess a certain degree of conductivity cannot, however, be ruled out.

Jupiter's inner magnetosphere, which extends up to a distance of twenty jovian radii, may be compared with the Earth's inner magnetosphere, which extends to about four Earth radii: the dipole field predominates and the intensity of the magnetic field, modelled on this basis, agrees well with observation. It should be noted, however, that unlike the Earth, there are perturbations caused by the presence of the satellites Io, Europa and Ganymede within this region.

In the middle magnetosphere, at radial distances between twenty and sixty jovian radii, the dipolar field is strongly perturbed by currents circulating in the plasma. In this region there is a disk at low latitudes, where the field is radial and has a weak (10^{-4} gauss), approximately constant, intensity. The field traps a low-energy plasma in which electrical fields circulate, forming a current "sheet" entrained by the planet's rotation, and which has a thickness of about one jovian radius; this current is too weak to perturb the inner magnetosphere's dipolar field.

In the outer magnetosphere the field is very irregular. Because of variations in the solar wind, the distance from the centre of Jupiter to the magnetosphere's limit in the direction of the Sun may vary from sixty to one hundred jovian radii. In the antisolar direction there is a magnetotail that has been detected by the Voyager probes as far out as the orbit of Saturn; this magnetotail therefore extends for several astronomical units (Fig. 7.26).

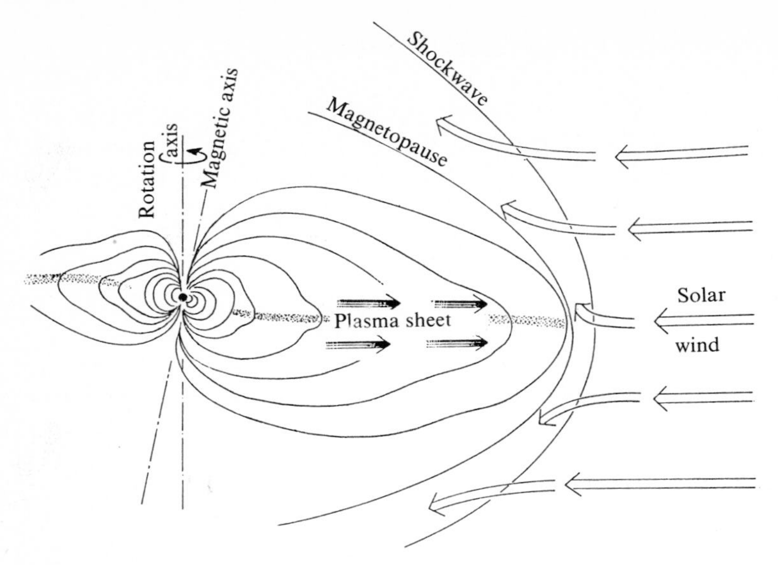

Fig. 7.26. The magnetosphere of Jupiter. [After J. Simpson et al., *Jupiter,* ed. by T. Gehrels (© University of Arizona 1976)]

As with the Earth, the jovian magnetosphere contains regions where very energetic particles, trapped by the magnetic field, accumulate. They are sometimes described as radiation belts. A remarkable feature, unique to Jupiter, is the interaction that occurs between the radiation belt and three of Jupiter's satellites, Io, Europa and Ganymede. The satellites absorb some of the energetic particles trapped in the radiation belt and modify the distribution of particles around their orbits. The bombardment of their surfaces and, in the case of Io, of the atmosphere, also causes them to act as sources of both atoms and molecules.

The Effects of Jupiter's Magnetic Field

Decimetric Emission. The first detection of Jupiter in the radio region was unexpected, and it was made by Burke and Franklin in 1955 at 22.2 MHz ($\lambda = 13.5$ m). Shortly afterwards Jupiter was detected at centimetric wavelengths, at three and four centimetres. Towards the end of the 1950s, the spectral region available extended towards longer wavelengths, first to 10.3 cm, then 21 cm, and then 68 cm. It soon became clear that the planet's thermal component (Fig. 7.27) was combined with another, non-thermal, component. The idea of *synchrotron radiation* was soon put forward, and this was the first indication of the existence of a magnetic field on Jupiter.

Since then, Jupiter's spectrum at centimetric and decimetric wavelengths has been the subject of a large number of studies. In particular, the use of long-base-line interferometry has enabled the desired degree of spatial resolution of Jupiter's disk to be achieved. The characteristic features of the decimetric radiation are:

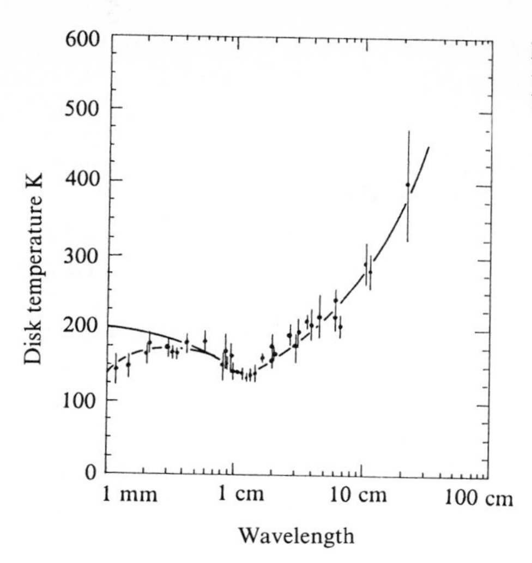

Fig. 7.27. Jupiter's centimetric, thermal radiation. [After G. Berge and S. Gulkis, *Jupiter*, ed. by T. Gehrels (© University of Arizona 1976)]

1. the source of the emission is a toroid encircling the planet, with the central plane of the toroid lying at an angle of 10° to the plane of the equator; the major diameter of the toroid is about two jovian radii;
2. the radiation is strongly linearly polarized, as the electrical field vector is parallel to the central plane of the toroid, there is little circular polarization;
3. the flux density varies by a factor of only two for a change in wavelength from 5 to 180 cm;
4. the emission is essentially constant with time, apart from an oscillation caused by the planet's rotation.

It is now generally accepted that Jupiter's decimetric radiation is caused by synchrotron radiation from high-energy electrons ($E_c \gtrsim 0.5$ MeV) trapped in Jupiter's dipole magnetic field.

Decametric Emission. Since its discovery in 1955, Jupiter's decametric radiation has been subject to continuous surveillance at between 39.5 MHz and 450 kHz, that is between wavelengths of about 7 m and 700 m. In 1964, Bigg demonstrated the correlation between decametric emission and the position of the satellite Io in its orbit. The decametric emission, which is very different from the decimetric emission, has the following characteristic features:

1. the emission is sporadic; it appears in the form of brief, intense *bursts*, with a very restricted range of frequencies; successions of bursts, or *storms*, are also observed, and these have durations of a few minutes to a few hours;
2. the radiation is strongly elliptically polarized, and has a sharp cut-off at its upper limit of 39.5 MHz;
3. the radiation arises in small, localized sources, close to the planet, at well-defined longitudes and at high latitudes;
4. the probability that bursts will occur is synchronized with the planet's rotation; this allows a longitude system (System III) to be precisely defined;

5. the probability that bursts will occur is also correlated with the phase of the satellite Io with respect to the planet-Earth line.

The small size of sources and the great intensity of the radio waves observed, meant that the idea of thermal radiation had to be rejected, as it would have led to unrealistic brightness temperatures, above 10^{12} K.

It is now accepted that the decametric emission arises from instabilities: population reversals in the plasma distribution within Jupiter's magnetic field. In particular, the cut-off frequency of 39.5 MHz may be interpreted as the electronic gyrofrequency: it corresponds to a magnetic field strength of 14 gauss, which is found at the level of Jupiter's ionosphere.

Ionospheric Phenomena. Jupiter's ionosphere, which lies immediately above the stratosphere, extends out to about 3000 km above the cloud tops. In this region, the Voyager probes detected auroral phenomena, that is non-thermal emission at visible and ultraviolet wavelengths, excited, as in the Earth's upper atmosphere, by the arrival of floods of charged particles accelerated along the magnetic field's lines of force.

Auroral phenomena have also been observed from Earth in the infrared: they occur as abnormally intense, non-thermal emissions at certain CH_4, C_2H_6 and C_2H_2 transitions that are confined to the disk of Jupiter itself. The profile of the emission lines of C_2H_6, obtained at very high spectral resolution, demonstrates their non-thermal origin.

Io's Plasma Torus. The interaction of Io with the radiation belts in Jupiter's magnetic field was detected in the 1960s from decametric observations; the existence of a hydrogen torus around Io's orbit was demonstrated by Pioneer 10 through the Lyman-α emission from the torus. More recently, measurements made from Earth, observations by the IUE satellite, and Voyager measurements have shown that this torus also contains sodium, and various ions, in particular H II and S II. They probably arise from interaction between the gases ejected by Io's volcanoes and the energetic particles around Jupiter. Voyager also measured the density of electrons within Io's torus.

7.3.2 Saturn's Magnetic Field

Unlike the case with Jupiter, and before the Pioneer and Voyager spacecraft encounters with Saturn, it was not possible to demonstrate that the planet had a magnetic field from radio observations from Earth. From the observations made close to the planet, Saturn's magnetic field is dipole, with the axis inclined by only 1° to the rotational axis, in contrast to Jupiter and the Earth. The equatorial displacement of the dipole is less than 0.01 saturnian radii (0.01 R_S), so the field has axial symmetry. The surface field intensity is 0.21 gauss. Saturn is distinctive in being surrounded by a ring of current that has a thickness of 5 R_S, lying between 8 and 16 R_S.

In the outer magnetosphere, which is comparable with that of the Earth, the magnetic field is the resultant of the fields from the dipole, the ring current, and contributions from the magnetopause and currents in the magnetotail; the last has

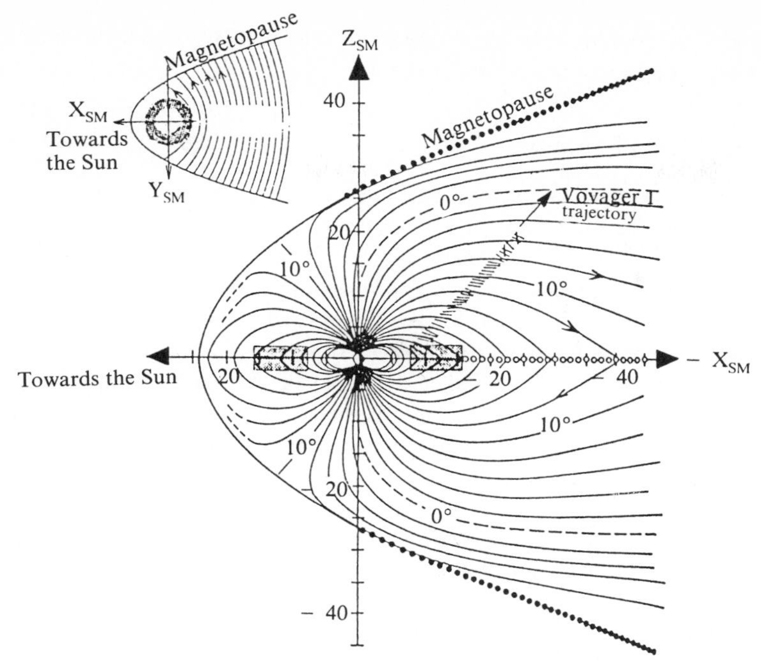

Fig. 7.28. The magnetic field of Saturn. [After A. Schardt et al., *Saturn,* ed. by T. Gehrels (© University of Arizona Press 1984)]

a diameter of 80 to 100 R_S (see Fig. 7.28). Unlike the Earth's, Saturn's external magnetosphere contains Titan's torus, which consists of hydrogen and nitrogen, and maintains the plasma density at between 10^{-2} and 5×10^{-1} ions/cm^3 and a temperature of 10^6 K. Some of the properties of the surfaces of Saturn's icy satellites may arise from their irradiation by particles accelerated in Saturn's magnetic field.

Although all the studies carried out from Earth failed to detect decametric emission, with the few decametric bursts picked up being unconfirmed, the Voyager mission showed that Saturn was also a radio source. Saturn's radio spectrum has three components: a non-thermal radiation (similar to the kilometric radiation emitted by the Earth's auroral zone, and to Jupiter's decametric radiation), noise at very narrow spectral bands, with a low-frequency continuum and finally, electrostatic discharges. Although the kilometric radiation was observed at great distances from Saturn, the continuum radiation and the discharges are trapped within the magnetosphere, and could be observed only from nearby. The storms at kilometre wavelengths show several sorts of periodicities: the first corresponds to the rotation period of Saturn's magnetic field itself (10 h 39.4 min); the second is approximately 66 hours and may be caused by the satellite Dione; finally Saturn's radio emission is closely correlated with solar-wind activity. Saturn's low-frequency continuum radiation, observed by Voyager, is trapped within the low-frequency cavity formed by the outer magnetosphere (see Chap. 4). The electrostatic discharges appear to be caused by radio bursts connected with the storm complex in Saturn's equatorial zone. Finally it should be noted that the absence of decimetric radiation from Saturn is explained by the Pioneer and Voyager measurements, which showed that the planet's radiation belts could not be the source of synchrotron radiation.

7.3.3 The Magnetic Field of Uranus

The existence of a magnetic field on Uranus was suggested in 1982, when measurements in the ultraviolet by the IUE satellite were interpreted as arising from auroral phenomena. However, this was by no means definite. Contrary to the case with Jupiter and Saturn, when the Voyager 2 probe approached Uranus in 1986 January, no emission was detected before the week of the fly-past. But the probe's magnetometer measured a very intense magnetic field (about 0.3 gauss, and therefore about 15 % stronger than Saturn's, and 15 % weaker than that of the Earth), and with a highly individual geometrical configuration. The magnetic axis is inclined 55° with respect to the planet's rotational axis; the magnetosphere resulting from such an inclination is extremely complex, and quite unlike those of any other Solar-System bodies.

The ultraviolet spectrometer on Voyager also indicated the existence of a halo on the side illuminated by the Sun, and that extends out to two planetary radii. It is the source of the ultraviolet emission detected earlier by IUE. In addition, Voyager detected an aurora on the night side of the planet, close to the magnetic pole.

7.3.4 Neptune's Magnetic Field

Before Voyager, no magnetic field had been detected on Neptune. According to theoretical models, however, Neptune's magnetic field was expected to be stronger than that of Uranus: the energy excess measured on Neptune could indicate the presence of convective motion in deep, conducting layers of the planet, where a magnetic field might arise.

The Voyager mission has now identified a magnetic field around Neptune, tilted 50° to the rotation axis. According to the first data reduction, its intensity is very asymmetrical: it reaches 1.2 Gauss in the southern hemisphere, but only 0.06 Gauss in the northern hemisphere.

8. Bodies Without Atmospheres in the Outer Solar System

After the invention of the telescope in the 17th century, it was discovered that the Earth was not the only planet to have a satellite. In 1610, Galileo observed the largest satellites of Jupiter and Huygens discovered Titan, the largest of Saturn's satellites in 1655. Gradually the number of satellites in the realm of the outer planets grew to about a dozen. Not until the 19th century, however, were the martian satellites, Phobos and Deimos, observed for the first time. It is only very recently, with the Pioneer and Voyager deep-space missions, that numerous new satellites have been discovered, and their main physical properties have been determined. What is striking is that this family of bodies orbiting the outer planets should show such an extraordinary diversity in their surface conditions, degree of internal activity, internal structure and evolution. Nevertheless it has been possible to establish some general laws regarding orbital parameters and macroscopic properties. For example, almost all the satellites have synchronous rotation, with equal periods of orbital revolution and sidereal rotation: they present the same face to the planet throughout their orbits. This implies that tidal forces have stabilized the rotation in an equilibrium state.

Our knowledge of the rings of particles surrounding the outer planets has progressed even more slowly. Although the rings around Saturn were discovered in the 17th century, it was not until 1977 that the rings of Uranus were detected. The Voyager probes then found a ring around Jupiter in 1979, and revealed the complexity of Saturn's rings in 1981, and of those of Uranus in 1986.

The rings and satellites associated with each of the outer planets form as many systems within the Solar System as a whole. Study of the satellites therefore makes constant reference to the evolutionary processes that have affected the inner planets, and the dynamics of the rings are referred to that of the protosolar nebula.

8.1 The Satellites of the Outer Planets

8.1.1 The Satellites of Jupiter

The Small Satellites

At present sixteen satellites of Jupiter are known (Table 1.2). Their discovery dates clearly show that there are two distinct families in size, and therefore in their properties. The satellites discovered by Galileo have diameters in excess of 3000 km, so they are similar, or somewhat larger than the Moon. The others discussed here are very small bodies: only *Amalthea*, discovered a century ago, is more than a hundred

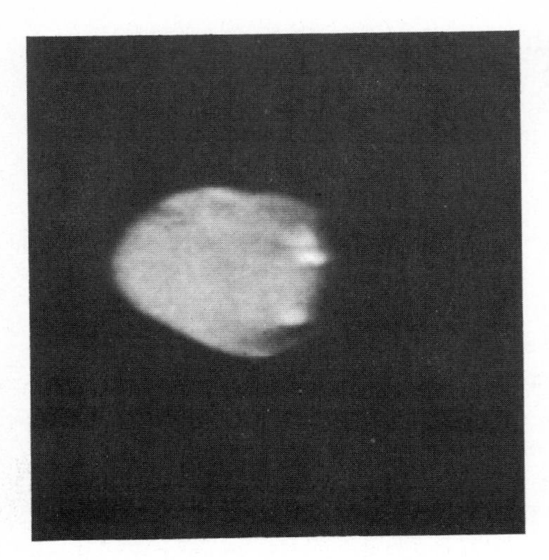

Fig. 8.1. Amalthea, as seen by Voyager 1 on 1979 March 4 from a distance of 425 000 km. (By courtesy of NASA)

kilometres in diameter. So the list of satellites is probably incomplete, because such small bodies are exceptionally difficult to detect. Neither is it possible to discover more than their existence and principal orbital parameters. Even the Voyager probes were unable to analyze the physical properties of any of them. The one characteristic common to all these objects is that of having too low a mass for them to have become spherical under the force of gravity: all are irregular in shape.

Voyager 1 passed closest to *Amalthea* (Fig. 8.1), at about 420 000 km. The body appeared very asymmetric (270 km by 160 km). It permanently turns the same face towards Jupiter, with its longest axis lying in the orbital plane. At the image resolution – about eight kilometres – impact craters can be distinguished. The reddish colour of the surface could be caused by the accretion of material ejected by Io, the closest of the Galilean satellites. We may expect the surface of Amalthea to have been altered by the effects of irradiation by particles accelerated in Jupiter's magnetosphere, in which Amalthea is permanently immersed.

Generally the small satellites are grouped in classes according to their orbital parameters. The four outer satellites, *Sinope, Pasiphae, Carme* and *Ananke*, have retrograde orbits with semi-major axes greater than twenty million kilometres and with high inclinations. They might be minor planets captured by Jupiter. The next group also includes four satellites: *Elara, Lysithea, Himalia* and *Leda*. Their orbits have semi-major axes between eleven and twelve million kilometres. Their inclinations are also high, but their motion is direct, not retrograde. Closer to Jupiter we find the group comprising the Galilean satellites. Even closer, no satellites other than Amalthea were known prior to the Voyager mission. We now know of *Thebe*, with an orbit outside that of Amalthea, and *Adrastea* and *Metis*, whose orbits are smaller. The last two satellites lie very close to Jupiter's system of rings, and are probably involved in its evolution.

The Galilean Satellites

The four largest satellites of Jupiter were discovered by Galileo in 1610. They were subsequently named *Io, Europa, Ganymede* and *Callisto*. Their dimensions put them in a class with the inner planets (Table 1.2). Their physical properties, on the other hand, as deduced from observations made by the Voyager probes, show that the predominant influence is that of the proximity of Jupiter, which is more than 10 000 times as massive. Although in the case of the Earth and the Moon we can speak of a double system, this certainly does not apply to Jupiter and its satellites.

This does not mean that these four objects are comparable. On the contrary, they show considerable differences, mainly caused by their different distances from Jupiter. For example, throughout their lifetimes all four have been bombarded by meteorites of all sizes, and probably at similar rates of flux. Yet the crater densities measured on their surfaces are completely different, and this reflects a difference in the degree of planetary activity that has affected the surfaces of these objects. For example, Callisto, the satellite farthest from Jupiter, has a surface covered in craters, a sign of an extinct body that has been inactive for several thousand million years. Ganymede has craters that are distributed in a non-uniform manner: certain regions are almost free of them. Europa, on the other hand, has a surface that is essentially smooth, whilst Io, the closest satellite to Jupiter, is subject to intense volcanism that completely remodels the surface on a time-scale of just a few thousand years.

Callisto was passed by Voyager 1 on 1979 March 6, at a minimum distance of 126 000 km, which gave an image resolution that reached two kilometres. Although we do not have complete photographic coverage of Callisto, the whole surface is remarkably uniform in appearance. The crater density is the same as the most "ancient" planetary bodies in the Solar System, i.e. those that have undergone little modification for at least three thousand million years, as with the Moon. However, there does seem to be a difference in the average diameter of the craters, which on Callisto do not exceed a hundred kilometres. Equivalent features to the large impact craters that gave rise to the giant lunar basins have not been observed, with the exception of the Valhalla basin, which has a diameter of more than 500 km (Fig. 8.2).

A second feature of Callisto's surface concerns the region around the Valhalla basin. This basin is surrounded by a series of concentric ridges, which extend out to about 1500 km from the centre of the basin. These ridges resemble a train of waves in some medium that is not completely rigid.

The lack of giant impacts and the presence of these ripples may be explained in the following manner. Giant impacts were common in the first few hundred million years of the Solar System's history, with numerous collisions between bodies of all sizes that arose with the growth of protoplanets (see Chap. 11). We may assume that at that epoch Callisto had a differentiated internal structure, where a surface crust covered a fluid mantle formed of silicates and ices. (Callisto has too low a density – 1.8 – to consist of just rocky material.) The heat released by high concentrations of radioactive elements in the mantle could have created a convective regime leading to flows of "dirty ice" covering the surface, rather like the way in which the terrestrial lithosphere is produced from the mantle. All traces of craters produced in the first few hundred million years would thus be effaced. The impact that created Valhalla

Fig. 8.2. Callisto, photographed by Voyager 1 from a distance of 350 000 km. The surface is covered with meteoritic-impact craters. The Valhalla basin, a large circular structure, is surrounded by large concentric rings. (By courtesy of NASA)

doubtless occurred later, when the crust had not completely solidified. Later still, the decrease in internal activity would have caused the resurfacing of Callisto's crust to cease. The absence of mountains on the surface of Callisto, where the only relief apart from the craters is these concentric ridges, bears witness to the cessation of all tectonic activity. Only meteoritic bombardment has modified the surface of Callisto over the last three thousand million years.

Ganymede was passed by Voyager 2 on 1979 July 8 at a minimum distance of 62 000 km, which produced a best image resolution of about a kilometre. Ganymede is the largest and most massive of the Galilean satellites, with a density essentially the same at that of Callisto. Ganymede differs from Callisto, on the other hand, in that its surface is heterogeneous both as regards colour and morphology: dark regions containing impact craters alternate with lighter regions where systems of faults of great extent can be seen (Fig. 8.3). Yet both appear to have been formed in a material consisting of a mixture of ice and rock.

In the dark regions, the great density of craters and their size distribution – which covers a very wide range of diameters – suggest that we are dealing with a very ancient surface, perhaps even more ancient than that of Callisto: it may be as old as four thousand million years. These regions also contain long folds that are sometimes cut by craters, which are therefore younger. These folds could have been caused by giant impacts that occurred very shortly after the formation of the satellite. The craters, which have now disappeared, would have been formed in a non-rigid medium, which would indicate primordial magmatic activity in a body containing a considerable proportion of ice.

The network of faults, or grooves, observed in the light regions on Ganymede have no equivalent on the surface of any other Solar-System body (Fig. 8.4). About one kilometre deep, they may attain several thousands of kilometres in length. They have been formed in a material that consists principally of ice and are relatively

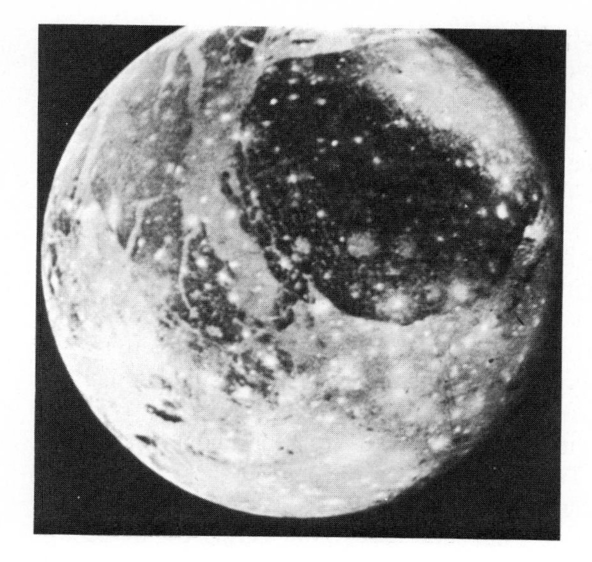

Fig. 8.3. Ganymede, seen by Voyager 1 from a distance of 2.6 million kilometres. (By courtesy of NASA)

young, as shown by the very low, impact-crater density. The correlation between the young age and the light colour could be explained in the same manner as that of the lunar regolith. We know that the progressive darkening of grains on the lunar surface is caused by irradiation by the solar wind and by micrometeoritic bombardment. The albedo contrasts on Ganymede could be explained similarly by the length of exposure of the surface to the interplanetary medium and to Jupiter's particle environment.

There are several possible explanations for the formation of these faults, the existence of which indicates that the internal structure was once differentiated. Around

Fig. 8.4. On this high-resolution image of Ganymede, taken from only 145 000 km, a network of grooves can be seen, which have depths of about a kilometre and are about ten kilometres apart. (By courtesy of NASA)

a silicate core, a partially fluid, icy-silicate mantle would have been the site of intense magmatic activity: the ascent of the magma towards the surface, by convection, would have caused stretching of the crust, which would have fractured, producing the faults. Magma would have welled up along the fracture zones, leading to their being filled by more recent, brighter ice.

An alternative theory is that the primitive crust was punctured by meteoritic impact. What might be called "plates" resulted, and these plates were carried along by the internal convection. Being of a greater density than the underlying mantle, they gradually sank into it, allowing lighter, brighter and more recent material to reach the surface. In this scheme, the faults would arise from movements of the crust that had been fractured by impacts: they would indicate that the tectonic activity was quite ancient, but which nevertheless presents striking analogies with the present-day terrestrial lithosphere.

The presence of ice on the surface of Ganymede has been known since 1971 from astronomical observations. This led to the suspicion that there might be a residual atmosphere of water vapour, or even of oxygen produced by the photodissociation of water. An experiment was therefore carried out during the fly-past of Voyager 1. It consisted of measuring the decrease in the intensity of light from a star as it was occulted by Ganymede. It produced an extremely low upper limit of 10^{-11} bar for the combined concentration of oxygen, water vapour, and carbon dioxide.

Europa, the smallest of the Galilean satellites, has a high density when compared with those of Callisto and Ganymede. Even in telescopic observations from Earth, however, its surface reveals the presence of ice: it is white in appearance, and it has a high albedo (0.6). Observations from space, by Voyager 1 from more than 700 000 km, and then by Voyager 2 on 1979 July 2 at a minimum distance of 206 000 km, confirms the presence of ice at the surface. In addition, at the resolution obtained, which reaches about four kilometres, practically no impact craters can be seen. This indicates that the surface has been recently renewed. The principal feature of Europa's surface is that it is covered by a network of extremely long fractures, intersecting in a complex arrangement (Fig. 8.5). Some resemble curved ridges and occasionally show repeating patterns over a distance of some hundreds of kilometres.

There is general agreement in considering that we are dealing with the fracturing of a superficial layer of ice. The very low number of craters detected (only three impact craters, about twenty kilometres in diameter, have been clearly identified), indicates that this ice has frozen at the surface very recently, and removed all trace of earlier impacts. This would also explain the high albedo, in accordance with the theory that the principal factor governing darkening of the surface is prolonged contact with the interplanetary medium. The origin of the fractures has not been clearly established. It could be the result of expansion forces, arising from magmatic activity in a convective, underlying mantle. According to other authors, it may be the result of tidal forces produced by Jupiter.

Io, the Galilean satellite closest to Jupiter, is the densest of the four. It is also the one that the Voyager probes approached most closely: Voyager 1 passed by on 1979 March 5 at a distance of only 20 000 km. The image-resolution approaches one kilometre. It was immediately apparent that the surface was rich in colour – red, orange, yellow and white – which completely distinguished it from any other

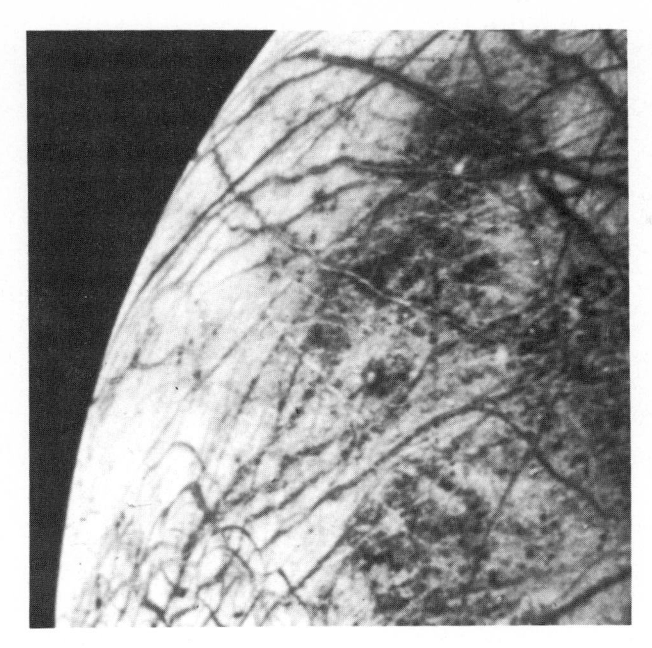

Fig. 8.5. Europa has a surface that is almost completely free from large impact craters. Great fractures, several thousand kilometres long, are the only type of structure that is observable. (By courtesy of NASA)

planetary surface. No impact craters could be detected, a sign of intense internal activity that gives rise to extremely rapid renewal of the surface. In contrast, numerous centres of volcanic activity were discovered. They appear as dark patches, a number of which resemble the calderas of terrestrial or martian volcanoes. These calderas are several tens of kilometres in diameter, and the area they occupy represents 5 % of Io's total surface area. Flows can be seen extending from some of these calderas, and some are several hundred kilometres in length (Fig. 8.6).

Mapping of the surface indicated Io's volcanic activity, which was confirmed by observations of the limb, where plumes more than 200 km high were identified. Using infrared spectrometry, it was possible to measure the regional variations in temperature and to identify the major components at the same time. Areas covered by recent lava flows could be identified by their temperature (about 300 K), higher than that of the surrounding areas (about 130 K). The various colours are attributed to the presence of sulphur compounds at different temperatures. The white areas correspond to sulphur dioxide SO_2, which, because of the very low surface temperature of Io, would be in solid form: ice, frost, or snow. The plumes are gaseous and primarily consist of sulphur dioxide, while the black, red and yellow colours correspond to sulphur compounds at decreasing temperatures.

Voyager 1 observed eight giant eruptions. When Voyager 2 arrived, four months later, the degree of activity shown by some of the volcanoes had changed markedly. Pele, which was the site of the most spectacular eruption, had ceased to be active. On the other hand, the plumes from the volcano Loki, observed in visible light, were 210 km high in July as against only 100 km in March. Neither of the probes detected

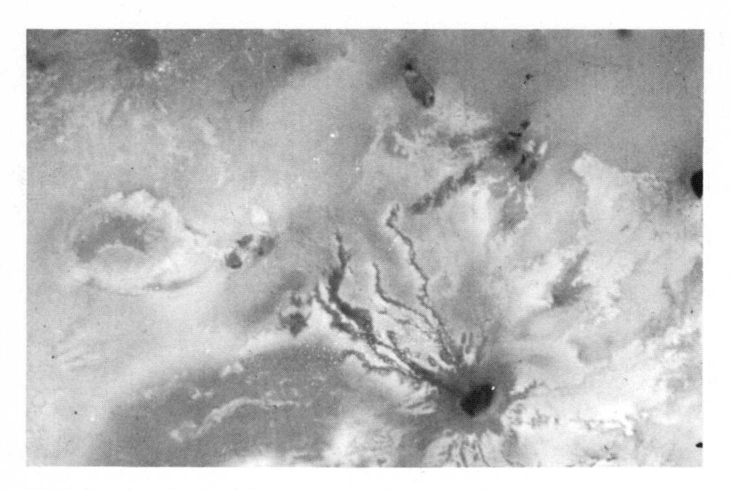

Fig. 8.6. Io's principal feature is active volcanism, as shown by the lava flows visible on this image obtained by Voyager 1 on 1979 March 4. (By courtesy of NASA)

eruptions reaching less than 70 km in altitude. It is estimated that in order to reach such a height the ejection velocity must be greater than 300 m/s, rising to 1000 m/s for eruptions reaching 280 km, the altitude of the plume from Pele. It has also been calculated that one of these volcanic eruptions ejects about 10 000 tonnes of material per second, which corresponds to more than one hundred thousand million tonnes per year. Because this material mainly consists of sulphur compounds, the condensation temperature of which is not very high, a significant fraction is deposited in liquid or solid form. Calculation of the mass of these deposits indicates that in just a few thousand years the whole surface of Io would be covered to a depth of several metres. The average time required for the surface of Io to be completely renewed is astonishingly short.

Apart from the Earth, Io is therefore the sole object known to be currently active volcanically. The energy sources of Io and the Earth are, however, completely different. In the latter, most of the energy required to maintain a high-temperature magma and to give rise to a convective regime within the mantle is derived from the radioactive disintegration of uranium, thorium and potassium. For Io, on the other hand, it has been calculated that the concentration of those elements required to sustain its volcanic activity would have to be about one hundred times their cosmic abundances, which is most improbable.

The mechanism proposed depends upon the close proximity of Jupiter. There is a considerable difference between the gravitational force exerted on the side of Io closest to Jupiter and that on the opposite side. This induces a mechanical deformation of the satellite, which takes on an almost ellipsoidal form, with its long axis pointing towards Jupiter. This deformation would be stable if the orbit were circular, because of the synchronous rotation of Io. But gravitational perturbations caused by the other Galilean satellites cause the orbit to be elliptical, around which Io's velocity is not constant (in accordance with Kepler's second law). As a result, the deformation of Io by Jupiter's differential gravitational attraction is not confined to specific zones throughout a revolution of Io. The mechanical relaxation results in a release of energy that is estimated at 10^{13} W. Violent tidal action is therefore the cause of Io's partial melting and volcanism. In support of this theory, it is observed that the volcanic regions are concentrated at low latitudes, i.e. just where one would expect the longitudinal oscillation of the equatorial bulges to release the maximum amount of energy.

This intense volcanic activity, together with the very low escape velocity (2.5 km/s, very similar to that of the Moon), would explain why Io has lost its primordial ices (water and carbon-dioxide), in contrast to the other Galilean satellites, which have lower densities. The surface of Io therefore mainly consists of sulphur dioxide. There is a very tenuous atmosphere, corresponding to this compound's phase equilibrium. At the subsolar point, where the temperature may reach 140 K, the saturated vapour pressure is of the order of 10^{-3} millibar. Close to the poles, where the temperature is below 100 K, this pressure falls to 10^{-8} millibar.

This atmosphere, to which the components ejected by the volcanism contribute, is subject to Jupiter's magnetic effects. Part of the atmosphere is ionized, primarily as a result of charge transfer through ion sputtering of Io's soil. This produces a plasma, which is entrained by Jupiter's magnetic field and co-rotates with a period of ten hours. Jupiter is therefore surrounded by a roughly toroidal volume of plasma at an average distance equal to Io's orbital radius, i.e. 5.9 jovian radii. The width of this plasma torus is of the order of one jovian radius. Its composition has been analyzed by UV spectrometry and by direct in-situ examination by the Voyager probes. The major components detected by these means are sulphur and oxygen, which confirm the composition of Io's surface and atmosphere, as well as that of the volcanic plumes. Its ionic and electronic densities are of the order of $1000\,\mathrm{cm}^{-3}$. The electron temperature is about 10^5 W. The UV luminosity of the torus corresponds to a radiative power of 10^{12} W.

The Trojan Minor Planets

We know that the solution to the three-body problem, obtained in the 18th century by the mathematician Joseph Louis Lagrange, predicts the existence of stable-equilibrium positions for a mass subject to the gravitational attraction of two distinct bodies. Lagrangian points therefore exist for every pair of objects such as the Earth-Moon and Earth-Sun pairs. An analogous situation arises for the Jupiter-Sun pair: two Lagrangian points lie on Jupiter's orbit, 60° ahead of and behind the Jupíter-Sun line (Fig. 1.2). Small bodies are in fact found in these regions of space, orbiting at the same velocity as Jupiter. They are known as the "Trojans" (they all have names of heroes of the Trojan War), and they may be minor planets captured in the early stages of the jovian system's formation.

8.1.2 The Satellites of Saturn

Following the Voyager probes' exploration in 1981, Saturn's system is known to consist of twenty-one satellites. The largest were discovered before the space age: first *Titan* by Huygens in 1655, then *Iapetus* by Cassini in 1671, followed by *Rhea, Dione, Tethys, Enceladus* and *Mimas*, and finally *Hyperion* and *Phoebe*. Twelve satellites were discovered between 1979 and 1981. The only satellite for which the physical properties could be obtained from Earth was Titan. This object indeed has the distinctive feature of being surrounded by an atmosphere, some components of which were detected spectroscopically before 1981. In this respect, *Titan* is more like *Triton* than like the other satellites of Saturn. Its features will therefore be described in Chap. 9.

Examination of Saturn's satellites by the Voyager probes allowed not just the orbital characteristics to be obtained, but also, for the first time, a considerable number of their main physical properties. Objects that were once just tiny dots on photographic plates now have topographic charts with resolutions approaching one kilometre in some cases.

The Large Icy Satellites

After Titan, the six largest satellites in Saturn's system are, in order of increasing distance from the planet, *Mimas, Enceladus, Tethys, Dione, Rhea,* and *Iapetus*. The first five have circular orbits with radii less than 550 000 km, that is less than ten saturnian radii, and the orbits all lie very close to Saturn's equatorial plane. Iapetus, on the other hand, moves in an orbit inclined at 14.7° to that plane, and at a distance of about sixty saturnian radii.

The density of these satellites is known, and it is only slightly greater than one (Table 1.2). This indicates that their basic composition is largely ice, and this is confirmed by the images of their surfaces.

Mimas, the smallest of this group, was passed by Voyager 1 at a distance of less than 90 000 km. We therefore have images with resolutions that attain two kilometres. Observation shows that the surface is saturated with meteoritic-impact craters. The largest of these is 130 km in diameter, about one-third of that of Mimas itself. It is about ten kilometres in depth, and has a central peak six kilometres high. This shows that under meteoritic bombardment the rheological behaviour of soils consisting of rock and that of soils consisting of ice are very similar. The size of the body responsible for creating the largest crater must have been about ten kilometres. Slightly larger and it would have completely fractured Mimas, which would not have been able to resist the impact. Traces of fractures that originated in this violent impact can be observed, but apart from these effects of the propagation of shock-waves produced by meteoritic impact, there are no signs of internal activity on the surface of Mimas.

Enceladus is the brightest of all the planetary bodies known to date: its surface reflects more than 90 % of incident light. As a result, we must be dealing with a surface consisting of extremely clean ice, free of any rocky particles. The satellite is also the coldest body: the average day-time temperature is about 70 K. Although little greater in size than Mimas, Enceladus is very different. Instead of a surface entirely covered in impact craters, regions can be seen that clearly suggest recent internal activity, like the youthful areas seen on Ganymede. In contrast to Ganymede, however, where the cratered areas are very ancient – and have, in other words, not been affected by internal activity for perhaps four thousand million years – the cratered terrain on Enceladus is only moderately old. The average diameter of the craters is low: no craters larger than 35 km in diameter have been detected. Overall,

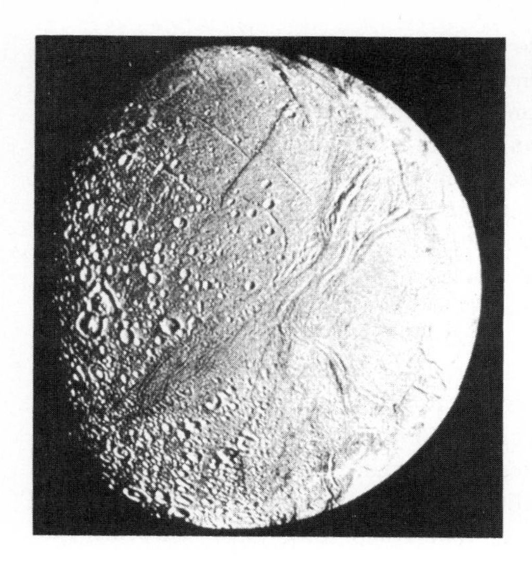

Fig. 8.7. Enceladus clearly shows two types of terrain, which are distinguished by their crater densities. Flows and faults, dating from the last thousand million years, are visible on the "young" terrain. (By courtesy of NASA)

it seems that this body's internal activity – which has very recently renewed whole areas of the surface, where not a single crater appears – has been a continuous process, covering a long period of its history.

Images of the surface of Enceladus show striations, cracks and faults in certain areas, as well as the paths of flows (Fig. 8.7); but on the other hand, no formations connected with volcanic activity can be seen. Interpretation of the surface structures requires deformations that could have been induced by magmatic activity in an underlying mantle. We do not know the source of the energy responsible for the complete or partial melting of this mantle. Bearing in mind the small size of Enceladus, radioactive disintegration does not appear to be sufficient, if the uranium and thorium abundances are within the range covered by other Solar-System bodies. Calculation of tidal effects caused by an orbital resonance with Dione (such as those operating on Io), show that the energy dissipated is far too low. Enceladus is therefore a body of very small size (about one-tenth of that of Ganymede) that nevertheless was the site of comparable internal activity in the past. To date we have no clear understanding of the mechanism responsible.

Tethys has a diameter twice that of Enceladus. There are no images with resolution better than five kilometres, but this is sufficient to discern two types of region, distinguished by their crater-density. The first type of area is very similar to the overall surface of Mimas: it is saturated with craters, one of which is exceptional in size. On Tethys this crater is 400 km in diameter, i.e. more than the diameter of Mimas itself. But this crater is shallower than the one on Mimas, and appears to have been degraded, not just by later impacts, but also equally by what were probably flows of ice: in a manner somewhat similar to the way in which the giant basins on the Moon became flooded by magma from the interior and gave rise to the lunar maria.

Within these areas of ancient terrain there are regions where the crater-densities are much lower, indicating recent rejuvenation of the surface. Signs of internal activity are similarly found in the presence of flows, resembling the stream-beds seen on Mars. In particular, a gigantic "canyon", more than 2000 km long, runs three-quarters of the way round Tethys: this Ithaca Chasma shows a marked resemblance to Vallis Marineris. The origin of the valley may well be completely different, however, and not caused by any ancient flow. It might, for example, be the scar remaining from very ancient deformation in response to tension in the upper crust. This could have been caused by expansion in the mantle, in accordance with the theory that the partially fluid mantle would gradually turn into ice as it cooled.

Dione, with a diameter essentially the same as that of Tethys, differs from it in having a slightly greater density and, in particular, in a greater variety of structures mingled with the impact craters. Significant contrasts in albedo between neighbouring areas may indicate the presence of rocky material mixed into the ice that dominates the surface, unless we are dealing with ice of different ages, where the lighter areas may have frozen later. The valleys and cracks form more complex networks than on Tethys, which is a reflection of the greater importance of past internal activity.

Rhea is the largest of the inner satellites, with a diameter of 1500 km. It shows less signs of internal activity than Dione, Tethys and Enceladus, however. This shows that the diameter of the body is not the predominant factor in determining the thermal regime with these satellites. The surface of Rhea appears different at high and medium resolutions. On a scale similar to the dimensions of the satellite itself, optical contrasts indicate that regions of the surface are covered by material of higher albedo, which might consist of recent ice. Overall, however, the surface appears to be uniformly covered with craters of all sizes. The crater diameter-distribution shows, according to some authors, that there were two types of impacting bodies. One, common to all the planetary bodies, corresponds to the meteoritic bombardment found throughout the Solar System. The other, specific to Rhea – but which might apply to other satellites of Saturn – might have arisen from a local source, following primordial collisions that took place in Saturn's system, and perhaps associated with the formation of the rings.

Iapetus is the outermost of the satellites in this group. Following the Voyager mission it still remains one of the most mysterious of Saturn's satellites. Despite a resolution that attained about twenty kilometres, the striking feature about Iapetus, which has been known from the time of the satellite's discovery by Cassini, remains unexplained: Iapetus has one dark hemisphere – the one turned towards Saturn – whilst the other one is light. The ratio between the albedos of these two types of terrain is of the order of 1:10. There are numerous craters in the light regions, but no details can be seen in the dark areas.

It is tempting to interpret the presence of these dark areas as having been caused by the deposit of some very absorbent material on a bright icy surface. The fact that one face is uniformly dark suggests that the deposit may have arisen from the permanent accretion of material through the motion of Iapetus in its orbit. Contradicting this interpretation, however, is the fact that the bottom of certain craters on the bright side are equally dark, which suggests volcanic deposits. Another observation has to be accounted for: there are no signs of any recent meteoritic impacts in the dark areas, although these ought to appear as bright ray-craters contrasting with the surrounding soil. Either the deposit is extremely thick or else it is constantly renewed. By eruptions or by accretion? The enigma remains.

The Smaller Satellites of Saturn

This group includes all the other objects in Saturn's system, with the exception of Titan and the six satellites that have just been discussed. They all have diameters less than that of Mimas. Fourteen are known at present, but we have no information about their densities or masses.

The two largest are the outermost, *Phoebe*, which orbits at thirteen million kilometres, and *Hyperion*, which is 1.5 million kilometres from Saturn. Phoebe has a quasi-spherical shape and a very dark surface. It seem to be an undifferentiated minor planet that has been captured by Saturn. Hyperion, on the other hand, is very irregular in shape, with dimensions of $400 \times 250 \times 200$ km. According to telescopic observations of its surface, which the Voyager images show to be heavily cratered, probably consists of ice. It may have been formed by the fragmentation of a more massive body through the effects of a slightly more violent impact than the one that created the largest crater on Mimas.

The other twelve satellites are close to Saturn. They have only been known since 1980. Some of these occupy the Lagrangian positions with respect to other satellites of Saturn, just as the Trojans occupy Jupiter's Lagrangian points. After the discovery of one of Dione's Lagrangian satellites in 1980, Voyager 2 found a second. These occupy the same orbit as Dione, one 60° ahead of, and one 60° behind the satellite. Voyager 2 also observed the two Lagrangian satellites of Tethys. These four satellites have sizes of about thirty to forty kilometres. Among the satellites discovered by Voyager, two small objects have the same orbit at 2.51 saturnian radii, and which they follow alternately approaching and receding from one another. Finally, the other small objects are either in the immediate vicinity, or even inside, the rings. They play a part in the gravitational confinement of the rings.

8.1.3 The Satellites of Uranus

Uranus has five principal satellites with diameters above 450 km. In order of increasing distance from Uranus they are: *Miranda, Ariel, Umbriel, Titania* and *Oberon*. The two most distant are the largest and were discovered by Herschel in 1787. Ariel and Umbriel were discovered later, in 1851, by Lassell. Miranda, the smallest and the closest to Uranus was discovered by Kuiper in 1948. All have orbits with very low eccentricities lying in the equatorial plane of Uranus. Only the orbit of Miranda departs slightly from this, as its inclination is 3.4°.

Voyager passed Uranus in 1986 January, and examined these satellites. They all have a density between 1.5 and 1.7, slightly higher than Saturn's icy satellites, and their colour is a brownish grey. Their low albedo (12 % in the case of Umbriel) might be the result of irradiation by particles accelerated in the magnetosphere of Uranus. Their surfaces consist of mixtures of water ice and carbon compounds, such as methane CH_4, and nitrogen compounds, such as ammonia NH_3. Laboratory experiments have shown that irradiation of such mixtures by protons at keV to MeV energies may induce polymerization of stable organic compounds with low albedo.

One of the principal discoveries by Voyager 2 was that of signs of geological activity on the surface of the satellites. Titania and Ariel have long depressions and

Fig. 8.8. Miranda. (By courtesy of NASA)

valleys; Ariel has areas of smooth surface, resembling mud-flows. On Oberon there is a mountain more than six kilometres high, and some of its impact craters are covered in a very dark deposit.

Miranda is the most spectacular of these objects (Fig. 8.8). Voyager 2 passed it at a distance of 28 000 km. Steep cliffs, in places more than five kilometres high, deep gorges, faults, and numerous valleys have all been detected on the surface. Some of the structures are extremely jagged or show abrupt changes in direction. No one expected to find geological activity of this extent on such a cold body. A possible origin for this activity has been sought in bombardment by meteoritic-type objects.

Voyager 2 also discovered several other small satellites inside the system of rings. The limited resolution, around five to ten kilometres, however, did not detect the objects very close to the rings that are presumably responsible for their confinement.

8.1.4 The Satellites of Neptune

The most massive of Neptune's satellites is *Triton*, the diameter of which is more than 2000 km. With the existence of an atmosphere it resembles *Titan*, the largest of Saturn's satellites. For this reason both will be discussed in Chap. 9.

A second satellite of Neptune is known: *Nereid*. It has a very highly eccentric orbit ($e = 0.75$), and its distance from Neptune varies from 1.39 to 9.73 million kilometres. The inclination of the orbit to Neptune's equatorial plane is 27°. Nereid's motion is direct and it completes one orbit in 359 days. Its apparent magnitude is 19, so it is hardly observable, and nothing is really known about its physical properties.

Its diameter is probably between 400 and 900 km. The Neptune flyby by Voyager 2, on 25 August 1989, has led to the discovery of six new satellites. According to preliminary analyses, the values of the semi-major axis a and the diameter d are the following for the 6 satellites (in km):

$$N1: \quad a = 117\,600 \;; \quad d = 420$$
$$N2: \quad a = 13\,600 \;; \quad d = 200$$
$$N3: \quad a = 52\,500 \;; \quad d = 160$$
$$N4: \quad a = 62\,000 \;; \quad d = 140$$
$$N5: \quad a = 50\,000 \;; \quad d = 90$$
$$N6: \quad a = 49\,200 \;; \quad d = 50 \quad .$$

8.2 Pluto and Charon

Pluto is the ninth planet in the Solar System, in order of increasing size of semi-major axis. But its orbit is highly eccentric (e = 0.25), and as a result it has been closer to the Sun than Neptune since 1979, and will remain so until 1998. This does not mean that the orbits of Pluto and Neptune intersect, because that of Pluto is steeply inclined (17°) to the ecliptic. If account is taken of the motion of the planets in their orbits, and of the fact that the period of Pluto (247.7 tropical years) is one-and-a-half times that of Neptune, it is possible to calculate that Pluto never comes closer to Neptune than 2.5 thousand million kilometres. The distance between Pluto and Uranus remains more than 1.6 thousand million kilometres, so all thoughts of a possible collision between these planets can be forgotten. The fact that the periods of Pluto and Neptune are in a simple ratio is the result of a dynamical resonance that has been discovered recently, and which corresponds to a stable state: any modification of the trajectory under the influence of an external perturbation would be followed by a return to the present state of equilibrium.

The search for Pluto was undertaken on the basis of calculations predicting its existence because of its gravitational effects on Neptune and Uranus. In fact, these assessments have proved to be inaccurate: the mass of Pluto is quite inadequate to cause the perturbations that were thought to exist, and which simply resulted from the imprecise orbital data available at the beginning of this century. Pluto was only discovered in 1930, at magnitude 15, at the end of a couple of decades of intensive searching, and thanks to the systematic examination of innumerable photographic plates. Seen from Earth, Pluto has an apparent diameter of less than a quarter of an arc-second. Because atmospheric turbulence produces a resolution that is only about one arc-second, Pluto appears as a point source.

It would seem that Pluto is covered in ices, mainly consisting of a mixture of solid water, methane and ammonia, and that it is surrounded by an atmosphere that contains methane. Assuming that the overall albedo is fairly high, a diameter of between 2000 and 3500 km is obtained. So we are dealing with a body that is cold, and very small in size, even less than the diameter of the Moon. This would make it the smallest of the planets in the Solar System. From the point of view of its physical characteristics, Pluto belongs neither to the inner planets not to the giant planets, but to the icy satellites of the latter. For this reason it has been suggested that Pluto might once have been a satellite of Uranus or Neptune, ejected from one of these systems by an encounter with a massive object, early in the Solar System's history. Such an event might equally explain the considerable eccentricity and inclination of Pluto's orbit. Another theory suggests that Pluto may belong to a family of icy minor planets, which accreted at large heliocentric distances. In this it would resemble *Chiron*, which has an eccentric orbit that lies between the orbits of Saturn and Uranus.

In 1978, Christy, using the 1.54-m reflector at the U.S. Naval Observatory, discovered a satellite of Pluto, *Charon*. It appeared as an elongation of the usual image of the planet in a particular direction. Since then speckle interferometry of Pluto has allowed certain characteristics of Charon to be obtained. It is probably an object about 800 km in diameter, orbiting Pluto at less than 20 000 km and in 6.39 days, so its period of revolution is equal to Pluto's rotational period. The relative sizes of Pluto and Charon mean that – as for the Earth and Moon – it can well be called a double planet, rather than a planet and satellite.

We can be certain that space exploration of this system will not occur for the next two or three decades. We must therefore, for the foreseeable future, remain content with spectroscopic analysis of the surfaces of these objects by telescopes on the ground or in Earth orbit.

8.3 The Search for a Tenth Planet

The American astronomer C.W. Tombaugh, who discovered Pluto in 1930, undertook a systematic search for distant planetary objects until 1943. He concluded that there were no objects brighter than the Earth out to 100 AU, and none brighter than Jupiter out to 470 AU, at least in areas close to the ecliptic. These studies have been continued since, and have remained without result. Only new comets and minor planets have been discovered. At present attempts are being made to detect the existence of any new planets by establishing the existence of perturbations in the orbits of comets, as well as in the trajectories of the interplanetary probes that are moving out beyond the orbits of Neptune and Pluto. So far this research, which just involves the American Pioneer probes, has been without any positive result.

It is obvious that these methods can only probe regions very close to the Sun. The limits of the Solar System are measured, however, not in tens of astronomical units, but in thousands: comets recede from the Sun to distances that are more than a thousand times greater than the distance separating Pluto and the Sun, and the closest star is three times as far again. The question of the existence of not just a tenth planet, but of a whole number of small objects that formed simultaneously with the known planets and were expelled to large heliocentric distances, remains completely open.

8.4 The Rings of the Outer Planets

8.4.1 The Formation of Planetary Rings

Rings consist of a myriad particles and small-sized blocks of material. Their movement around each of the outer planets takes place on orbits that all lie very close to the planets' equatorial planes. They therefore form extremely flat disks, the main properties of which are deduced from the sunlight that they scatter. The Voyager observations of Saturn's ring system showed that the dynamical properties of rings are very complex, and certainly cannot be explained simply in terms of just the gravitational potential of the central body. It is essential to include the effects of the small satellites that are embedded in the outer rings. It is, however, fairly simple to account for the existence of rings, and for the fact that they are confined to the immediate vicinity of planets, i.e. that their outer radii are less than the semi-major axis of the orbit of the principal satellites.

In the gravitational field of a planet, a satellite is subjected to stronger forces on the side facing the planet. This differential force is known as the tidal force. It acts against the cohesive forces within the satellite, which include the satellite's internal gravitational forces and the non-gravitational inter-molecular forces, the mechanical strength of the material. As a satellite approaches the planet that it is orbiting, the differential effect increases: mechanical distortions, or tides, begin to occur. There is a limiting distance at which cohesion is no longer possible. This limit L_R, is known as the Roche limit (see Sect. 3.2.2), and may be expressed simply as a function of the planetary radius R and of the densities ϱ_P and ϱ of the planet and satellite: $L_R = 2.5\,R(\varrho_P/\varrho)^{1/3}$. When an object approaches a planet to a distance less than L_R, it is violently fragmented, giving rise to smaller bodies. Their inelastic collisions decrease the average size of the particles, which gradually spread out to populate a ring of material.

This version of the origin of rings by fragmentation of satellites that fall below the Roche limit is not the only one possible. We can imagine another theory whereby the existence of rings would result from the opposite process, namely the impossibility of satellites growing by the accretion of particles in the region below the Roche limit. When the giant planets were formed, the dynamical evolution of the clouds of gas and dust under the gravitational influence of the central protoplanets was accompanied by the condensation of large amounts of material in the equatorial planes – if we are to believe numerical models that take inelastic collisions into account. As a result, the planets would not have accreted all the initial material, which would have been concentrated into disks of pulverized material. Satellites would have formed through the effects of collisions at low relative velocities, giving rise to blocks of material some millimetres to a few metres across. Subsequently, gravitational attraction between the blocks would have accelerated the process, leading to the formation of satellites that were at least a few tens of kilometres across. The growth of these satellites would, however, have been limited by the differential attraction of the central body. These tidal effects would abort the formation of satellites out to distances close to the Roche limit. This would explain why today we find a distinction between satellites, far from the planet, and rings, at closer distances.

At present it is difficult to chose between these two models for the origin of rings. It does seem, however, that the differences between the properties of the rings of Jupiter and Saturn, as far as we know them today, favour the second interpretation. The rings of Jupiter are, in fact, considered to consist of mineral grains of relatively high density, whilst those of Saturn mainly contain ice. This may be explained if the luminosity of Jupiter was higher than that of Saturn. Jupiter would have prevented the condensation of icy grains out to very large distances. Being in the form of vapour, the volatile compounds would have dispersed into the interplanetary medium. Only refractory grains (primarily silicates) would remain to populate the present-day rings. But the opposite would apply to Saturn, where condensation of ice grains was permitted by the lower planetary temperature.

8.4.2 Jupiter's Rings

Jupiter's rings were discovered by Voyager 1 on 1979 March 4. The photographic discovery came from a long exposure of a stellar field slightly offset from the planet itself, and which was intended to detect possible small satellites within the orbit of Amalthea. The edge of the rings appeared as a whitish trail (Fig. 8.9), details of which show that it arose from the superimposition of six segments, each of which corresponded to the equatorial plane. During the long exposure Voyager 1 has oscillated slightly, probably because of the effect of the local magnetic field, but had been corrected by the 3-axis pointing system. The tracks of star images in the field showed this motion: each trail consisting of a jagged line with six periodic waves. Three months later, Voyager 2 arrived in the vicinity. It had been programmed to obtain a better image of the ring by observing it under good conditions of visibility from outside the ring plane (Fig. 8.10).

The principal difficulty in observing these rings from Earth is that they only extend to about 53 000 km above the top of the clouds in Jupiter's atmosphere. They

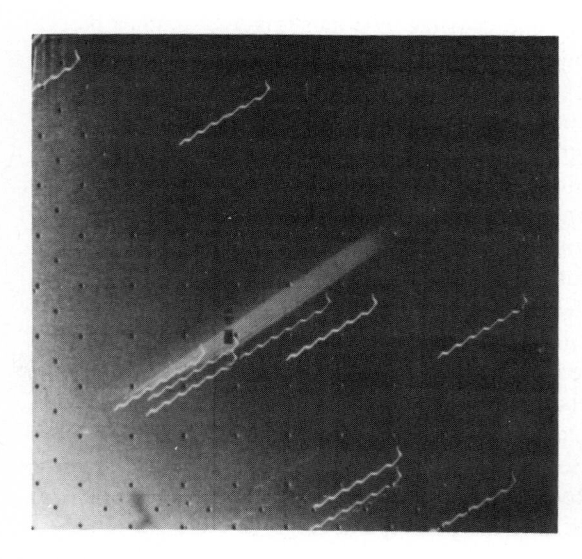

Fig. 8.9. The discovery photograph of Jupiter's ring. (By kind permission of NASA)

Fig. 8.10. Jupiter's ring as seen by Voyager 2. (By kind permission of NASA)

are therefore less than two jovian radii from the planet and the sunlight that they scatter is very faint in comparison with the light reflected by Jupiter. It is only in very narrow spectral regions, at which the atmosphere is dark, that the rings have subsequently been detected from Earth.

Voyager 2 detected three principal components to Jupiter's system of rings. The outer ring is the brightest, although its optical thickness is very low at about 3×10^{-5}. It extends over a width of 6000 km, from a sharp outer border at 1.81 R_J (R_J is the radius of Jupiter) and a more diffuse limit at 1.72 R_J. Within this there is a more tenuous zone, with an optical thickness of 7×10^{-6}, from 1.81 R_J down to the top of Jupiter's atmosphere. An even less dense halo surrounds the whole system. The thickness of the rings does not appear to exceed a few tens of kilometres.

Seen by back-lighting, that is in forward-scattered light, the rings are twenty times as bright as they are in reflected light. This is attributed to the predominance of particles with sizes below about ten micrometres. Such particles have a very short lifetime in Jupiter's environment, both as a result of dynamical forces (the Poynting-Robertson effect and electromagnetic forces) and of erosion (by ionic and micrometeoritic bombardment, and by collisions). Jupiter's rings much therefore be continually replenished by new particles. Possible sources are through meteoritic bombardment of the nearby satellites (*Adrastea* and *Metis*), from micrometeorites trapped in Jupiter's magnetic field, or even by particles ejected from Io's volcanic eruptions. Another theory for the origin of these particles is that they arose from the erosion by collision of primitive blocks of rock, and that these were proto-satellites that were unable to grow because they occurred inside the Roche limit.

8.4.3 Saturn's Rings

Galileo discovered that Saturn was accompanied by intriguing objects. At first thought to be two satellites symmetrically placed with respect to the planet, these objects did not show any apparent relative movement and their luminosity varied from year to year. In 1654, forty-four years later, Huygens suggested that they could be explained in terms of a solid ring of material lying in Saturn's equatorial plane: the variation in the appearance being caused by changes in the inclination of the equator with respect to the Earth. Shortly afterwards, in 1675, Cassini discovered that this ring was not uniform, but was split into two rings (the A and B rings) by a division which has since carried his name. A inner ring, C, was discovered in 1850. Laplace showed, in 1785, that a solid ring would not be stable against tidal forces. Maxwell later published the idea that the rings consisted of individual grains in differential rotation. This theory received observational confirmation at the end of the 19th century, thanks to measurements of the rotational velocities by means of the Doppler-Fizeau effect. The rotational periods of the particles depend on their distance from Saturn, in accordance with the dictates of Kepler's third law. It is only in the last twenty years that models have succeeding in describing the origin and evolution of the rings, in particular the effects of collisions between particles, like those occurring in the Solar System as a whole.

The arrival of the Voyager probes completely changed studies of the rings, because of the resolution of the images obtained, and the possibility of observing the rings by both scattered and reflected light, under conditions not possible from Earth. More and more rings were discovered, both inside ring C, where the existence of a ring D was confirmed, and outside ring A (Fig. 8.11). Ring E, the most distant,

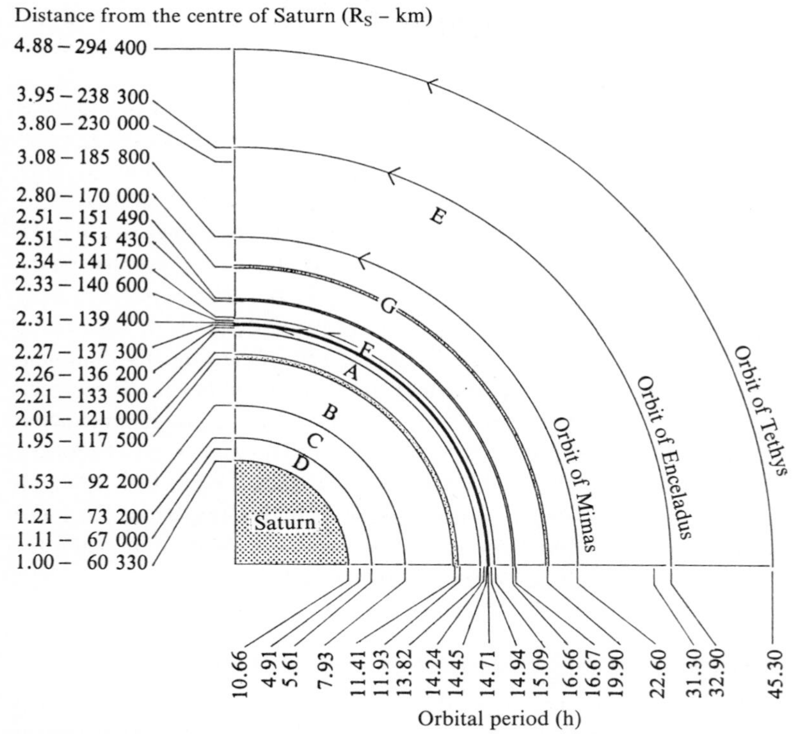

Fig. 8.11. The rings of Saturn. [After A. Brahic, "Saturn", in *Le grand Atlas Universalis de l'Astronomie* (Encyclopedia Universalis 1983). By courtesy of the publisher]

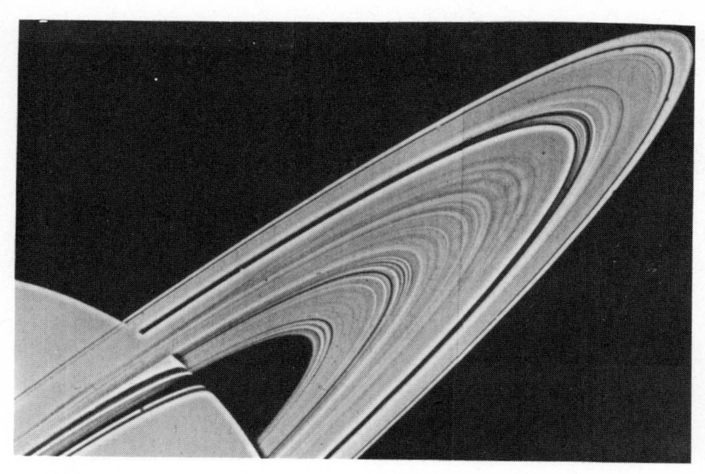

Fig. 8.12. Saturn's rings as observed by the Voyager 1 probe. (By kind permission of NASA)

stretches beyond the Roche limit. Ring F, on the other hand, is only 4000 km outside ring A. Ring G lies between them. Whilst the optical thicknesses of rings A and B are close to one, that of ring E is only some 10^{-7}.

The most spectacular results are not the discovery of new rings but concern their very complex structure. With the high resolution of the images, the rings no longer appeared as zones of relatively homogeneous brightness, but as thousands of fine concentric rings, clearly distinct optically (Fig. 8.12). The same applied to the Cassini division, which is not free of particles but is also filled with a multiplicity of rings, which are visible only under appropriate lighting conditions. Although it had previously been generally agreed that the rings were essentially circular and of uniform density, strongly eccentric rings were detected, together with braided rings and local fluctuations in density, as well as a whole set of structures that did not show circular symmetry. In particular, radial structures (or "spokes" were observed, dark in reflected light but bright in scattered, which evolved rapidly with time. With lengths that could exceed 10 000 km, they appeared in a few minutes and lasted for some tens of hours at the most. They may consist of minute (sub-micron) particles elevated out of the plane of the rings by magnetic storms.

Just as some rings lie beyond the Roche limit, so small satellites have been discovered lying inside it. With the exception of these few larger objects, the blocks forming the rings are smaller than a kilometre in size. This dimension is not just the maximum dimension of the blocks themselves but also the maximum thickness of the disk formed by the rings: the dynamical processes occurring in the rings are highly efficient in confining the particles to a single plane, and this is particularly evident when one compares the thickness of the disk with its diameter, which is of the order of 300 000 km.

The size distribution of the particle diameters is still poorly known, but it would appear that it ranges from a micrometre to a kilometre. As in the case of Jupiter's rings the presence of tiny particles, whose lifetime in Saturn's environment is short, indicates the existence of suitable reservoirs and repeated replenishment.

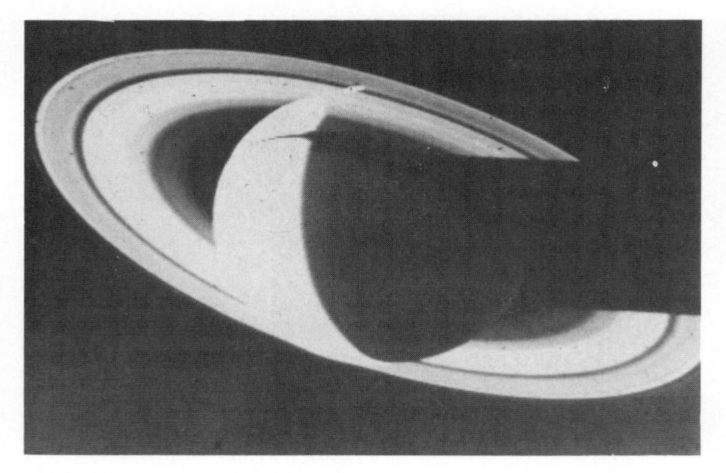

Fig. 8.13. Saturn and its system of rings as seen by Voyager 1 as the probe was receding from the planet. (By kind permission of NASA)

More generally, the rings, which probably form a structure that is stable on a long time-scale similar to that of the planets themselves, are subjected to dynamical evolution on a very short time-scale. In particular, gravitational interactions with the small satellites and the larger blocks are responsible for temporary variations in local density, irregularities in the structure, and spiral density waves. It has been possible to show that the overall effect of the interactions, operating on particles that are subject to repeated collisions, finally leads to stable confinement of the particles.

The chemical composition of the particles can be determined, although only in approximate qualitative terms, from the optical properties and their variations with phase angle. First, the high albedo favours particles covered with a shell of ice, agreeing with the low temperature of the rings – which varies from 70 to 90 K according to the degree of insolation – and also with the low density of Saturn's satellites. Second, important differences in colour appear between the various rings, although it has not yet been possible to link these with the chemical composition of the grains. On the other hand these variations indicate that transport of particles from one ring to another is not a dominant process, which confirms the high degree of stability in Saturn's system of rings (Fig. 8.13).

8.4.4 The Rings of Uranus

The rings of Uranus were discovered on 1977 March 10, during a stellar occultation by the planet. These have since been studied by the same means, and the first direct photograph was taken in 1984 from the Las Campanas Observatory in Chile, using a very sensitive camera which is scheduled to be fitted later to the focus of the Space Telescope. A total of nine narrow rings were identified, lying between 42 000 and 52 000 km from the centre of the planet. They are known, in increasing order of distance from Uranus, first by the numbers 6, 5, and 4, and then by the Greek letters α, β, η, γ, δ and ε. Each is less than 100 km in width, and eight of them are less than 10 km. Their albedo is particularly low, being less than 5 %.

When Voyager 2 flew past Uranus on 1986 January 24, numerous details were obtained. Two small satellites were discovered, one on each side of the principal ring, ε, which thus appears similar to the F ring of Saturn, discovered by Voyager 1 in 1981 December. This ring ε has an average width of thirty-six kilometres. It appears to consist mainly of large-sized particles, i.e. larger than a centimetre and perhaps reaching or surpassing one metre. This distinguishes it from Saturn's rings.

Voyager 2 also discovered 2 new rings: the tenth, very narrow, lying between rings ε and δ, and the eleventh, more diffuse, within ring 6, which was the closest to Uranus of the rings previously known.

The very low albedo of the material in the rings may be caused by the presence of organic polymers, which might result from the irradiation of icy particles containing hydrocarbons by magnetospheric winds.

8.4.5 Neptune's Rings

Several observations of stellar occultations carried out since 1984 seemed to indicate the existence of a ring surrounding Neptune. One difficulty in the interpretations of the experiments was that the ring thus detected did not appear to be complete, but fragmentary: it might be just an arc. It was suggested that its width was 20 000 km and its radius 55 000 km. A spectacular confirmation of these results has been provided by Voyager 2's flyby of Neptune. In fact, the arcs do belong to entire rings; the density of matter is significantly larger in the parts of the rings (the "arcs") that had been detected from the Earth. Voyager 2 has detected 2 main, narrow rings, with a width of about 10–15 km, and 3 other, fainter rings, wider and probably composed of smaller particles.

9. Satellites with Atmospheres

Apart from Io, where the atmosphere of sulphur sporadically emitted by the volcanoes is not stable, there are only two satellites, Titan and Triton, that are definitely surrounded by atmospheres. These two objects apparently have little in common. The dense atmosphere of Titan has been known to astronomers for a long time. On the other hand, recent observations of Triton have revealed the possible existence of a tenuous nitrogen atmosphere around the globe. This hypothesis has been confirmed by the Voyager 2 observations.

9.1 Titan

After Ganymede, Titan is the largest satellite in the Solar System, distinctly larger than the planet Mercury. It was in 1908 that visual observations of a limb-darkening effect by Comas Sola first indicated the probable presence of a dense atmosphere on Titan. In 1944, Kuiper identified methane in the atmosphere, using near-infrared spectra. Beginning in 1970, Titan was the subject of an intensive campaign of observations from Earth. At the same time, in 1970, on the basis of thermochemical models, Lewis suggested that nitrogen was possibly present in Titan's atmosphere. The interpretation of the spectra remained ambiguous, however. The surface of Titan cannot be observed directly, because of the dense layer of clouds, brownish-orange in colour, which covers the disk. Although the pressure at the top of the cloud layer could be estimated from the spectroscopic measurements, great uncertainty remained about the physical conditions on the surface. The Voyager mission in 1981 allowed this information to be determined, as well as the atmospheric composition of Titan and details of its environment.

9.1.1 Titan's Neutral Atmosphere

The Chemical Composition of the Atmosphere

Apart from methane, several molecules were detected on Titan before the Voyager fly-by, thanks to infrared spectroscopy. These were C_2H_6, CH_3D, and C_2H_2. After the Voyager observations the following were added to the list: N_2 – which forms by far the major component – observed in the ultraviolet (Fig. 9.1), then H_2, HCN, C_3H_8, C_2H_4, C_3H_4, C_4H_2, HC_3N, C_2N_2, and CO_2, these components being identified by the IRIS infrared spectrometer (Fig. 9.2). The relative abundances of the

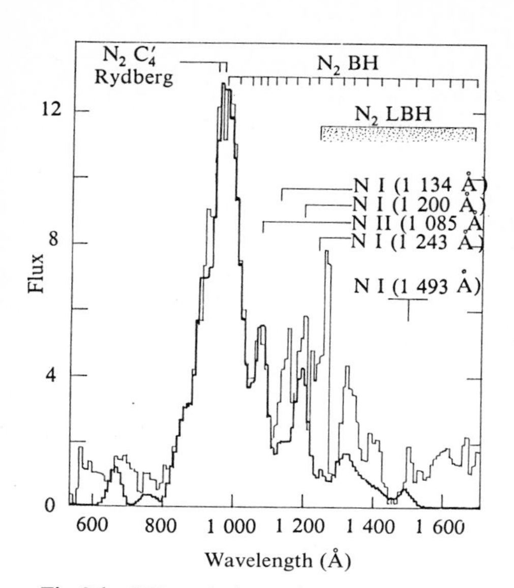

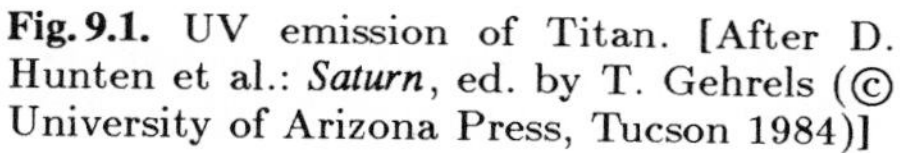

Fig. 9.1. UV emission of Titan. [After D. Hunten et al.: *Saturn*, ed. by T. Gehrels (© University of Arizona Press, Tucson 1984)]

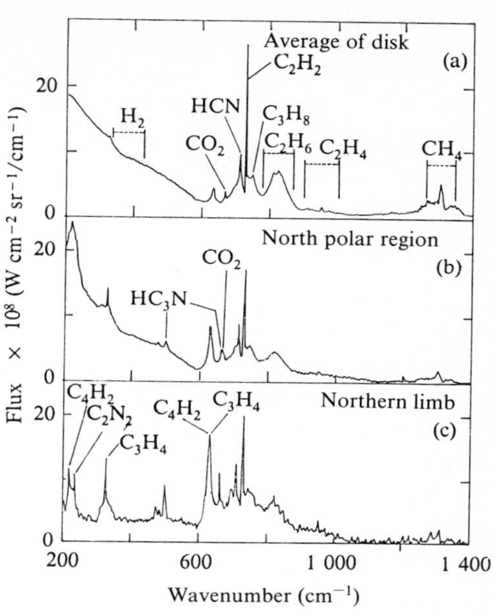

Fig. 9.2. Infrared spectra of Titan, as recorded by the Voyager IRIS experiment. [After R. Samuelson et al.: Journal of Geophysical Research **88**, 8709 (1983)]

gases are indicated in Table 9.1. In addition, by comparison between the Voyager results in the infrared and radio observations, is has been possible to determine indirectly that argon is present, and also its relative abundance. The mean molecular mass of Titan's atmosphere, as determined by the Voyager radio-occultation experiment, is 28.6. It is therefore necessary to assume that a heavier gas than nitrogen is present; argon seems to be the only heavy gas with a saturated vapour pressure sufficient to explain the mean molecular mass. The A/N_2 ratio would then be of the order of 0.15. Finally, after the Voyager fly-past, another molecule, CO, has been observed from Earth in near-infrared spectra.

The question of the origin of Titan's atmosphere raises an interesting problem, because it touches upon the mechanisms by which accretion occurred at great distances from the Sun, at the time of the formation of the Solar System. At very low temperatures certain molecules condense as *clathrates* (see Sect. 9.1.3), of the form (x,nH_2O). This is particularly true for CH_4, N_2, CO, and A. The presence of all these molecules on Titan suggests that the atmosphere is probably secondary, i.e. outgassed from the body and the clathrates of which it is formed. The theory is reinforced by measurements of the density of the solid body of Titan, which is remarkably low (1.9 g/cm^3), which is an indication of the presence of an important amount of ice in its chemical composition.

After methane, the four most abundant hydrocarbons are, in order: C_2H_6, C_3H_8, C_2H_2, and C_2H_4; the abundances appear to vary slightly with latitude. All these compounds condense at around the temperature minimum. The five less-abundant

Table 9.1. Chemical composition of Titan's atmosphere (after D. Hunten et al., *Saturn,* T. Gehrels, ed., Tucson, University of Arizona Press, © 1984)

Molecule	Spectral region (μm)	Year of first detection	Abundance
N_2	0.1	1981	0.65–0.98
A	indirect detection	1981	0–0.25
CH_4	0.8	1944	0.02–0.10
H_2	0.6	1975	2×10^{-3}
C_2H_2	13.7	1975	2×10^{-6}
C_2H6	12.1	1975	2×10^{-5}
C_2H_4	10.5	1975	4×10^{-7}
CH_3D	8.6	1975	
C_3H_4	15.8, 30.8	1981	3×10^{-8}
C_3H_8	13.4	1981	4×10^{-6}
HCN	14.0	1981	2×10^{-7}
C_2N_2	42.9	1981	10^{-8}–10^{-7}
HC_3N	20, 15.1	1981	10^{-8}–10^{-7}
C_4H_2	45.4, 15.9	1981	10^{-8}–10^{-7}
CO_2	15.0	1982	1.5×10^{-9}
CO	1.6	1983	6×10^{-5}–1.5×10^{-4}

hydrocarbons show strong variations in concentration with latitude, maximum occurring towards the north pole. This effect, which is probably seasonal, is not fully understood at present. On the other hand CO_2 appears to be evenly distributed in latitude.

One specific feature of Titan's atmosphere is that it is in a state of permanent evolution. Methane is liberated from the surface, part of which is probably liquid; some of the methane must return to the pools of liquid in the form of precipitation, whilst some escapes upwards. The latter is subjected to photodissociation by the solar ultraviolet flux, which gives birth to the various hydrocarbons observed; these products descend to cooler atmospheric layers where they condense and fall back onto the surface as precipitation. Only ethane C_2H_6 and propane C_3H_8 probably remain liquid, mixed in with the liquid methane.

The Thermal Structure of the Atmosphere

Results from the Voyager radio occultation have enabled the thermal structure of the lower atmosphere to be determined. In particular they allowed conditions at the surface to be established: a pressure of 1.5 ± 0.2 bar and a temperature of about 94 K. The temperature profiles determined by this method depend, however, on the mean molecular mass selected (in the present case, $m = 28$). Such profiles are valid between the surface and the 10-mbar pressure level, but beyond that the inaccuracy inherent in the initial choice of conditions becomes too great. Complementary information about the thermal structure is provided by the IRIS infrared-spectrometer experiments. The profile derived from the two experiments is shown in Fig. 9.3.

The IRIS experiment also allowed the evolution of the thermal structure as a function of latitude to be studied. This structure is symmetrical with respect to the equator, which is about 3 K warmer than the poles. Temperature differences between

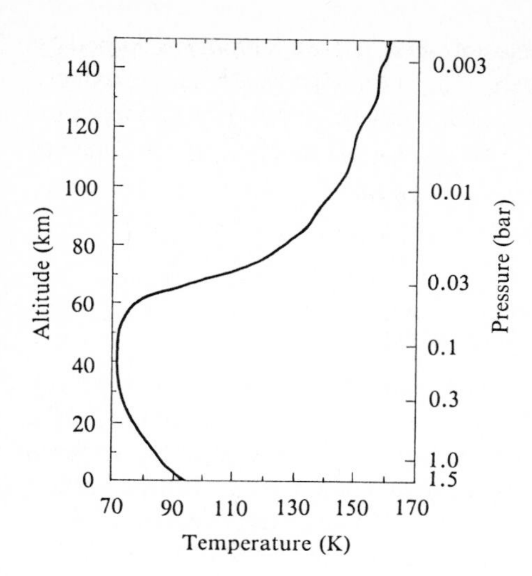

Fig. 9.3. The thermal profile of Titan's atmosphere. [After R. Samuelson: Icarus **53**, 364 (1983)]

day and night are small, of the order of 1 K. But a slight asymmetry between the two hemispheres should be noted, the southern hemisphere being cooler by 1 K. This difference is perhaps related to the asymmetry observed in the visible, where the southern hemisphere is slightly brighter. These effects may be caused by clouds of different thicknesses in the northern and southern hemispheres (Fig. 9.4).

At the surface, the temperature gradient is equal to the adiabatic gradient for a nitrogen atmosphere; it then declines to zero at the tropopause. It would appear that Titan's thermal structure is governed by radiative equilibrium. Recently, Samuelson has been able to account for the observed structure in a satisfactory way by using a simple radiative-transfer model. He separates the incident solar flux into two components, one (in the ultraviolet) is absorbed by the atmosphere and contributes to its

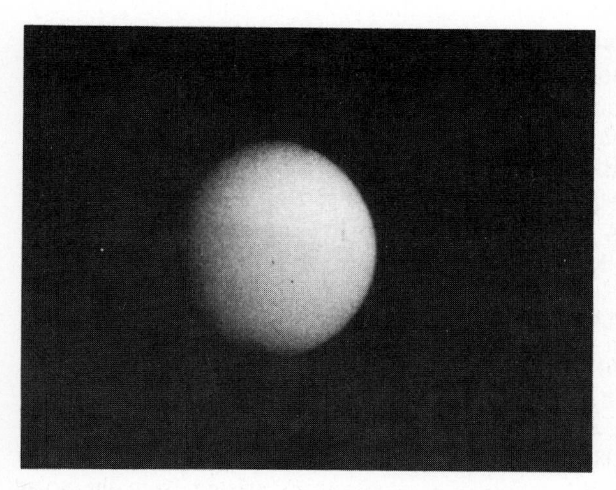

Fig. 9.4. Titan, as imaged by Voyager 1. [B. Smith et al.: "Encounter with Saturn: Voyager 1 Imaging Science Results", Science, Vol. 212, 1981 April 10 (© AAAS 1981)]

heating, and the other (in the visible and near infrared) is preferentially absorbed by the surface after scattering in the atmosphere. The ultraviolet component produces the temperature-inversion effect, whilst the visible and near-infrared component produces the greenhouse effect. According to this model, 10 to 30% of the incident solar flux reaches the surface of Titan.

9.1.2 Titan's Upper Atmosphere

Composition and Structure of the Upper Atmosphere

The structure of Titan's upper atmosphere was determined by means of the UVS (ultraviolet spectrometer) experiment on the Voyager probe. The observation of a stellar occultation, which was measured through different filters corresponding to the dissociation and ionization energies of various atmospheric components, allowed vertical distribution profiles for these to be determined. In particular, the occultation light-curve revealed the presence of several absorbing layers in the upper atmosphere, which are doubtless of molecular origin. From the measurements, the altitude of the homopause has been determined (3500 km), as well as the value of the *eddy diffusion coefficient, K*, which is of the order of $10^8\,cm^2\,s^{-1}$, significantly higher than for the giant planets. In the lower stratosphere the coefficient K decreases to $10^3\,cm^2\,s^{-1}$, which confirms that Titan's atmosphere is very stable at the temperature inversion. The occultation experiment showed that CH_4 and C_2H_2 were the principal absorbants; the loss of CH_4 must therefore be compensated by photolysis; whence the evolutionary model described in Sect. 9.1.1.

Another remarkable result obtained by the UVS spectrometry is the upper limit determined for the Ne/N_2 and A/N_2 ratios: 0.01 and 0.06 respectively, at an altitude of 3900 km. However, this result does not contradict the values given in Table 9.1 concerning Titan's atmospheric composition. In fact, using the atmospheric models deduced from all the measurements, the upper limits measured imply that at the surface the Ne/N_2 and A/N_2 ratios should be less than 0.002, 0.6 and 0.3 respectively. The absence of neon is therefore definitely confirmed, whilst the presence of argon is very likely. It only remains to determine whether the argon is present as argon-36 or argon-40: the reply to this question will give a clue as to its origin. If the argon has been outgassed from Titan, where it was trapped at the time of the formation of the Solar System, then it should be primordial and thus mainly in the form of ^{36}A, with ^{38}A amounting to about 20%, and argon-40 should be absent. We should remember that most of the ^{40}A in the atmospheres of the inner planets arises from radioactive disintegration of ^{40}K.

Interaction with the Magnetosphere

We have already mentioned Voyager's discovery of ultraviolet emission from nitrogen (Fig. 9.1). Apart from being the first direct indication of the presence of nitrogen as an atmospheric component, this result has another important consequence. The presence of N^+ emission at 1085 Å as well as the Rydberg bands of N_2 at 958 and 981 Å cannot be explained by the action of photoelectrons arising from the solar ultraviolet flux. These emissions are doubtless caused by the presence of energetic electrons from the magnetospheric plasma, which dissociate the N_2 molecules and partially ionize the individual N atoms that are liberated. More generally, it can be shown that the energy necessary to excite the whole of Titan's ultraviolet spectrum between 600 and 1400 Å is about five times as great as the amount of energy provided by the solar ultraviolet flux. It is possible to estimate the amount of energy received by Titan's atmosphere through its interaction with the magnetospheric plasma; this seems to be sufficient to explain the observed emissions.

Another consequence of the impact of the electrons that result from this interaction concerns the escape of nitrogen atoms. It can be shown that in Titan's *thermosphere*, the temperature is sufficiently high to allow H and H_2 to escape. This is the origin of the *hydrogen torus* that lies around Titan's orbit and extends a considerable distance out into space. Voyager observations of its Lyman-α emission enabled it to be established that this torus extends from 8 to 25 saturnian radii (the orbit of Titan itself lies at 20 R_S). The bombardment by electrons present in the magnetospheric plasma is the cause of the dissociation of N_2 and of the escape of N atoms, which also go to populate the torus. There is also the creation of N_2^+ ions, which play an important part in Titan's photochemistry.

Titan's Aeronomy and the Formation of Complex Molecules

The photochemistry of Titan, like those of the giant planets, has been the subject of numerous studies. That of Titan is particularly interesting, because it may lead to the formation of complex molecules, thereby creating a medium that could be favourable to the building of prebiotic compounds.

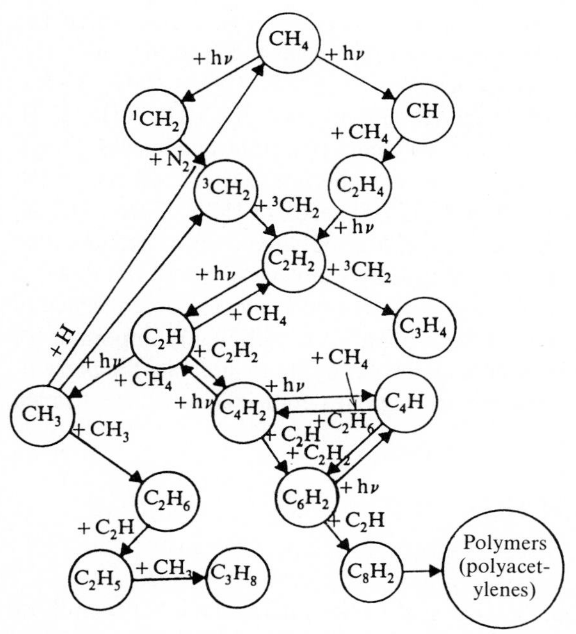

Fig. 9.5. The photochemistry of CH_4 in Titan's atmosphere. [After D. Strobel: Planetary and Space Science **30**, 839 (© Pergamon Journals Ltd. 1982)]

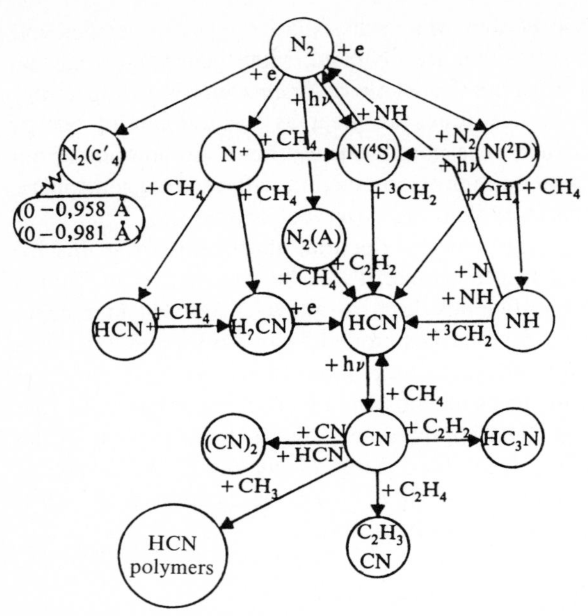

Fig. 9.6. The photochemistry of N_2 in Titan's atmosphere. [After D. Strobel: Planetary and Space Science **30**, 839 (© Pergamon Journals Ltd. 1982)]

Titan's photochemistry is based partly on photochemistry of hydrocarbons that result from the dissociation of CH_4, and partly on the dissociation of N_2. The dissociation of CO and CO_2 also plays a part. All these mechanisms have been studied in detail, particularly by Strobel. The CH_4 photochemistry leads to the formation in significant quantities of the following components (see Fig. 9.5): C_2H_4, C_2H_2, C_3H_4, C_2H_6, C_3H_8, C_4H_2, all of which are observed, and additionally: CH_2, CH_3, C_2H, C_4H, C_6H_2 and C_8H_2, leading to chains of polymers (the polyacetylenes $C_{2n}H_2$). From the dissociation of N_2 by energetic magnetospheric electrons, we obtain the N^+ ion, which itself dissociates CH_4 (Fig. 9.6). This leads to HCN, C_2N_2, HC_3N, all of which are observed, and thence to the $(HCN)_n$ polymers, up to adenine, one of the four bases involved in the structure of the DNA molecule. Could life develop on the surface of Titan? There are numerous problems (low temperature, absence of aqueous media, etc.) but the question has been posed. It is sufficiently important for numerous laboratory studies to have been undertaken, trying to simulate the present-day environment on Titan and to analyze the complex molecules that are formed under these conditions.

9.1.3 The Interior and Surface of Titan

The Clathrates

The physical characteristics of Titan's solid surface are intermediate between those of Ganymede and Callisto (see Table 9.2). The fraction of ice in Titan should be of the order of 0.3 to 0.45, according to whether the silicates are hydrated or not. The

Table 9.2. Comparison of Callisto, Titan and Ganymede

	Callisto	Titan	Ganymede
Mass (10^{26} g)	1.075	1.345	1.482
Radius (km)	2410 (±10)	2575 (±0.5)	2638 (±10)
Mean density (g cm^{-3})	1.83 (±0.02)	1.881 (±0.005)	1.93 (±0.02)
Fraction in ice (by mass) (Hydrated silicates)	0.34	0.30	0.27
Fraction in ice (by mass) (Anhydrous silicates)	0.475	0.45	0.42

Hypotheses: $\bar{\varrho}$ ice = 1.2 g cm^{-3}
$\bar{\varrho}$ (hydrated silicates) = 2.5 g cm^{-3}
$\bar{\varrho}$ (anhydrous silicates + iron) = 3.5 g cm^{-3}
(After D. Hunten et al., *Saturn,* T. Gehrels, ed., Tucson, University of Arizona Press, © 1984)

similarity in the physical conditions does not mean, however, that the composition of the interiors of the three satellites is identical. Indeed, other ices of the type ($NH_3 \cdot H_2O$) or ($CH_4 \cdot H_2O$) have densities very close to that of water. Moreover, Titan formed at a very low temperature. Below 100 K, clathrates are liable to appear. A clathrate is a crystalline network – in this case H_2O – that contains cavities within which foreign molecules such as CH_4, N_2, CO, A, etc. are trapped. We may therefore assume that the current composition of Titan's atmosphere is defined by the nature of the molecules that were trapped within the clathrates. Theory shows that Titan's present atmosphere corresponds to equilibrium, at about 200 K, for a clathrate containing comparable abundances of N_2 and CH_4. We can imagine that Titan went through a phase at this temperature during which it had seas consisting of water and ammonia beneath an atmosphere rich in CH_4 and containing some N_2, perhaps produced by photolysis of NH_3. As the temperature declined, clathrates formed, trapping most of the methane and leaving behind, in qualitative terms, the atmosphere that we see today.

Many questions remain unanswered. The existence of a deep sea of pure methane, in particular, is difficult to explain, because according to the CH_4–N_2 phase diagram, it would imply the presence of large quantities of liquid nitrogen, the origin of which is difficult to envisage. On the other hand, the sea of CH_4 could have been enriched in ethane C_2H_6, which would modify the phase diagram and would meet the difficulty. Ethane could arise quite naturally from photolysis of CH_4. It should be noted that the permanent nature of this photolysis would appear to indicate that clathrates containing CH_4 continuously decompose, creating a permanent flux of gaseous CH_4 into the upper atmosphere (see Sects. 9.1.1 and 9.1.2).

Possible Modes of Formation

Three types of formation may be envisaged:

1. accretion at low temperature (i.e. corresponding to negligible heating);
2. accretion with heating, in the presence of a dense gaseous phase (in this case the energy arising from the accretion should be lost by convection via the dense atmosphere);
3. accretion with heating, but without a dense gaseous phase.

Table 9.3. Models of the formation of Titan

Composition of the body	Rocks $+ H_2O$	Rocks $+ H_2O$ $+ NH_3$	Rocks $+ H_2O + NH_3$ + volatile layer (CH_4)	Rocks $+ H_2O + NH_3$ + clathrates or methane
Accretion at low temperature	Final state like Callisto; atmosphere not explained	H_2O–NH_3 volcanism; atmosphere not explained	Atmosphere possible, but CH_4/N_2 unexplained	CH_4 degassing; N_2/CH_4 too low
Accretion at high temperature without gases	Final state like Ganymede; atmosphere not explained	H_2O–NH_3 volcanism + ocean; atmosphere not explained	Atmosphere possible. Difficult to explain production of N_2 from NH_3	Production of N_2 insufficient; CH_4/N_2 unexplained
Accretion at high temperature with gases	Final state like Ganymede; atmosphere not explained	NH_3–H_2O atmosphere + ocean; CH_4 not explained	Production of N_2 possible; adequate CH_4 abundance	massive atmosphere; adequate production of N_2

(After D. Hunten et al., *Saturn,* T. Gehrels, ed., Tucson, University of Arizona Press, © 1984)

In each of these three cases, we can envisage either homogeneous accretion (rocks and ices mixed from the start) or heterogeneous accretion (the formation of a rocky core and then the accumulation of ices). It is very unlikely that accretion could have occurred at a low temperature; it would have required a very slow mechanism, contradicting the general theory for the formation of satellites, which suggests rapid accretion. In the same way, it is probable that accretion, as for the Galilean satellites, was heterogeneous. Even if the accretion was initially homogeneous, differentiation would doubtless have occurred during the course of the body's evolution. But we are still reduced to theorizing. Table 9.3 gives, for example, a whole series of possible formation and evolutionary models; one of these is illustrated in Fig. 9.7.

Finally it may be said that Titan's peculiarities and its unique nature among Solar-System bodies make it one of the most fascinating objects to be explored in future. Numerous fundamental questions remain unanswered: where does the

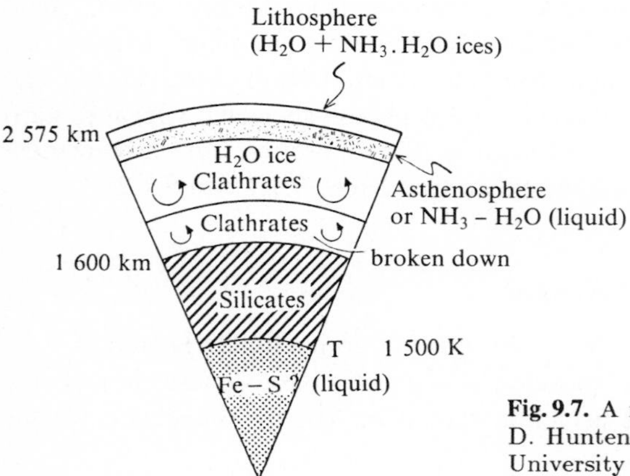

Fig. 9.7. A model of the interior of Titan. [After D. Hunten et al.: *Saturn*, ed. by T. Gehrels (© University of Arizona Press, Tucson 1984)]

nitrogen in the atmosphere come from? What is the nature of the surface, liquid or solid? What are the mechanisms and peculiarities of the atmospheric organic chemistry? It is probable that thorough study of Titan by a space mission, some time early in the 21st century, will provide the replies to a number of questions about the mechanisms governing the accretion, formation and evolution of solid bodies at great distances from the Sun.

9.2 Triton

Triton is the largest of Neptune's satellites. Because of its mass and the low temperature of its surroundings, it should be able to retain an atmosphere of heavy gases; this theory, proposed a long time ago, has been the subject of much interest since the recent discoveries concerning Titan's atmosphere.

Early spectroscopic research concentrated on methane, and dated from 1970. Until the end of the 1970s, results were negative. Then in 1978 and 1979 several authors detected weak absorption at 8900 Å, 1.7 μm and 2.3 μm, which they attributed to gaseous methane. However, CH_4 ice also shows absorption bands at the same wavelengths, so it is difficult to be specific about the state in which the methane exists. An important contribution was made by Cruikshank and Apt in 1984. They recorded the spectrum of Triton between 0.8 and 2.5 μm – in which they detected six CH_4 bands – and compared it with different solid and gaseous absorption models. This research showed that the methane is primarily in solid form; variations with time in the intensity of the bands suggest, on the other hand, that the CH_4 ice is not uniformly distributed over the surface of Triton.

The spectroscopy of Triton in the near-infrared range has provided another result. An unexpected absorption band at 2.16 μm, detected by Cruikshank and his colleagues, has been attributed to liquid, or even solid, nitrogen (Fig. 9.8). As the saturated vapour pressure of nitrogen is larger than that of methane by a factor of 10 to 100, these authors suggested the presence of a nitrogen atmosphere around Triton. The infrared spectrum of Triton also revealed the presence of possible water ice on its surface.

The Voyager 2 observations, at the time of Triton's flyby on 26 August 1989, have confirmed the existence of a nitrogen atmosphere, but they have also demonstrated that this atmosphere is very tenuous. The UV emission bands of nitrogen, previously observed on Titan, have been identified on Triton by the UVS experiment, between 1000 and 1100 Å. Moreover, the UV occultation experiment has revealed a nitrogen atmosphere extending over 800 km, with a pressure of 10 microbars at the surface. The methane was detected only in the vicinity of the surface, with a partial pressure of 1 nanobar. According to first estimates, the surface temperature is 37 K; it decreases over the first 50 km, and then increases again to reach about 100 K at a few hundred km. The Voyager images have detected a haze at an altitude of 50 km, probably composed of frozen hydrocarbons and possibly nitriles, photodissociation products of methane and nitrogen.

As Triton's atmosphere is very tenuous, the surface of the satellite has been investigated in detail by the imaging experiment of Voyager. The albedo is surpris-

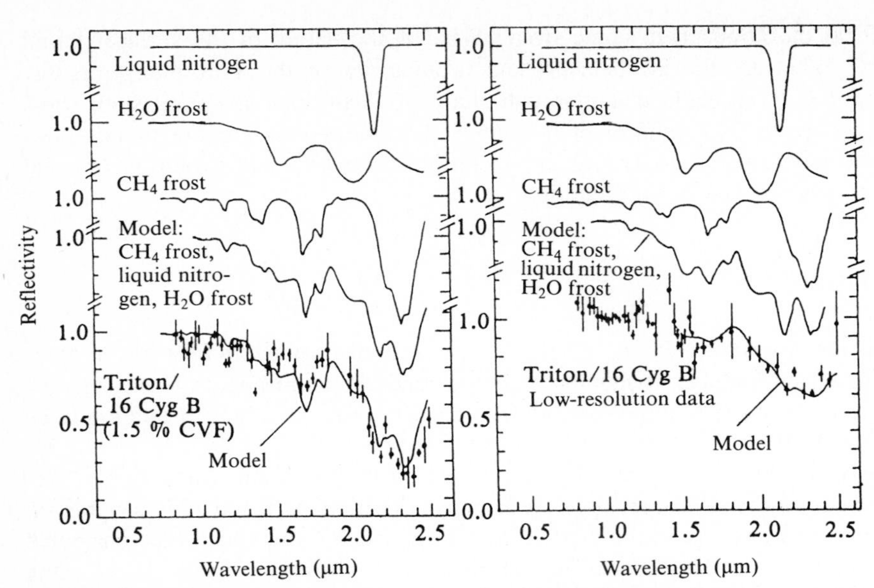

Fig. 9.8. Spectra of Triton in the near infrared. In both cases the best fit is obtained by a model that includes H_2O frost, CH_4 frost, and liquid nitrogen. [After D. Cruikshank et al.: Icarus **58**, 293 (1984)]

ingly high, and reaches 0.9 in some areas; it is probably due mostly to methane or nitrogen ice. There is little relief, with no pattern detected above a height of a few hundred meters. There are few craters, which indicates a relatively young surface. The Voyager images have revealed traces of flows, and craters filled with frozen liquid – which are indications of recent volcanic activity – based on water, nitrogen and methane. This implies that Triton's temperature had to be higher in the past. This volcanic activity could still be present in the form of geysers, as possibly indicated by some images; but this has not yet been confirmed.

10. The Comets

Our knowledge of comets dates back to antiquity. Some are, or have been, among the most spectacular objects in the Solar System, and they have been the subject of innumerable studies. The mysteries surrounding their origin, the often unpredictable nature of their apparitions, their movements – which remained inexplicable for a very long time – and finally their changes in appearance, all contributed to the legends, and even superstitions, that built up around them. From the most ancient times, chronicles tell us that on numerous occasions comets were seen as threatening portents of catastrophes (wars, famines and epidemics), or of the deaths of important persons. These superstitions die hard, and persisted long after the orbits and periodic return of comets were explained by Halley at the beginning of the 18th century, using the equations of celestial mechanics. Despite having lost much of their mystery, comets nevertheless remain fascinating objects for research to scientists. Comets appear to be unaltered remnants of the Solar System's primitive material, the analysis of which has unique interest for the study of the physical and chemical conditions prevailing in the primordial solar nebula.

Fig. 10.1. Comet Mrkos, photographed in 1957

10.1 The Nomenclature of Comets

The number of comets observed is growing rapidly, which is why the International Astronomical Union (IAU) has adopted a precise system of nomenclature. When a comet is observed for the first time, or is recovered, it is given a provisional designation, consisting of the year followed by a letter, awarded chronologically and in alphabetical order (1985a, 1985b, etc.). After the orbital elements have been calculated, and after a delay of about two years, this designation is replaced with another, definitive one. This consists of the year of the comet's perihelion passage, followed by a number, in roman numerals, corresponding to the chronological order of perihelion passage (1985I, 1985II, etc.). Periodic comets therefore receive a new designation at each perihelion passage.

Apart from these numbers, comets are also called by the name of their first discoverer, or discoverers. In a few specific cases, the name of the person who first calculated the orbit is used; this is the case for Halley and Encke, among others. Finally, comets discovered by the infrared satellite IRAS carry the name IRAS.

For periodic comets with periods less than 200 years, the name of the discoverer is preceded by P/; if several comets have the same name, they are followed by a number. This applies, for example, to P/Temple 1 and P/Temple 2. Within this group there are the comets of the same type as Halley (20 years $< P < 200$ years) and Jupiter's family of comets (<20 years), the latter being so called because their orbits are significantly influenced by Jupiter. The most recent catalogue of comets is that of Marsden (1986). The number of apparitions catalogued is 1187, corresponding to 748 individual comets, 135 of which are short-period comets; 85 of them have been observed twice or more. Table 10.1 gives a list of the best-known periodic comets.

Table 10.1. The best-known periodic comets

No.	Name	P Sidereal period (year)	q Perihelion distance (AU)	Q Aphelion distance (AU)
1	Encke	3.305	0.340 66	4.097 0
2	Grigg-Skjellerup	5.100	0.993 38	4.932 1
3	Tempel 2	5.270	1.369 39	4.687 6
4	Honda-Mrkos-Pajdusakova	5.279	0.578 87	5.485 1
5	Neujmin 2	5.43	1.338 18	4.840 5
6	Brorsen	5.46	0.589 85	5.613 2
7	Tempel 1	5.498	1.496 86	4.733 6
8	Tuttle-Giacobini-Kresak	5.56	1.151 92	5.125 6
9	Tempel Swift	5.68	1.152 56	5.217 3
10	Wirtanen	5.875	1.256 06	5.255 6
11	D'Arrest	6.228	1.164 00	5.606 4
12	Du Toit-Neujmin-Delporte	6.306	1.677 05	5.149 7
13	De Vico-Swift	6.31	1.624 28	5.207 3
14	Pons-Winnecke	6.362	1.254 22	5.612 6
15	Forbes	6.397	1.533 07	5.358 8
16	Kopff	6.430	1.571 99	5.343 3
17	Schwassmann-Wachmann 2	6.509	2.142 19	4.830 3
18	Giacobini-Zinner	6.517	0.993 99	5.983 6

Table 10.1 (continued)

No.	Name	P Sidereal period (year)	q Perihelion distance (AU)	Q Aphelion distance (AU)
19	Wolf-Harrington	6.551	1.621 79	5.380 3
20	Churyumov-Gerasimenko	6.588	1.298 49	5.730 3
21	Biela (nucleus A)	6.62	0.860 59	6.188 5
22	Tsuchinshan 1	6.64	1.492 51	5.570 6
23	Perrine-Mrkos	6.72	1.272 21	5.847 6
24	Reinmuth 2	6.736	1.941 05	5.192 3
25	Johnson	6.763	2.195 66	4.957 1
26	Borrelly	6.765	1.316 46	5.837 2
27	Tsuchinshan 2	6.80	1.774 95	5.403 2
28	Harrington	6.802	1.582 45	5.597 5
29	Gunn	6.803	2.444 93	4.735 6
30	Arend-Rigaux	6.838	1.444 37	5.761 3
31	Brooks 2	6.878	1.840 04	5.393 0
32	Finlay	6.953	1.095 88	6.190 1
33	Taylor (nucleus B)	6.97	1.951 72	5.349 1
34	Holmes	7.05	2.155 13	5.195 9
35	Daniel	7.09	1.661 66	5.722 2
36	Shajn-Schaldach	7.27	2.227 49	5.276 1
37	Faye	7.388	1.609 77	5.976 5
38	Ashbrook-Jackson	7.425	2.284 21	5.327 9
39	Whipple	7.440	2.468 63	5.153 5
40	Harrington-Abell	7.581	1.773 77	5.944 6
41	Reinmuth 1	7.631	1.994 74	5.757 2
42	Oterma	7.88	3.387 82	4.532 2
43	Arend	7.984	1.846 91	6.142 1
44	Schaumasse	8.18	1.195 94	6.924 2
45	Jackson-Neujmin	8.39	1.427 75	6.829 0
46	Wolf	8.43	2.506 05	5.777 2
47	Comas Solà	8.55	1.768 77	6.592 8
48	Kearns-Kwee	9.01	2.228 58	6.430 1
49	Swift-Gehrels	9.23	1.353.85	7.443 7
50	Neujmin 3	10.57	1.976 33	7.657 5
51	Klemola	10.97	1.765 91	8.107 5
52	Gale	10.99	1.182 91	8.704 3
53	Väisälä 1	11.28	1.866 13	8.192 8
54	Slaughter-Burnham	11.62	2.543 36	7.716 3
55	Van Biesbroeck	12.41	2.409 69	8.310 0
56	Wild	13.29	1.980 42	9.242 0
57	Tuttle	13.77	1.022 93	10.464 9
58	Du Toit 1	14.976	1.293 97	10.857 7
59	Schwassmann-Wachmann 1	15.03	5.447 49	6.731 5
60	Neujmin 1	17.93	1.542 57	12.158 2
61	Crommelin	27.89	0.743 36	17.649 8
62	Tempel-Tuttle	32.91	0.981 73	19.537
63	Coggia-Stephan	38.84	1.595 29	21.344
64	Westphal	61.86	1.254 02	30.03
65	Olbers	69.47	1.178 66	32.62
66	Pons-Brooks	70.98	0.773 88	33.51
67	Brorsen-Metcalf	71.93	0.484 89	34.10
68	Halley	76.09	0.587 21	35.33
69	Herschel-Rigollet	155	0.748 49	56.94

(After *L'Encyclopédie Scientifique de l'Univers,* Bureau des longitudes, Gauthier-Villars, 1979.)

10.2 Cometary Orbits and the Problem of Their Origin

As we have seen earlier (Sect. 1.1.1), the path of a comet – like that of any body in the Sun's gravitational field – is a conic section, which may be described by its *orbital elements* (see Fig. 1.1):

1. the time T of perihelion passage
2. the distance q of the comet at perihelion
3. the argument of perihelion ω
4. the longitude Ω of the ascending node
5. the inclination i of the orbit
6. the eccentricity e of the conic section.

An initial determination of the orbit of a new comet is, in principle, possible with just three astrometric measurements of the comet, if these are well-spaced in time. As most new comets have quasi-parabolic orbits, calculations are initially made assuming that e is equal to one. With further observations of the object, the parameters are refined by using the least-squares method; then when measurements extending over a period of several months are available, departures from the parabolic curve become obvious, and the eccentricity can be calculated; then planetary perturbations have to be taken into account. Cometary paths and coming apparitions can thus be predicted several decades in advance. In the case of well-known periodic comets, the positions can be calculated to a precision of several seconds of arc; for example, P/Halley was rediscovered on 1982 October 2 at magnitude 24, less than 10 arc-seconds from its theoretical position, calculated from observations of all its previous apparitions. Table 10.2 gives a list of periodic comets, whose perihelion passage ("return") will take place before the year 2000.

Taking the effects of planetary perturbations on cometary orbits into account has also allowed the initial cometary orbits to be calculated, before these were modified by interactions with the planets. This was how Oort, in 1950, by examining about twenty long-period comets, came to the conclusion that a vast cometary reservoir exists, lying at 50 000 to 100 000 AU from the Sun. More recent work by Marsden, which involves more than two hundred comets, confirms this result. It is a remarkable fact that where sufficiently precise calculation has been possible, all the initial paths have proved to be very elongated ellipses. Comets do not, therefore, enter from interstellar space, but instead belong to the Solar System, although their source is a very distant region, known as the *Oort Cloud* (Fig. 10.2). According to Oort's theory, generally accepted today, the Oort Cloud may contain some 10^{11} comets. Perturbations caused by nearby stars may cause 5 to 10 % of the total cometary population to be ejected from the cloud. When a comet from the Oort Cloud enters sufficiently close to be subject to planetary perturbations, it may be turned into a short-period comet. This means that we can never expect to see more than a minute fraction of the total cometary population.

How can we explain the presence of comets in the Oort Cloud? At present there is no definite answer to this question. Two theories have been put forward; in the first, comets were formed close to the planets, some tens of astronomical units from the Sun. They would have been subsequently ejected out to the present-day

Table 10.2. Periodic comets whose return is expected before 2000

		$T_{per.}$	ω	Ω	i	q	e	P
Reinmuth 1	1988 May	10.0	13.02	119.15	8.14	1.869 3	0.503 0	7.29
Finlay	1988 June	5.9	322.21	41.74	3.65	1.094 4	0.699 5	6.95
Tempel 2	1988 Sept.	16.7	191.04	119.12	12.43	1.383 4	0.544 4	5.29
Longmore	1988 Oct.	12.2	195.70	15.00	24.39	2.409 0	0.341 4	7.00
du Toit 1	1988 Dec.	25.9	257.05	21.84	18.69	1.273 5	0.787 9	14.7
Tempel 1	1989 Jan.	4.4	178.98	68.33	10.54	1.496 7	0.519 7	5.50
d'Arrest	1989 Feb.	4.0	177.07	138.80	19.43	1.292 1	0.624 6	6.39
Perrine-Mrkos	1989 Mar.	1.4	166.56	239.39	17.82	1.297 7	0.637 8	6.78
Tempel-Swift	1989 Apr.	12.1	163.42	240.31	13.17	1.588 4	0.539 1	6.40
Churyumov-Gerasimenko	1989 June	18.8	11.37	50.36	7.11	1.299 6	0.630 3	6.59
Pons-Winnecke	1989 Aug.	19.9	172.32	92.75	22.27	1.261 0	0.633 5	6.38
Gunn	1989 Sept.	24.9	196.94	67.87	10.37	2.471 6	0.314 4	6.84
Brorsen-Metcalf	1989 Sept.	28.9	129.73	310.84	19.33	0.478 1	0.972 0	70.6
Lovas	1989 Oct.	9.2	73.62	341.72	12.20	1.679 6	0.614 1	9.08
du Toit-Neujmin-Delporte	1989 Oct.	18.6	115.35	188.31	2.85	1.715 4	0.501 7	6.39
Schwassmann-Wachmann 1	1989 Oct.	26.2	49.86	312.12	9.37	5.771 8	0.044 7	14.9
Gehrels 2	1989 Nov.	3.7	183.55	215.52	6.67	2.348 3	0.409 8	7.94
Clark	1989 Nov.	28.3	208.93	59.07	9.50	1.555 8	0.501 3	5.51
Kopff	1990 Jan.	20.3	162.82	120.30	4.72	1.585 1	0.543 0	6.46
Tuttle-Giacobini-Kresak	1990 Feb.	8.0	61.57	140.88	9.23	1.068 0	0.655 7	5.46
Sanquin	1990 Apr.	3.3	162.84	181.81	18.72	1.813 7	0.663 3	12.5
Schwassmann-Wachmann 3	1990 May	18.8	198.78	69.26	11.41	0.936 3	0.694 0	5.35
Tritton	1990 July	7.2	147.75	299.95	7.05	1.435 7	0.580 7	6.34
Honda-Mrkos-Pajdusakova	1990 Sept.	12.7	325.76	88.64	4.23	0.541 2	0.821 9	5.30
Encke	1990 Oct.	28.5	186.25	334.04	11.94	0.330 9	0.850 2	3.28
Johnson	1990 Nov.	19.0	208.28	116.67	13.66	2.312 6	0.366 1	6.97
Kearns-Kwee	1990 Nov.	23.2	131.84	315.03	9.01	2.215 4	0.486 7	8.97
Wild 2	1990 Dec.	17.3	41.54	135.60	3.25	1.577 9	0.541 0	6.37
Taylor	1990 Dec.	31.2	355.59	108.18	20.55	1.950 4	0.465 7	6.97
Russell 1	1991 Jan.	4.0	333.74	225.80	17.72	2.173 4	0.438 7	7.62
Swift-Gehrels	1991 Feb.	22.1	84.83	313.73	9.25	1.354 7	0.691 7	9.21
Wolf-Harrington	1991 Apr.	4.9	186.98	254.17	18.47	1.607 8	0.539 1	6.51
Haneda-Campos	1991 Apr.	8.6	305.40	67.21	4.93	1.224 9	0.640 2	6.28
Van Biesbroeck	1991 Apr.	25.0	134.15	148.44	6.62	2.401 0	0.552 7	12.4
Arend	1991 May	25.2	47.07	355.50	19.93	1.850 0	0.537 0	7.99
Harrington-Abell	1991 May	31.9	139.18	336.60	10.25	1.760 7	0.542 1	7.54
Tsuchinshan 1	1991 Aug.	30.8	22.76	96.19	10.50	1.497 5	0.576 4	6.65
Wirtanen	1991 Sept.	21.1	356.19	81.60	11.67	1.083 0	0.652 3	5.50
Arend-Rigaux	1991 Oct.	3.0	329.06	121.45	17.89	1.437 8	0.600 1	6.82
Faye	1991 Nov.	15.6	203.96	198.88	9.09	1.593 4	0.578 2	7.34
Kowal 2	1992 Jan.	1.2	189.62	247.01	15.84	1.499 5	0.567 8	6.46
Chernykh	1992 Jan.	28.0	263.20	129.75	5.08	2.356 3	0.593 8	14.0
Giacobini-Zinner	1992 Apr.	14.6	172.52	194.68	31.83	1.0340	0.706 6	6.61
Tsuchinshan 2	1992 May	20.3	203.15	287.59	6.72	1.782 2	0.504 4	6.82
Kowal 1	1992 June	1.9	180.49	28.07	4.36	4.673 6	0.234 4	15.1

Table 10.2 (continued)

		$T_{per.}$	ω	Ω	i	q	e	P
Grigg-Skjellerup	1992 July	24.9	359.27	212.64	21.10	0.994 6	0.664 4	5.10
Smirnova-Chernykh	1992 Aug.	2.9	88.83	76.84	6.63	3.572 7	0.147 0	8.57
Neujmin 2	1992 Aug.	20.3	214.90	307.17	5.38	1.265 5	0.587 9	5.38
Wolf	1992 Aug.	28.4	162.30	203.44	27.48	2.427 5	0.405 6	8.25
Daniel	1992 Aug.	31.6	11.01	68.37	20.13	1.649 7	0.551 9	7.06
Schuster	1992 Sept.	6.6	355.72	49.92	20.13	1.539 2	0.589 6	7.26
Giclas	1992 Sept.	18.2	276.54	111.80	7.28	1.846 1	0.493 8	6.96
Gale	1992 Dec.	17.0	215.44	59.26	10.73	1.213 6	0.758 1	11.2
Schaumasse	1993 Mar.	5.0	57.46	80.38	11.84	1.202 3	0.704 8	8.22
Forbes	1993 Mar.	15.8	310.64	333.59	7.16	1.449 5	0.567 6	6.14
Holmes	1993 Apr.	10.6	23.23	327.34	19.16	2.176 7	0.410 4	7.09
Vaisala 1	1993 Apr.	29.2	47.36	134.40	11.60	1.783 0	0.634 7	10.8
Slaughter-Burnham	1993 June	19.9	44.08	345.75	8.15	2.542 6	0.503 6	11.6
Ashbrook-Jackson	1993 July	13.9	348.68	1.97	12.49	2.316 0	0.395 0	7.49
Gehrels 3	1993 Nov.	12.1	227.19	243.38	1.06	3.383 4	0.175 4	8.31
Neujmin 3	1993 Nov.	13.3	146.98	149.77	3.99	2.001 5	0.586 0	10.6
Sahjn-Schaldach	1993 Nov.	16.0	216.52	166.21	6.08	2.344 4	0.387 7	7.49
West-Kohoutek-Ikemura	1993 Dec.	25.8	359.97	83.48	30.54	1.576 8	0.543 3	6.42
Schwassmann-Wachmann 2	1994 Jan.	24.1	358.15	125.62	3.76	2.070 3	0.398 8	6.39
Encke	1994 Feb.	9.4	186.28	334.02	11.94	0.330 9	0.850 2	3.28
Kojima	1994 Feb.	18.6	348.50	154.15	0.89	2.400 0	0.392 4	7.85
Tempel 2	1994 Mar.	16.8	194.86	117.58	11.98	1.483 5	0.522 4	5.48
Bus	1994 June	13.8	23.83	181.53	2.57	2.184 4	0.373 2	6.51
Tuttle	1994 June	27.0	206.71	269.85	54.69	0.998 0	0.824 1	13.5
Kohoutek	1994 June	29.0	175.93	268.91	5.91	1.785 4	0.496 0	6.67
Reinmuth 2	1994 June	29.8	45.92	295.43	6.98	1.893 0	0.464 1	6.64
Tempel 1	1994 July	3.3	178.87	68.32	10.55	1.494 2	0.520 2	5.50
Wild 3	1994 July	20.5	179.23	71.94	15.46	2.300 2	0.365 8	6.91
Harrington	1994 Aug.	23.7	233.45	118.59	8.66	1.572 2	0.561 3	6.78
Brooks 2	1994 Sept.	1.5	198.00	176.20	5.50	1.843 0	0.491 0	6.89
Russell 2	1994 Oct.	28.8	249.48	41.75	12.04	2.279 2	0.399 1	7.39
Borrelly	1994 Oct.	31.8	353.30	74.70	30.30	1.365 0	0.623 0	6.89
Whipple	1994 Dec.	22.3	201.88	181.79	9.93	3.093 8	0.258 7	8.53
de Vico-Swift	1995 Apr.	4.8	1.86	358.34	6.08	2.145 0	0.430 8	7.31
Finlay	1995 May	4.6	323.47	41.42	3.67	1.035 8	0.710 3	6.76
Clark	1995 May	31.2	208.81	59.07	9.50	1.552 5	0.502 1	5.51
d'Arrest	1995 July	27.2	178.04	138.30	19.53	1.346 0	0.614 0	6.51
Tuttle-Giacobini-Kresak	1995 July	28.5	61.67	140.82	9.23	1.065 2	0.656 4	5.46
Tempel-Swift	1995 Aug.	30.5	163.31	240.30	13.18	1.583 7	0.540 0	6.39
Reinmuth 1	1995 Sept.	3.4	13.25	119.08	8.13	1.873 6	0.502 5	7.31
Schwassmann-Wachmann 3	1995 Sept.	22.3	198.76	69.28	11.42	0.933 0	0.694 8	5.34
Jackson-Neujmin	1995 Oct.	6.6	200.35	160.02	13.48	1.381 0	0.661 4	8.24
Longmore	1995 Oct.	9.4	195.79	14.96	24.40	2.399 1	0.343 0	6.98

Table 10.2 (continued)

		$T_{per.}$	ω	Ω	i	q	e	P
Perrine-Mrkos	1995 Dec.	7.2	166.56	239.91	17.84	1.292 8	0.638 6	6.77
Honda-Mrkos-Pajdusakova	1995 Dec.	30.1	324.98	89.49	4.11	0.532 8	0.824 1	5.27
Pons-Winnecke	1996 Jan.	2.7	172.30	92.74	22.30	1.255 9	0.634 4	6.37
Churyumov-Gerasimenko	1996 Jan.	18.4	11.34	50.36	7.11	1.300 0	0.630 2	6.59
du Toit-Neujmin-Delporte	1996 Mar.	5.9	115.26	188.25	2.85	1.719 7	0.500 8	6.39
Denning-Fujikawa	1996 June	4.7	337.41	35.76	9.18	0.788 2	0.818 3	9.03
Comas Sola	1996 June	11.0	45.79	60.17	12.82	1.846 0	0.568 0	8.83
Kopff	1996 July	2.2	162.75	120.29	4.72	1.579 5	0.544 1	6.45
Gunn	1996 July	24.2	196.79	67.86	10.38	2.461 9	0.316 3	6.83
Tritton	1996 Nov.	2.9	147.63	299.97	7.04	1.436 5	0.580 5	6.34
Wirtanen	1997 Mar.	14.5	356.37	81.52	11.71	1.063 5	0.656 8	5.46
Boethin	1997 May	1.1	20.02	16.07	4.99	1.158 1	0.774 7	11.7
Wild 2	1997 May	7.0	41.67	135.56	3.25	1.582 4	0.540 2	6.39
Encke	1997 May	23.5	186.28	334.01	11.92	0.331 4	0.850 0	3.28
Gehrels 2	1997 Aug.	7.6	192.93	209.81	6.25	1.998 0	0.464 0	7.20
Haneda-Campos	1997 Aug.	14.4	306.94	66.02	4.94	1.267 3	0.631 9	6.39
Grigg-Skjellerup	1997 Sept.	3.8	359.34	212.61	21.09	0.996 8	0.663 9	5.11
Wolf-Harrington	1997 Sept.	29.4	187.16	254.04	18.51	1.581 8	0.544 0	6.46
Johnson	1997 Oct.	31.9	207.97	116.68	13.67	2.308 4	0.367 2	6.97
Taylor	1997 Dec.	15.8	355.41	108.17	20.55	1.947 9	0.466 1	6.97
Neujmin 2	1998 Jan.	7.3	214.94	307.11	5.37	1.270 1	0.586 8	5.39
Tempel-Tuttle	1998 Feb.	27.4	172.52	234.56	162.48	0.975 8	0.905 6	33.2
Tsuchinshan 1	1998 Apr.	19.4	22.74	96.15	10.50	1.495 8	0.576 6	6.64
Klemola	1998 May	1.9	154.54	174.85	11.10	1.754 5	0.641 3	10.8
Kowal 2	1998 May	12.9	194.26	242.74	14.40	1.345 8	0.600 8	6.19
Arend-Rigaux	1998 July	13.2	330.55	121.05	18.30	1.368 3	0.611 6	6.61
Russell 1	1998 Aug.	25.3	333.90	225.72	17.71	2.180 9	0.437 9	7.64
Lovas	1998 Oct.	11.7	74.52	339.29	12.23	1.692 0	0.613 0	9.14
Giacobini-Zinner	1998 Nov.	23.8	172.55	194.70	31.86	1.033 7	0.706 6	6.61
Harrington-Abell	1998 Dec.	7.1	139.29	336.58	10.27	1.749 1	0.543 6	7.50
Tsuchinshan 2	1999 Mar.	8.6	203.24	287.45	6.71	1.770 4	0.506 4	6.79
Faye	1999 May	6.5	204.99	198.63	9.05	1.656 8	0.568	37.52
Forbes	1999 May	8.5	310.83	333.48	7.16	1.449 3	0.567 5	6.13
Arend	1999 Aug.	3.0	49.00	354.68	19.17	1.916 8	0.530 3	8.24
Giclas	1999 Sept.	1.9	276.47	111.76	7.28	1.844 8	0.494 2	6.97
Tempel 2	1999 Sept.	8.4	194.99	117.54	11.98	1.481 7	0.522 8	5.47
Kearns-Kwee	1999 Sept.	17.5	127.53	312.32	9.34	2.337 4	0.476 7	9.44
Schuster	1999 Dec.	16.8	355.84	49.91	20.13	1.549 6	0.588 0	7.29
Wild 1	1999 Dec.	27.4	167.99	357.83	19.93	1.960 9	0.649 7	13.2

(After B. Marsden and E. Roemer, *Comets*, L. Wilkening, ed., Tucson, University of Arizona Press, © 1982)

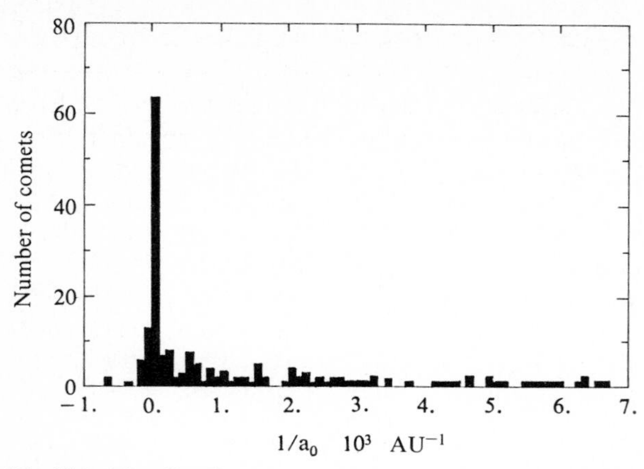

Fig. 10.2. The distribution of comets as a function of $1/a_0$ (a_0 being the semi-major axis of the ellipse). [After P. Weissman: *Comets*, ed. by L. Wilkening (© University of Arizona Press, Tucson 1984)]

Oort Cloud, as a result of perturbations by Jupiter, Saturn and Uranus. In the second hypothesis, comets were formed where they are found nowadays by accretion of cometoids, during the collapse of the primordial solar nebula. The very low density of matter at great heliocentric distances means that this second hypothesis is very improbable.

10.3 Measurements of the Brightness of Comets

A unit frequently used in cometary physics is the *magnitude*, defined so as to measure the integrated flux from astrophysical objects in white light or in different standard filters, in the visible and infrared regions of the spectrum. The two filters most often used are the B and V filters, centred on 4400 Å and 5500 Å respectively, with a pass-band of 1000 Å. the corresponding magnitudes being defined as follows:

$$m_{\mathrm{V}} = -2.5 \log\left(\int_\lambda V_\lambda f_\lambda d\lambda\right) - 13.74 \tag{10.1}$$

$$m_{\mathrm{B}} = -2.5 \log\left(\int_\lambda B_\lambda f_\lambda d\lambda\right) - 12.97 \tag{10.2}$$

where the integrals represent the flux received from the object, integrated over the V and B filters' passbands respectively: f_λ is the flux from the object, V_λ and B_λ are the respective transmissions of the V and B filters. The integrals are expressed in erg cm^{-2} s^{-1}.

In dealing with comets, two types of magnitude may be considered: total magnitudes (which include all the radiation phenomena caused by perihelion passage), and nuclear magnitudes, which only involve the nucleus. Far from the Sun, in the absence of outgassing, the total magnitude of a comet reduces to its nuclear magnitude.

If one knows the nuclear magnitude of a comet (which mainly depends on the size and albedo of the nucleus) at a given point, it is easy to calculate its nuclear magnitude at any other position: the nuclear flux, observed from the Earth, varies as r^{-2} and Δ^{-2}, where r is the heliocentric distance and Δ the geocentric distance. This gives the following relationship:

$$m_{\mathrm{n}} = m_0 + 5.0 \log(\Delta) + 5.0 \log(r) \tag{10.3}$$

where Δ and r are expressed in AU; m_0 is the magnitude that the nucleus would have at a distance of 1 AU, in the absence of outgassing. The relationship is valid for any filter, and even without any filter: one is then measuring the integrated flux over the whole visible range. For example, for Comet Halley, we have, integrating over the whole visible range, $m_0 = 13.8$.

The variation in a comet's total magnitude is more difficult to calculate precisely. When the comet becomes active, the observed flux has two components: emission from the gas itself, by fluorescence, and a contribution from dust, which is just scattered sunlight. In order to describe the variation in a comet's total magnitude as a function of r and Δ, one has to resort to various empirical formulae, derived from observations and the coefficients of which vary from one comet to the next.

The light-curve is frequently asymmetric (with respect to perihelion passage). This doubtless arises from inhomogeneities in the surface of the nucleus. As this does not present the same face to the Sun before and after perihelion, it may well produce an asymmetry in the curve of the production rate (see below).

10.4 The Physics of Comets

Comets pass most of their lives at great distances from the Sun, in a region that is very cold and empty, preserving them in their initial state. This is why they are considered to be important witnesses of the conditions under which they were formed. Unfortunately, when comets are far from the Sun they are inaccessible for observation. We are only able to study them when they pass perihelion, if this is at no more than a few astronomical units. Far from the Sun, a comet just consists of a nucleus; when it approaches the Sun, the ices on the surface sublime, leading to the loss of gas and dust. A coma is then seen to appear, and this grows as the comet approaches the Sun. If the comet is sufficiently active, two tails may form, one wide and curved, consisting of dust, and the other narrow and straight, caused by the plasma created by the ionization of the gas by the solar ultraviolet flux.

It is the scattering of the solar visible radiation that is responsible for the appearance of comets as we generally think of them; the nucleus itself often cannot be observed, as it is hidden by the shell of dust and gas surrounding it. In order to determine the nature of comets, i.e. the composition of their nuclei, one is forced to observe secondary phenomena, such as the ions, radicals and daughter molecules, all of which are produced by the dissociation of parent molecules ejected from the nucleus itself. Starting with the last links in the chain, we have to work back to the source.

This is difficult for several reasons. First, the parent molecules derived from the nucleus are themselves very hard to observe: these molecules' most intense transitions are at infrared or millimetric wavelengths, which are always difficult to observe. What can be identified in profusion are radicals and ions that have very intense electronic transitions in the visible and ultraviolet regions, and thanks to which we have a large amount of cometary data. But there are numerous dissociation products and multiple dissociation and ionization reactions; despite numerous attempts, to this day it has not been possible to derive a parent-molecule composition that accounts for all the secondary products observed and their respective abundances.

Table 10.3. Atoms and molecules observed in comets by UV and visible spectrometry

Atom or molecule	Characteristic band or line	Transition	Year of first detection
H I	1 216 Å	$^2P^0 \rightarrow {}^2S$	1972
	6 563 Å	$^2D \rightarrow {}^2P^0$	1974 and 1975
O I	1 304 Å	$^3S^0 \rightarrow {}^3P$	1972
[O I]	2 972 Å	$^1S \rightarrow {}^3P$	1980
	6 300 Å	$^1D \rightarrow {}^3P$	1958
	5 577 Å	$^1S \rightarrow {}^3P$	1984
C I	1 561 Å	$^3D^0 \rightarrow {}^3P$	1974
	1 657 Å	$^3P^0 \rightarrow {}^3P$	1974
[C I]	1 931 Å	$^1D \rightarrow {}^1P^0$	1976
C II	1 335 Å	$^2D \rightarrow {}^2P^0$	1976
S I	1 814 Å	$^3P \rightarrow {}^3S^0$	1980
Na I	3 303 Å	$^2P^0 \rightarrow {}^2S$	1967
	5 890 Å	$^2P^0 \rightarrow {}^2S$	1967
K I	7 665 Å	$^2P^0 \rightarrow {}^2S$	1967
	4 044 Å	$^2P^0 \rightarrow {}^2S$	1967
Ca II	3 934 Å	$^2P^0 \rightarrow {}^2S$	1967
Cr I	3 579 Å	$^7P^0 \rightarrow {}^7S$	1967
Mn I	4 031 Å	$^6P^0 \rightarrow {}^6S$	1967
Fe I	3 441 Å	$^5P^0 \rightarrow {}^5D$	1967
	3 570 Å	$^3G^0 \rightarrow {}^5F$	1967
	3 581 Å	$^5G^0 \rightarrow {}^5F$	1967
	3 720 Å	$^5F^0 \rightarrow {}^5D$	1967
	3 749 Å	$^5F^0 \rightarrow {}^5F$	1967
	3 813 Å	$^3P^0 \rightarrow {}^5F$	1967
	3 816 Å	$^3D^0 \rightarrow {}^3F$	1967
	3 820 Å	$^5D^0 \rightarrow {}^5F$	1967
	3 860 Å	$^5D^0 \rightarrow {}^5D$	1967
	4 046 Å	$^3F^0 \rightarrow {}^3F$	1967
Ni I	3 381 Å	$^1P^0 \rightarrow {}^1D$	1967
	3 446 Å	$^3D^0 \rightarrow {}^3D$	1967
	3 458 Å	$^3F^0 \rightarrow {}^3D$	1967
	3 462 Å	$^5F^0 \rightarrow {}^3D$	1967
	3 525 Å	$^3P^0 \rightarrow {}^3D$	1967
	3 566 Å	$^1D^0 \rightarrow {}^1D$	1967
	3 619 Å	$^1F^0 \rightarrow {}^1D$	1967
Cu I	3 248 Å	$^2P^0 \rightarrow {}^2S$	1967
C_2	2 313 Å	$D^1\Sigma_u^+ \rightarrow X^1\Sigma_g$	1980
	5 165 Å	$d^3\Pi_g \rightarrow a^3\Pi_u$	1882
	7 715 Å	$A^1\Pi_u \rightarrow X^1\Sigma_g^+$	1981

Table 10.3 (continued)

Atom or molecule	Characteristic band or line	Transition	Year of first detection
$^{12}C^{13}C$	4 745 Å	$d^3\Pi_g \rightarrow a^3\Pi_u$	1943 and 1964
C_3	4 040 Å	$\tilde{A}^1\Pi_u \rightarrow \tilde{X}^1\Sigma_g^+$	1882 and 1951
CH	3 889 Å	$B^2\Sigma^- \rightarrow X^2\Pi_r$	1940
	4 315 Å	$A^2\Delta \rightarrow X^2\Pi_r$	1938
CN	3 883 Å	$B^2\Sigma^+ \rightarrow X^2\Sigma^+$	1882
	7 873 Å	$A^2\Pi_i \rightarrow X^2\Sigma^+$	1948
CO	1 510 Å	$A^1\Pi \rightarrow X^1\Sigma^+$	1976
CS	2 576 Å	$A^1\Pi \rightarrow X^1\Sigma^+$	1980
NH	3 360 Å	$A^3\Pi_i \rightarrow X^3\Sigma^-$	1941
NH_2	5 700 Å	$\tilde{A}^2A_1 \rightarrow \tilde{X}^2B_1$	1943
OH	3 090 Å	$A^2\Sigma^+ \rightarrow X^2\Pi_i$	1941
CH^+	4 225 Å	$A^1\Pi \rightarrow X^1\Sigma^+$	1941
CN^+	2 181 Å	$f^1\Sigma \rightarrow a^1\Sigma$	1980
	3 185 Å	$c^1\Sigma \rightarrow a^1\Sigma$	
CO^+	2 190 Å	$B^2\Sigma^+ \rightarrow X^2\Sigma^+$	1976
	3 954 Å	$B^2\Sigma^+ \rightarrow A^2\Pi_i$	1950
	4 273 Å	$A^2\Pi_i \rightarrow X^2\Sigma^+$	1910 and 1950
CO_2^+	2 890 Å	$B^2\Sigma_u^+ \rightarrow X^2\Pi_g$	1976
	3 509 Å	$\tilde{A}^2\Pi_u \rightarrow \tilde{X}^2\Pi_g$	1950
H_2O^+	6 198 Å	$\tilde{A}^2A_1 \rightarrow \tilde{X}^2B_1$	1974
N_2^+	3 914 Å	$B^2\Sigma_u^+ \rightarrow X^2\Sigma_g^+$	1910
OH^+	3 565 Å	$A^3\Pi_i \rightarrow X^3\Sigma^-$	1950
S_2	2 800 Å	–	1983

(After S. Wykoff, *Comets,* L. Wilkening, ed., Tucson, University of Arizona Press, © 1982)

It has now been shown that H_2O is the predominant parent molecule. Even before Comet Halley's return in 1986, several factors indicated this:

1. the abundances of H and OH, which were obviously equal and much higher than those of any other components;
2. the detection of H_2O^+;
3. the presence of forbidden transitions of oxygen.

The H_2O molecule itself had been looked for at radio wavelengths, but its detection was doubtful. It has now been unambiguously identified in Comet Halley, both from Earth and from space probes. Before 1986, there was great uncertainty about the nature and relative abundances of other parent molecules, and therefore about the composition of the nucleus itself. This is why the experiments carried out in situ near Comet Halley by the Giotto and VEGA probes are invaluable for cometary physics. Several parent molecules have been detected (see Table 10.4), in particular CO_2 (by infrared spectrometry from VEGA), with a CO_2/H_2O ratio of the order of 10^{-2}, and HCN (from the ground, at millimetric wavelengths), with a HCN/H_2O ratio of the order of 10^{-3}. In addition, the spacecraft showed the presence of carbonaceous molecules in the immediate vicinity of P/Halley's nucleus.

Table 10.4. Parent molecules detected in Comet Halley

Molecule	Wavelength of observation (μm)	Band or transition	Relative abundance
H_2O	2.7	ν_3	1
CO_2	4.3	ν_3	0.02
CO	4.7		0.05
H_2CO	3.6	ν_1, ν_5	0.04
OCS	4.8		0.007
HCN	1.1		0.001
Saturated hydrocarbons	3.4	C–H	–
Unsaturated hydrocarbons	3.3	C–H	–

10.4.1 The Nucleus

Size, Rotation and Mass

Sizes of cometary nuclei may be estimated in two ways. The oldest method employs photometry at visible wavelengths of distant comets. The flux observed is then the reflected solar flux; knowing the distance and making some assumptions about the albedo, one can deduce the diameter of the nucleus. Generally an albedo of between 0.05 and 0.50 is taken, analogous to that of other bodies in the Solar System. The measurements of Comet Halley made in situ indicate that the albedo is very low, of the order of a few per cent. The second method employs radar echoes; it is more precise, but can be used only on comets close to the Earth. These two methods have led to estimates of between a kilometre and ten kilometres for cometary diameters. A comet's nucleus is thus a small object, which explains the difficulties that arise with direct observation (Fig. 10.3).

The rotation period of the nucleus is an equally difficult parameter to determine. Theoretically, it should be possible to obtain it from the variations in the light-curve, as with asteroids. But as the object is small and, by necessity, distant (otherwise it

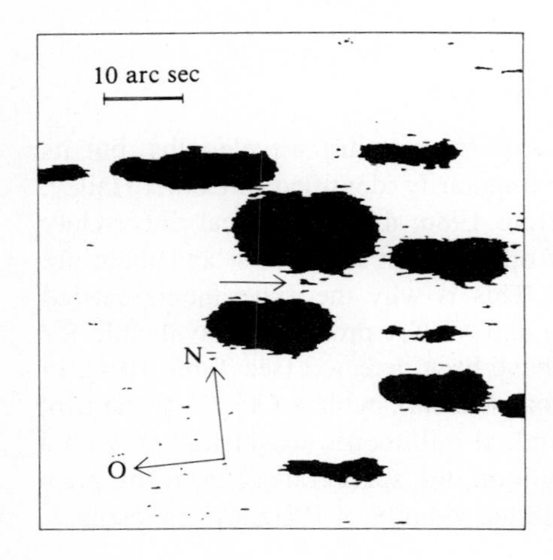

Fig. 10.3. One of the first observations of P/Halley at its most recent return. This photograph was taken in 1982 by using an electronographic camera on the Canada-France-Hawaii Telescope. [B. Sicardy et al.: *Astronomy and Astrophysics* **121**, L4 (1983). By courtesy of the publishers]

would be hidden within the coma), results are uncertain. Similarly, it is difficult to obtain an definite result from a light-curve, even a periodic one, because the function is not a simple sine-wave. Nevertheless this parameter is a very important one, as is the orientation of the rotation axis with respect to the Sun: if the nucleus is heterogeneous this has a great influence on the rate of outgassing and its variation with time.

The mass of the nucleus is yet another parameter, determination of which is very imprecise. It is far too small to have any detectable effects on the paths of the planets that perturb its orbit. Occasionally the nuclei of comets break-up under the effect of internal forces; this happened to Comet Biela. In such a case, it is possible to obtain an estimate of the mass from the fragments' mutual perturbations. Values thus found range from 10^{13} to 10^{19} g. Another indirect estimate may be obtained by measuring the delay in reaching perihelion caused by non-gravitational forces. Before perihelion, the mass of the gas and dust ejected towards the Sun before perihelion passage tends to produce a reaction that delays the comet in its orbit. From the delay at perihelion, obtained by astrometry, and by estimating the mass of gas ejected, obtained from the light-curve, information about the mass of the nucleus, and thus its density, may be obtained. For Comet Halley, this method gave relatively low densities (0.2 to 0.5 g/cm^3).

The Structure of Cometary Nuclei

Before 1950, the model of a cometary nucleus that was most commonly accepted was that of a cloud of particles that had formed as a result of some interstellar accretion mechanism. Nevertheless there were objections to this model; it was unable to account for the high percentage of volatile components for example, or for the existence of non-gravitational forces (which were confirmed by observations of the paths of certain comets, in particular that of Comet Halley). In 1950, these objections were analyzed in detail by Whipple, who proposed the model that thenceforward became famous as the "dirty snowball" model. In this, the cometary nucleus is described as being an aggregate of ices and dust particles, produced by condensation of primordial material. This concept, which is able to account for all of the observed phenomena, is the one current today. Information drawn from the space-probe missions to P/Halley indicate that the nucleus is very dark. Most of it may be covered by a dark material, possibly a carbon compound (Fig. 10.4). The surface enrichment in this material may be the result of preferential vaporization of H_2O ice at previous perihelion passages. Since 1950, numerous authors have refined the "dirty snowball" model. The mechanism for the formation of a cometary nucleus may be explained in the same way as that of more massive bodies, by the accumulation of cometoids, which have a mass distribution of the type:

$$n(m) \propto m^{-5} \tag{10.4}$$

This structure explains, in particular, how large parts of a comet, in contact with the nucleus over just small portions of their surface, may break away through vaporization, causing the fragmentation of nuclei.

Accretion of a nucleus by collisions between grains of ice can occur only at low temperatures, and with low relative velocities (< 0.05 km/s). The resulting density

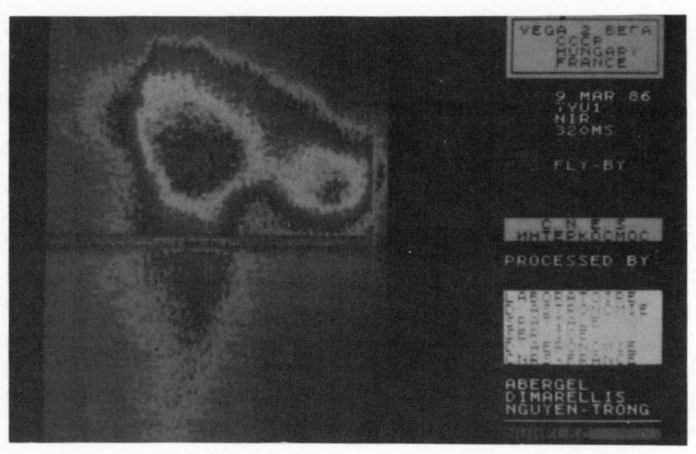

Fig. 10.4. Image of the central region of Comet Halley, obtained at a distance of 8200 km by the VEGA 2 probe on 1986 March 2. Two bright regions, which probably represent jets, can be clearly distinguished. The nucleus is elongated, with dimensions of 15 × 7.5 × 7.5 km. Its albedo is about 0.04. (By courtesy of CNES-Intercosmos)

is below 0.5 g/cm^3. The central pressure P_c, the gravity g, and the escape velocity V_e may be calculated:

$$P_c = \frac{2}{3}\pi G\bar{\varrho}^2 R^2 \qquad (10.5)$$

$$g = \frac{4}{3}\pi G\bar{\varrho} R \qquad (10.6)$$

$$V_e = [\frac{8}{3}\pi G\bar{\varrho}]^{1/2} R \qquad (10.7)$$

where G is the gravitational constant, ϱ the mean density, and R the radius. Calculations show that for a radius of less than ten kilometres gravity is too low for the nucleus to become compacted.

Dust grains are trapped within the icy matrix: this is shown by the analysis of the swarms of micrometeorites that the Earth encounters when it crosses the orbits of short-period comets (see Sect. 11.3). It is also possible that gaseous molecules other than those of H_2O are trapped in a matrix of H_2O ice in the form of hydrated clathrates; these might include as much as 18 % of foreign molecules within their cavities. In this case, the sublimation of these molecules would be governed by the sublimation of the H_2O ice.

To sum up, we may say that cometary nuclei are porous and fragile aggregates of ices and dust, which have little cohesion and are therefore liable to break apart.

What happens to a cometary nucleus after multiple passages close to the Sun? The answer to this question is still uncertain. The ices slowly sublime and dust particles are ejected; it is possible that some residue persists, the remnant of any non-volatile material trapped in the icy matrix of the initial nucleus, or the central part of a nucleus that was initially covered in ice. Some asteroids may be the inert remnants of comets, and this idea is now seriously entertained by numerous researchers.

10.4.2 The Coma

Because cometary nuclei are so difficult to observe, we have to turn to what we can analyze easily: the dust and gas ejected from the nucleus (Fig. 10.5). For this we need to understand the mechanisms governing sublimation, expansion within the coma, dissociation and ionization of gases; once again many problems remain without a solution.

Fig. 10.5. Comet West, photographed in 1976

Sublimation from the Nucleus

As a slowly-rotating cometary nucleus gradually approaches the Sun, its surface temperature slowly rises. If this surface is covered by a volatile material, the equilibrium state is described by the following equation:

$$\Theta_0 \frac{1-a}{r^2} = \sigma T^4 + Z(T) \cdot L(T) \quad . \tag{10.8}$$

The left-hand side of this equation expresses the solar flux absorbed by the nucleus: a is the albedo of the nucleus, Θ_0 is the solar flux at a distance of one astronomical unit, and r is the heliocentric distance. The right-hand side expresses the energy emitted by the surface: part thermal emission (σ being Stefan's constant, and T the temperature), and part the product of the latent heat of sublimation $L(T)$ and the gas production rate $Z(T)$.

Table 10.5. Gas production rates for different types of ice

Snow controlling vaporization	Z_0 (10^{18} mol cm^{-2} s^{-1})	T_0 (K)	T_1 (K)	d (AU)
Nitrogen	14.3	40	35	77.6
Carbon monoxide	13.0	44	39	62.5
Methane	10.6	55	50	38.0
Formaldehyde	5.0	90	82	14.1
Ammonia	3.7	112	99	9.7
Carbon dioxide	3.5	121	107	8.3
Cyanic acid	2.3	160	140	4.8
Ammonia	2.7	213	193	2.6
Clathrate	1.9	214	194	2.5
Water	1.7	215	195	2.5

(After A. Delsemme, *Comets,* L. Wilkening, ed., Tucson, University of Arizona Press, © 1982)

Z_0 is the vaporization rate at the subsolar point on a perfectly absorbing nucleus at a distance of 1 AU from the sun.

T_0 is the temperature for a stationary state of Z_0 (subsolar point, nucleus without rotation).

T_1 is the effective mean temperature of a rotating nucleus ($Z_1 = 1/4\ Z_0$).

d is the heliocentric distance beyond which the vaporization rate becomes negligible (less than 2.5% of the solar flux is used for vaporization).

It is theoretically possible to solve (10.8) by successive approximations, thus deriving T and Z, for different types of ice. This has been done by Delsemme and his results are summarized in Table 10.5. In particular, it allows the heliocentric distance at which degassing will commence to be calculated, and the variation in the rate of degassing as a function of heliocentric distance. When the degassing mechanism predominates (the term σT^4 on the right-hand side of (10.8) becoming negligible), the production rate becomes proportional to r^{-2}. Comparing these theoretical curves with observed light-curves for a dozen comets, it has been found that in new comets there is reasonable agreement with the water-ice sublimation curve. H_2O ice is therefore the predominant form at the surface of cometary nuclei. We shall see later that other arguments, obtained from analysis of the gases emitted by cometary nuclei, confirm this conclusion.

The Expansion of Gas in the Coma

From the theory developed earlier it is possible to estimate the velocity of the molecules that sublime. At the instant that they leave the nucleus they have a velocity V_0 such that (Delsemme 1982):

$$\tfrac{1}{2}\overline{V} \leq V_0 \leq \tfrac{2}{3}\overline{V} \quad . \tag{10.9}$$

$\overline{V}$ is the average velocity of molecules at temperature T:

$$\overline{V} = \left[\frac{8kT}{\pi m}\right]^{1/2} \tag{10.10}$$

m being the mass of the molecule and k Boltzmann's constant.

The molecules are subsequently accelerated radially by collisions until they attain sonic velocity:

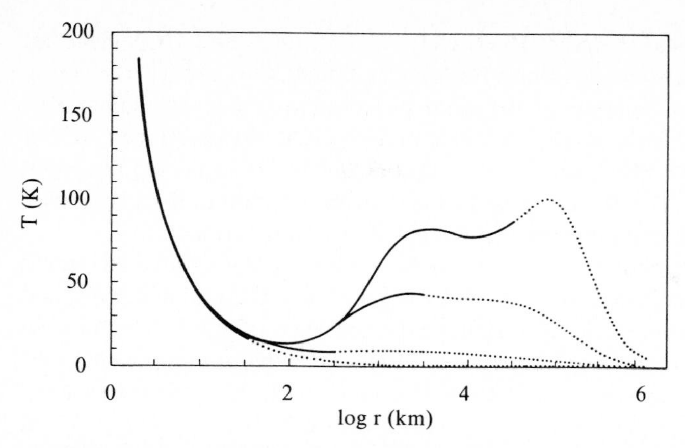

Fig. 10.6. The temperature distribution within the inner coma, in the absence of dust. The dotted curves indicate passage to free molecular flow. The production rates are (from top to bottom) $10^{27}\,s^{-1}$, $10^{28}\,s^{-1}$, $10^{29}\,s^{-1}$, and $10^{30}\,s^{-1}$, respectively. (D. Bockleé-Morvan and J. Crovisier, 1987, ESA SP-278, 235)

$$V_s = \left[\frac{\gamma k T}{m}\right]^{1/2} \tag{10.11}$$

where γ is the ratio of specific heats C_p and C_v. For water:

$$V_s = 0.71\overline{V} \quad . \tag{10.12}$$

The gases continue to expand adiabatically until they attain a limiting velocity, which in the case of water, is of the order of 0.8 km/s. Under the action of solar radiation, however, other processes intervene in the coma: photoexcitation, photodissociation, and photoionization of the molecules. Photolytic heating is effective in the region where collisions predominate, i.e. out to radii of the order of 10^3 km for gas production rates of $10^{29}\,s^{-1}$. Beyond this distance the flow becomes a "free-molecular" flow and the idea of a gas temperature becomes meaningless. It should be noted that close to the nucleus, within the first one hundred kilometres, expansion of the gas causes an abrupt drop in temperature, down to about 20 K according to some models; it then rises after some hundreds of kilometres (see Fig. 10.6).

The Parent Molecules

Being produced by direct sublimation of ices from the nucleus, the parent molecules provide, when they can be identified, one key to the chemical composition of comets. For a long time they were poorly known. The reason is simple: spectroscopic observations of comets have, until now, been primarily confined to the visible, ultraviolet and radio regions. The parent molecules do not have any strong spectral signatures in the visible and ultraviolet regions, where the transitions mainly correspond to dissociation and ionization energies. The observable transitions are molecular rotation and rotation-vibration bands in the infrared and sub-millimetre regions. These spectral regions are difficult to observe from the ground. Infrared and millimetric astronomy is in the process of rapid development, however, and recently considerable progress has been made at these wavelengths: H_2O, CO_2, H_2CO, OCS and HCN have all been detected by these methods in Comet Halley (see Table 10.4). Apart from these results, there are a certain number of marginal observations made on Comet Kohoutek, which was particularly bright, in 1973, and on IRAS-Araki-Alcock, which came exceptionally close to the Earth in 1983. Parent molecules suspected are H_2O, NH_3, HCN and CH_3CN. The S_2 molecule has also been detected in the ultraviolet, at low abundances, in the central region of Comet IRAS-Araki-Alcock.

The most abundant parent molecule, as we have seen, is H_2O. Once suspected from radio observation, indicated as probable by the nuclear sublimation curves, and confirmed by measurements made of P/Halley, the presence of the H_2O molecule is also confirmed by the relative abundances of H and OH, major components (see later), as well as by the presence of H_2O^+. The other molecules identified have mixing ratios of a few per cent or even less. Hydrocarbons, both in saturated and non-saturated form may be another type of parent molecule that is present. If these hydrocarbons are present in gaseous form, their mixing ratio may amount to a few tenths of that of H_2O. As regards the CO molecule, it appears that part may be degassed from the nucleus and part produced by the dissociation of CO_2, which might explain the presence of CO_2^+; it should be noted that CO is observed in the UV, and that its abundance varies greatly from comet to comet. In trying to work backwards to the initial composition of parent molecules, we have to examine the dissociation and ionization products overall, with their measured relative abundances, taking every possible photochemical process into account.

Modelling the spatial distribution of the parent molecules is necessary for this work. Assuming a constant expansion velocity V, the density $n(r)$, at a distance r from the nucleus is:

$$n(r) = \frac{Q}{4\pi V r^2} \tag{10.13}$$

Q being the production rate in mol/s. For an active comet, the gas production rate may be of the order of 10^{30} molecules/s, at a heliocentric distance of 1 AU.

After a certain time t, known as the lifetime, a parent molecule is dissociated, and its density beyond a sphere of radius $V \cdot t$ decreases very rapidly. Haser's model takes this distribution into account:

$$n(r) = \frac{Q}{4\pi V r^2}\, \mathrm{e}^{-r/Vt} \quad . \tag{10.14}$$

Table 10.6 gives the lifetimes of certain parent molecules.

Table 10.6. Lifetimes of several possible parent molecules

Molecule	Lifetime (s)	Possible dissociation products
H_2O	8.3×10^4	OH, H, O
CO	1.5×10^6	C, O
HCN	7.7×10^4	CN, CH, NH
H_2CO	3.5×10^3	CO, CH
CO_2	5.0×10^5	CO
NH_3	5.5×10^3	NH_2, NH

Daughter Molecules, Radicals and Ions

Numerous spectroscopic observations in the ultraviolet and visible regions have allowed a more or less complete analysis of the dissociation products found in comets (see Table 10.3). The excitation mechanism is, in general, resonant fluorescence, the source of the pumping being the solar flux (see Sect. 2.2.1); in a few cases non-resonant fluorescence is observed (forbidden oxygen transitions, and the OH line at radio wavelengths). In some cases (particularly for CN and OH) a very strong Swings effect is observed: as the solar ultraviolet spectrum is very irregular, the fluorescence rate varies abruptly as a function of the Doppler shift caused by the motion of the comet with respect to the Sun. The same effect is found in observations of doubled OH lines at radio wavelengths.

In the visible, the strongest cometary emissions are those of C_2, CN and CH; C_3, NH_3 and certain forbidden transitions of $O(^1D)$ are also encountered. Ions observed in the visible are CO^+, H_2O^+ as well as $N_2{}^+$ and CH^+ (Fig. 10.7). In the ultraviolet, where IUE carried out detailed studies, the strongest species are OH (at 3080 Å), H (in the Ly-α line), CS, CO, as well as H, O, C, S, and C^+, $CO_2{}^+$, CO^+ and perhaps CN^+ (Fig. 10.8).

Among these neutral species, the most abundant elements are H and OH. H is observed at great distances from the nucleus in the outer coma, with velocities of the order of 8 km/s, in agreement with what is expected if H is obtained by the dissociation of OH. The OH radical, less abundant than H, has a scale-length $V \cdot t$ of the order of 10^5 km; its fluorescence rate varies, in accordance with theory, as the square of the heliocentric distance. Several estimates of the H/OH ratio tend towards a value of two; this agrees with the calculated value if H and OH primarily originate with the dissociation of H_2O. H is simultaneously a primary dissociation

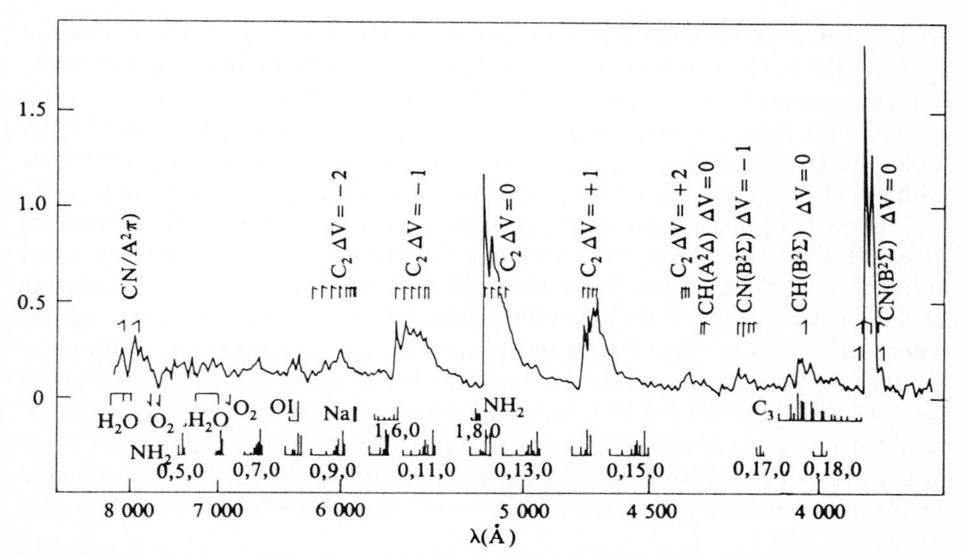

Fig. 10.7. The spectrum of Comet Kohoutek in the visible, obtained in 1973. [After M.A. Hearn: *Comets*, ed. by L. Wilkening (© University of Arizona Press, Tucson 1984)]

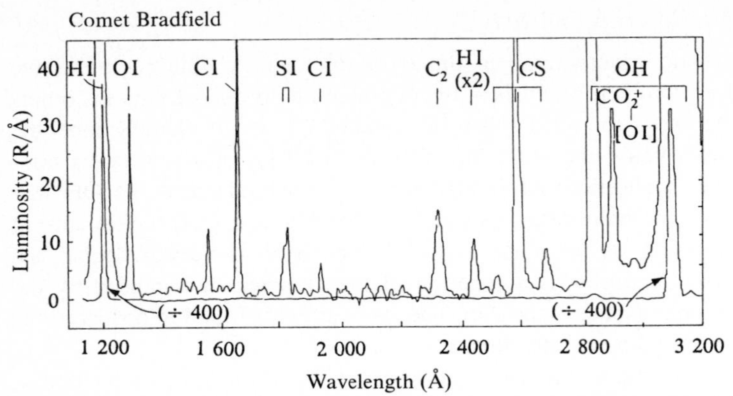

Fig. 10.8. The ultraviolet spectrum of Comet Bradfield (1979 X). This spectrum was obtained by the IUE satellite. [After P. Feldman: *Comets*, ed. by L. Wilkening (© University of Arizona Press, Tucson 1984)]

product (with a velocity of 20 km/s) following the $H_2O \rightarrow OH + H$ reaction, and a secondary product (with a velocity of 8 km/s) following the $OH \rightarrow O + H$ reaction.

Atomic oxygen is notable in that it may be observed in forbidden transitions in the visible, at 6300 Å–6364 Å (^1D–^3P) and at 5577 Å (^1S–^1D). These lines appear to be caused by the formation of O in an excited state (^1D or ^1S), by dissociation of a parent molecule, the atom later returning to the ground state ^3P. It remains to be established which molecules may be parents for $O(^1D)$. Naturally the primary candidate is H_2O; but calculation predicts that only 10 % of H_2O should be dissociated in the reaction:

$$H_2O \rightarrow H_2 + O(^1D) \qquad (10.15)$$

whilst some measurements appear to indicate an $O(^1D)/H_2O$ ratio of the order of 35 % (Festou). The existence of a second parent molecule therefore appears to be required; one possible candidate is CO_2, detected in P/Halley.

Another radical is of particular interest: CN, observed at 3880 Å and 7870 Å with a scale-length of 3×10^5 km, one of the presumed parent molecules for which, with a scale-length of 2×10^4 km, might be HCN, again detected in P/Halley.

There is complete uncertainty as to the origin of the other radicals that are observed: C_2, CH, and C_3 – which is more concentrated towards the nucleus – NH and NH_2 – also localized in the centre. Table 10.7 indicates the relative abundances of these components and their presumed scale-lengths. It should be noted that the parent molecules for C_3 and NH_3 must have very short scale-lengths, of the order of 1000 km. The presence of the CO molecule is not clearly explained; the CO/OH ratio, as well as the CO_2^+/OH ratio, varies considerably from one comet to another: sometimes absent, CO may sometimes even reach an abundance half that of OH, as in the case of Comet Kohoutek. As has been mentioned earlier, CO appears to be both a parent molecule and a daughter molecule, probably arising from dissociation of CO_2.

Investigation of the composition of the parent molecules from observation of the dissociation and ionization products has been the subject of exhaustive studies.

Table 10.7. Relative abundances of gaseous cometary components

Component	Relative abundance M(x)/M(OH)	Lifetime τ(x)/τ(OH)	Production rate Q(x)/Q(OH)
C_2	0.4 %	1	0.4 %
CN	0.4 %	3	0.13 %
CH	0.005 %	< 0.1	$\geqslant 0.05$ %
NH	?	4	?
C_3	0.007 %	0.5	0.015 %
NH_2	0.01 %	$\leqslant 0.1$?	$\geqslant 0.1$ %

(After M. A. Hearn, *Comets,* L. Wilkening, ed., Tucson, University of Arizona Press, © 1982)

In particular, photochemical models have been developed – involving about one hundred species – and several hundred photochemical reactions. Some workers, starting with a qualitatively "interstellar" composition of H_2O, CO_2, CH_4 and NH_3, have managed to reproduce reasonably well the observed abundances of C_2 and CN, but not those of C_3 and NH_3, which are, respectively, underestimated and overestimated in calculations. The NH_2 abundance obtained may be lessened if one reduces NH_3 in favour of N_2. At present, however, no model is capable of accounting for the overall abundances of radicals and ions that are observed.

Isotopic Ratios in Comets

Determination of isotopic ratios in comets is of particular interest for understanding the conditions under which they were formed. Indeed, the isotopic ratios, which are assumed to be equal to those of the primordial nebula during the comets' formation phase, are liable to have been modified to a greater or lesser extent by fractionation, which is sensitive to the prevailing physical and chemical conditions, particularly the temperature. This is especially true of the D/H ratio, where a deuterium enrichment factor of as much as ten may be observed.

Before observation of P/Halley in 1986, the only isotopic ratio in comets measured from the ground was the $^{12}C/^{13}C$ ratio, determined from the (1–0) bands of C_2 and $^{12}C^{13}C$ in the visible region. The $^{12}C^{13}C$ bands merged with an NH_2 band, however, which introduced a considerable degree of uncertainty. The various measurements seemed to indicate a $^{12}C/^{13}C$ ratio slightly higher than the value (89) measured for the Solar System. Measurements obtained for Comet Halley, both for the gas (from CN and the (0–0) band of C_2), and for the grains (from mass spectrometry by the space probes) appear to confirm a solar value, or one slightly enriched in ^{12}C. If this enrichment is real, it might be explained by a fractionation effect through an ion-neutral-atom exchange reaction, leading to enrichment in $^{12}C^+$ and therefore in ^{12}C, but this remains merely a hypothesis.

Observations of P/Halley by mass spectrometry have allowed new isotopic ratios to be determined: $^{34}S/^{32}S$ is equal to 0.045 ± 0.010, agreeing with the terrestrial value; similarly the $^{18}O/^{16}O$ ratio is identical to the terrestrial value (500). A particularly interesting result is the determination of the D/H ratio, from mass-spectrometry measurements of the values of H_2DO^+ and H_3O^+. The D/H ratio obtained lies between 0.6×10^{-4} and 5×10^{-4}. It is distinctly higher than the ratio suggested for the early solar nebula ($\sim 2 \times 10^{-5}$), which is also found in Jupiter and Saturn (see Sect. 7.1.4); and it approaches the values found for objects that have obtained their hydrogen from volatile molecules (terrestrial planets, meteorites, and Titan).

10.4.3 Cometary Dust

It has been known for a long time that dust is an important component in comets, primarily because of the wide, curved tails, otherwise known as "type II" tails, which develop around perihelion. In certain particularly massive comets a dust coma has been detected at distances as great as seven astronomical units, theoretically too far for sublimation of water to take place; some authors have suggested sublimation of CO_2 might be responsible for activity beginning at great heliocentric distances.

The Kinematics of Dust Tails

Analysis of isophotes of a dust tail has been dealt with in detail by Finson and Probstein. Cometary dust is ejected from the nucleus with a given velocity V, and undergoes acceleration through two forces acting in opposite directions: solar gravitational attraction and radiation pressure. The ratio of these two accelerations is conventionally described as $1 - \mu$, and we have

$$1 - \mu = C(\varrho d)^{-1} \qquad (10.16)$$

where ϱ is the density and d the diameter of the particle.

From this it is possible to calculate the density of particles in the dust tail; when the initial velocity V is zero the calculation is particularly simple. The isophotes of a dust tail then represent a two-dimensional solution of two distinct parameters:

1. the distribution of particle sizes varies along each *synchrone* (all the particles emitted by the nucleus at a given instant);
2. the production rate for particles of a give size varies along *syndynes* (all the particles having a given size), as a function of the time they were ejected.

Figure 10.9 shows the distribution of dust in a cometary tail with synchrones and syndynes, as well as how a dust tail is formed. In practice, the ejection velocity of the dust particles is not zero, and this leads to broadening of the tail. Observations of this broadening have enabled the ejection velocity to be estimated at 0.3 km/s on average, or about one third of the velocity at which gas is emitted. Subsequently the particles are accelerated by the gases until they attain their final velocity (which is of the order of a km/s). This happens within the first one hundred kilometres.

Beyond 10^4 km, the neutral particles are influenced only by radiation pressure and solar gravitation:

$$F_{\mathrm{rad}} = \frac{e}{c} \pi s^2 \frac{\Theta}{4\pi r^2} \qquad (10.17)$$

$$F_{\mathrm{grav}} = \frac{GM}{r^2} \left(\frac{4}{3} \varrho \pi s^3 \right) \qquad (10.18)$$

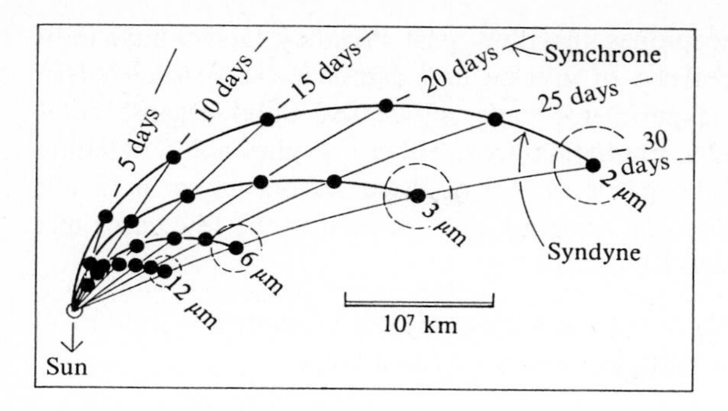

Fig. 10.9. Distribution of dust in the cometary tail. Particles of a given size are aligned along syndynes; particles emitted at the same time are aligned along synchrones

where e is the scattering efficiency of the particles, ϱ is their density, R their mean radius, Θ the solar flux at one astronomical unit, r the heliocentric distance, and M the mass of the Sun. From this we may derive – see (10.16) –

$$\frac{F_{\text{rad}}}{F_{\text{grav}}} \propto \frac{1}{\varrho s} \quad . \tag{10.19}$$

While the scattering efficiency does not depend on the radius, the radiation pressure becomes more effective the smaller the radius of the particles; in observed dust tails the ratio $F_{\text{rad}}/F_{\text{grav}}$ normally varies between 0.1 and 1.

Apart from the particles observed in the dust tail, cometary nuclei shed large particles ($s \gtrsim 30\,\mu$m) which are observed in antitails, the existence of which is simply a geometric effect. They are particles that are ejected with velocities that are close to zero, and which lie in the comet's orbital plane. When the Earth crosses this orbital plane, this collection of particles, seen from the side, appear to form a very narrow tail pointing towards the Sun.

Composition of the Particles

Preliminary information comes from infrared spectroscopy of comets: in many cases the characteristic emissions of silicates at 10 μm and 18 μm are seen superimposed on the comet's thermal continuum spectrum. In some cases, the signature of H_2O ice is seen at 3 μm. In addition, observation in the visible of the spectrum of Comet Ikeya Seki, which grazed the Sun in 1968, revealed emissions from numerous metals (Ti, V, Cr, Mn, Fe, Co, Ni, and Cu). These emissions had been produced by vaporization of the dust in the vicinity of the Sun. Measured abundances were in accordance with solar abundances.

Added to these spectroscopic observations there are a few results from in-situ observations:

1. analysis of stratospheric dust, which is assumed to be of cometary origin, and which has a composition close to that of type C1 chondrites;
2. mass-spectrometry and infrared-spectrometry measurements made on board

the Giotto and Vega spaceprobes that flew past P/Halley, and which indicate the simultaneous presence of silicate and carbonaceous particles (see Sect. 10.4.4). Laboratory experiments have shown that solid organic compounds may be formed by irradiating ices, either by ultraviolet radiation (Greenberg), or by energetic particles. All of these factors argue in favour of a "primitive" composition for cometary dust, at least in the few cases that have been studied (see Sect. 10.4.4).

Nevertheless the idea that there may be various classes of comets that differ in the average composition of their particles cannot be ruled out.

Dimensions of the Particles

Here again, the information that we have comes from two different sources. On the one hand, polarimetry and photometry of comets and of the zodiacal light, as a function of phase angle, in both the visible and infrared regions, provide constraints on the sizes and nature of the particles, by virtue of Mie theory. However, observations rarely allow us to separate the two parameters. In general, the observations are compatible with the so-called Sekanina-Miller distribution; where a is the radius of the particles and ϱ their density (in g/cm^3):

$$\begin{aligned} n(a) &= 0 \quad , \quad \varrho a < 0.45\,\mu\text{m} \\ n(a) &= (2a - 0.9)a^{-5} \quad , \quad 0.45\,\mu\text{m} < \varrho a < 1.3\,\mu\text{m} \\ n(a) &= a^{-4.2} \quad , \quad \varrho a > 1.3\,\mu\text{m} \quad . \end{aligned} \tag{10.20}$$

On the other hand, direct measurements, made when the Vega and Giotto probes flew through the coma of P/Halley, provide the flux and size of the particles encountered as a function of distance from the nucleus. In particular, this demonstrated the existence of a high proportion of particles with very low masses ($< 10^{-13}$ g), at large distances from the nucleus in the direction of the Sun, showing that they are largely unaffected by radiation pressure.

Dust Production Rates

Estimates of the rate of production, based on observations of comets at visible wavelengths, are difficult to make because they depend strongly on the scattering index and size of the particles. On the other hand, the mass of the particles may be estimated independently of their size by simply measuring the infrared flux; this is valid for particles whose size is less than the wavelength employed. Production rates measured to date vary from 10^4 g/s (for short-period comets, where activity has slowed down) to 10^6 g/s (for long-period comets starting to be active). As regards the ratio between the gas and dust production rates, it has been possible to estimate this in several cases by analyzing the dust tail by the Finson-Probstein method. The first ratios measured were of the order of 0.7. There are large uncertainties, however, and the possibility that this ratio varies from comet to comet cannot be discounted.

10.4.4 Cometary Material and the Interstellar Medium

One of the most remarkable results of the exploration of Comet Halley has been the discovery of carbonaceous material within the nuclear region of the comet. Numerous indications, both direct and indirect, confirm this:

1. the very low albedo (0.04) – which means that the comet is one of the darkest objects in the Solar System – and the high temperature (more than 300 K) of part of the surface, undoubtedly inactive, are probably caused by the presence of a layer of refractory carbon at least partially covering the nucleus;
2. the detection of large quantities of carbonaceous particles that are "primitive" in composition, the so-called "CHON" particles (Fig. 10.10);
3. the discovery of formaldehyde polymers $(H_2CO)_n$, as well as that of H_2CO itself;
4. the spectral signature at 3 μm of hydrocarbons that are in either a gaseous or solid form (Fig. 10.11).

All these indications reinforce the idea of a close link between cometary material and interstellar dust.

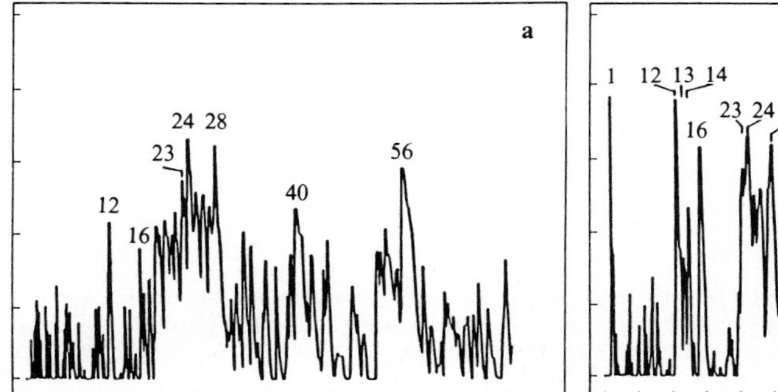

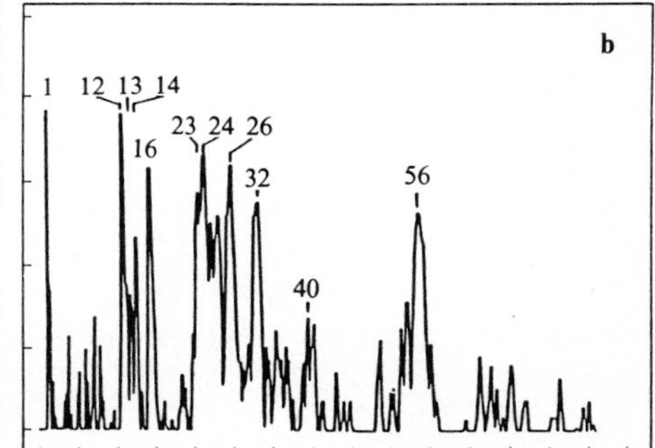

Fig. 10.10a,b. Spectra of two different types of particles obtained by the PUMA mass-spectrometer on board Vega 1 (from Halley's Comet). **(a)** a particle with a composition close to that of type C1 carbonaceous chondrites; **(b)** a primitive type of carbonaceous ("CHON") particle. Most of the particles show spectra of type **(b)**. [After J. Kissel et al.: Nature **321**, 280 (1986)]

We know that the signature at 3 μm is also observed in the interstellar medium. This has two components. The first, observed in emission at 3.3 μm close to UV sources, has been interpreted as the signature of large molecules of polycyclic aromatic hydrocarbons ("PAH"), temporarily heated by individual UV photons. The second component, observed in absorption at 3.4 μm is dense molecular clouds, has been attributed to the refractory mantles of organic particles consisting of saturated hydrocarbons. This second signature is also observed in the laboratory in the organic residue remaining after the irradiation of various ices (H_2O, CO, CO_2, CH_4) by a UV source or by a beam of energetic particles. In Comet Halley, the two signatures at 3.3 and 3.4 μm are observed in emission (Fig. 10.11). They may be caused by the same materials as in the interstellar medium (PAH, refractory organic material) or

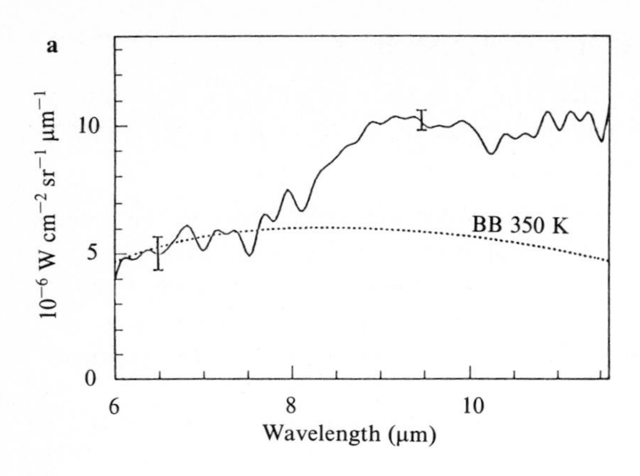

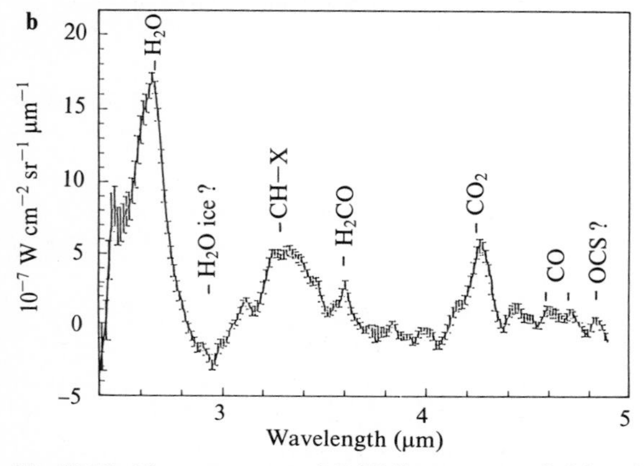

Fig. 10.11. The spectrum of P/Halley as recorded by the IKS experiment on board Vega 1, on 1986 March 6, in the 6–12 μm band (*top*) and 2.5–5 μm band (*bottom*). The wide emission beyond 8 μm **(a)** is caused by silicate grains. The emission at 3 μm **(b)** marked CH-X, is attributed to saturated or unsaturated hydrocarbons in either solid or gaseous form (M. Combes et al., Icarus **76**, 404, 1988)

simply by free hydrocarbon molecules – excited by infrared fluorescence like the other parent molecules – that are present in both saturated and unsaturated forms.

The existence of a refractory organic material on the nucleus or in its immediate neighbourhood is to be expected, given the comet's history. In the interstellar medium the refractory material appears to arise from the irradiation of particles of ice that contain carbon by UV radiation or energetic particles. With a comet, we may expect the surface of the nucleus to be irradiated by solar UV radiation and cosmic rays. As, however, the surface of the comet is completely renewed at each perihelion passage, there must be some doubt as to whether the exposure time of the particles at the surface is sufficient for the irradiation to be effective. Another possibility is that the cometary material was irradiated in depth at the beginning of the Solar System, before the cometary nucleus was formed. This theory is possible if, as suggested by theories of stellar evolution, the Sun passed through a "T-Tauri" stage, which typically shows very intense solar winds.

As it is possible to draw analogies between the irradiation mechanisms affecting both cometary and interstellar material, it is interesting to see if similarities also exist between their elementary compositions. The composition of interstellar material may be investigated by infrared spectroscopy of ices (H_2O, NH_3, CO, etc.) and of solid particles (PAH, carbon and silicate particles, etc.). From all these data one can reasonably conclude that the elements present in solid form (C, O, N, Si, etc.) have abundances comparable to their cosmic values, whilst for the hydrogen condensed in the grains, the H/O ratio typically varies between 1.5 and 2.2. For Comet Halley, the same exercise may be carried out, using the abundances measured for the parent molecules (see Table 10.4), making an assumption about the origin of the hydrocarbons responsible for the emission at 3 μm. It is then found that the cometary H/O ratio is also 1.5 to 2. The relative abundances of C, O, and N are also compatible with cosmic abundances, some of these elements being trapped in cometary particles. It will be seen, therefore, that it is possible to describe an elementary composition that applies to both cometary material and to interstellar dust. The "primitive" origin of comets, which has been suspected for a long time, is therefore definitely confirmed.

10.5 The Interaction of Comets with the Solar Wind

Comets are bodies with masses too low for them to have their own magnetic fields. However, their presence in the interplanetary medium locally modifies the magnetic field existing in the interplanetary medium (see Sect. 4.3.2). The situation is rather similar to that of Venus: a shockwave deflects the lines of force in the solar wind and heats the latter, at least when the comet is sufficiently close to the Sun for it to be surrounded by a coma and an ionosphere. When the comet is far from the Sun, and its production rate is negligible, the interaction between the comet and the solar wind is comparable to that of the Moon. As a comet's nucleus is very small, there is practically no modification of the surrounding solar wind (Fig. 10.12). The first stage in the interaction of the solar wind with a comet comes when the former penetrates the halo of hydrogen atoms, the velocity of which may be of the order of 10 km/s

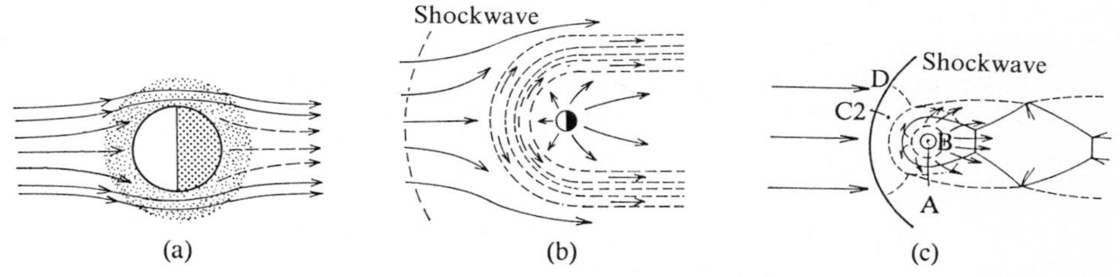

Fig. 10.12. Three types of interaction between the solar wind and a comet. (1) Lunar-type interaction (weak interaction), found in low-activity comets. (2) Venus-type interaction, corresponding to intermediate-activity comets. (3) Solar-wind-type of interaction (strong interaction), which applies to very active comets. [After W. Ip, W. Axford: *Comets,* ed. by L. Wilkening (© University of Arizona Press, Tucson 1984)]

(see Sect. 4.3.2). This velocity is negligible in comparison with that of the solar wind, which encounters the obstacle presented by the comet at about 400 km/s. This interaction creates a shockwave, some 10^5–10^6 km in front of the comet's nucleus. The charged particles are unable to move across the magnetic-field vector $\boldsymbol{B}$, but instead spiral round the lines of force. As a result, cometary ions are captured on the solar wind's lines of force. The solar wind thus gains mass, and is progressively braked and slowed down as far as a contact surface where the solar-wind pressure and the coma's dynamic pressure are equal. Within this contact surface, neither the magnetic field nor the plasma are very strong.

The closer to the nucleus, the more the solar-wind magnetic-field lines are compressed, causing an increase in the magnetic field. They then drape round each side of the nucleus, forming two lobes of opposite polarity. Between these two lobes, the magnetic field ought to fall to a value close to zero, and one ought to observe a current sheet, where the plasma density is very high. This model, by Biermann and Alfvén, has been confirmed by measurements made by the ICE probe on Comet Giacobini-Zinner (see below).

The structure of the ion tail of a comet often shows a series of pairs of symmetrical, ionized rays, with diameters of 10^3 to 10^4 km, that are inclined at low angles to the central axis of the tail (see Fig. 10.13). MHD models suggested that the existence of these ionized rays was correlated with the existence of tangential discontinuities in the interplanetary magnetic field. An ionized ray would be caused by narrow filaments where the density is high and the field weak, separated by regions with the opposite characteristics. Examples are known of ion tails that consist of a large number of these narrow, ionized bundles.

The last stage in the formation of these ionized rays is their evolution towards the central ionized tail, and then their amalgamation with the latter (Fig. 10.14). Presumably cometary ions, swept up by the solar wind from the outer coma, form the fast particles in the ionized rays, whilst the plasma originating from the inner coma, with a lesser velocity, is the principal component of the central ion tail.

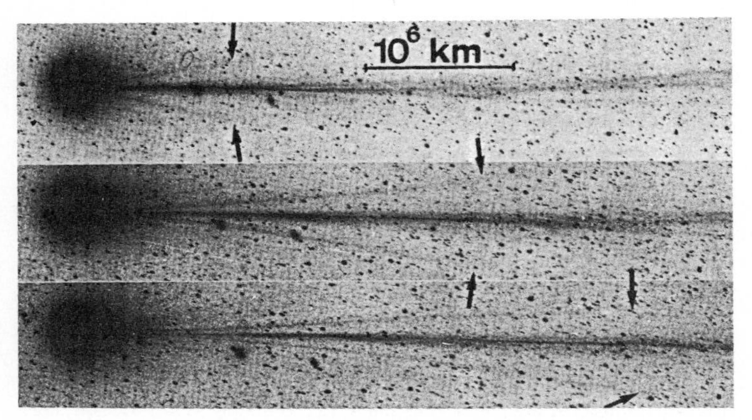

Fig. 10.13. Changes in a symmetrical pair of ionized rays (Comet Kobayashi-Berger-Milon). [W. Ip, W. Axford: *Comets*, ed. by L. Wilkening (© University of Arizona Press, Tucson 1984)]

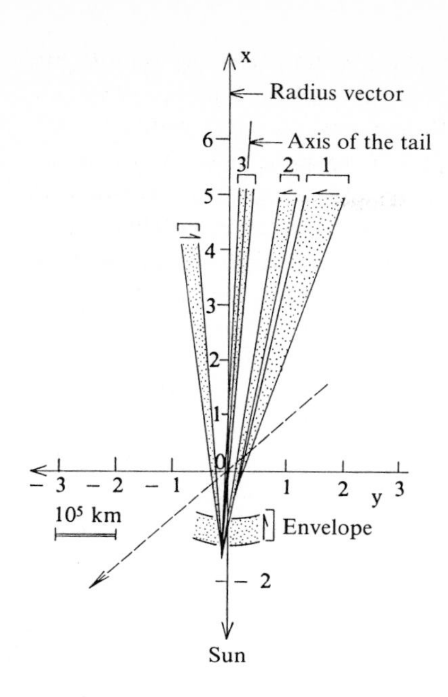

Fig. 10.14. The movement of four ionized rays observed in Comet Morehouse in 1908. [After W. Ip, W. Axford: *Comets*, ed. by L. Wilkening (© University of Arizona Press, Tucson 1984)]

Apart from these ionized rays, cometary plasma tails show a large variety of structures: waves, knots, helices, etc., which vary both in intensity and in their position in space, on time-scales that are of the order of a few hours. Another characteristic phenomenon of ion tails, observed on numerous occasions over a long period, consists of an apparent disconnection of the plasma tail behind the coma, followed by the tail reforming from ionized rays that converge towards its position. This effect arises when the nucleus crosses a sector boundary in the solar wind's $\boldsymbol{B}$ magnetic field, i.e. when it crosses a region where the magnetic field changes sense. A new plasma tail is then observed, whilst the old tail, of opposite polarity, becomes detached.

The first mission designed to study the interaction between a comet and the solar wind in situ, was the ICE mission, which flew past Comet Giacobini-Zinner on 1985 September 11 at a minimum distance of 7800 km. This was originally the ISEE-3 probe, launched by NASA several years previously to study the solar wind. Its path was modified to allow it to encounter the comet. As a result of its original purpose, the probe carried only instruments designed to investigate plasmas.

ICE began to detect energetic particles (with energies of several tens of keV to a few MeV) at a distance of several million kilometres from the comet. The probe did not find a distinct shock, but turbulence was encountered between about 150 000 km and 20 000 km. Closer in, the probe crossed the two "lobes", where the magnetic field is draped round the comet and where the intensity was of the order of 60 γ. The magnetic field was of opposite polarity on the two sides of the *neutral sheet*, where it was essentially zero – the actual value was 5 γ, whereas that of the ambient interplanetary medium was 8 γ). Within the lobes the electron density followed a R^{-2} law (where R is the distance from the centre of the comet), and rose rapidly in the neutral sheet, reaching a value of 10^3 electrons/cm^3. Beyond the neutral sheet, the structure was essentially symmetrical. The ICE probe also detected ions in the mass-range 14–33: the most abundant were H_2O^+, followed by a group at masses 23–24, which can be identified as Na^+ and Mg^+, and then the CO^+ group.

Initial results from the Giotto and Vega probes gave comparable information about Comet Halley. Unlike the ICE probe, these probes did not cross the neutral sheet. A magnetic field of about $60\,\gamma$ was detected; the mass-spectrometers found that the H_2O^+ group was very strong (the H_3O^+ ion being about ten times more abundant than H_2O^+), and many other ions were present.

11. Interplanetary Dust, Micrometeorites and Meteorites

Apart from radiation and particles of solar or galactic origin, interplanetary space contains dust and rocky bodies of all sizes. Their collisions with the Earth cause meteorite falls and showers of shooting stars. Scientific study of this extraterrestrial material, which has gone on for centuries, has still not provided a definite answer to the question of their origin: are they derived from cometary nuclei or from minor planets?

It would seem, however, that we can ascribe a cometary origin to the smallest interplanetary particles, with sizes close to a micrometre. On the other hand, the largest, that is those with masses greater than a kilogramme, which are the source of the meteorites that fall on Earth, derive from asteroids. Both numerically and in total mass, most of the material reaching the Earth consists of small interplanetary particles: their number decreases very rapidly as their size increases. This is why bodies with dimensions reaching or exceeding a hundred metres are not generally considered to form part of the interplanetary medium. They are classified as asteroidal debris, but it is not possible to set a rigid, lower boundary to this in size.

The lifetime of interplanetary particles, whatever their size, is much less than their "planetary" lifetimes. All derive from *parent bodies* from which they have been lost only recently in comparison with their age, which is at least equal to that of the Solar System. The condensation of the parent bodies goes back to the earliest phases in the evolution of the early solar nebula, during the cooling that occurred about 4.55 thousand million years ago.

The smallest of these parent bodies, with sizes not exceeding a few kilometres, were able to radiate away to space the energy that they accumulated in their accretion, because of the high ratio between their surface area and their volume. These objects therefore remained at a sufficiently low temperature for there to be no internal fusion, which as we know, produces mineralogical *differentiation* through recrystallization. As a result, cometary nuclei and type-C asteroids are the most primitive bodies in the Solar System, as they have not undergone any major metamorphic episode since their initial accretion. In addition, we cannot rule out the idea that some of the grains present in the proto-solar nebula, which had condensed in the atmospheres of other stars, may have survived the nebula's later thermal evolution. They may have been incorporated, without any major alteration in their physical, chemical or isotopic properties, into small-sized planetary bodies. Overall then, at least part of the solid extraterrestrial material is a sample of the primitive material from which the Solar System as a whole was formed. Study and analysis of these samples has become a new discipline uniting astrophysics and planetology.

11.1 The Mass-distribution

The mass-distribution of interplanetary particles at a heliocentric distance of one astronomical unit (Fig. 11.1) has been obtained directly by satellite measurements, down to a mass of about 10^{-6} g, as well as by observation of micrometeorite impact craters on lunar samples and metal foil exposed to space, either on the surface of the Moon, or in Earth orbit. For larger masses it has been derived from observations of meteors: meteors with masses less than one gramme may be detected by the radar echo from the ionized trail that they leave in passing through the atmosphere. For still larger masses, recourse is made to photographic observations. Finally, for the very largest masses, the flux is deduced from the size-distribution of craters on the surface of the Moon. These data enable us to describe the overall mass-distribution of the interplanetary-particle flux in terms of three relationships of the form $N(m) = Am^{-\gamma}$, where $N(m)$ is the sum total of all particles with a mass greater than m, per unit surface area, per unit time. As can be seen from the curve in Fig. 11.1, three mass-regions may be defined, corresponding to three different pairs (A, γ): for $m < 10^{-7}$ g, $N(m) = 1.45\,\mathrm{m}^{-0.46}$; for $10^{-7}\,\mathrm{g} < m < 1$ g, $N(m) = 9.16 \times 10^{-6}\,\mathrm{m}^{-1.213}$; and finally for $m > 1$ g, $N(m) = 10^{-5}\,\mathrm{m}^{-1.3}$. In all three cases, $N(m)$ is expressed in grains per cm^3 and millions of years, per 2π steradians.

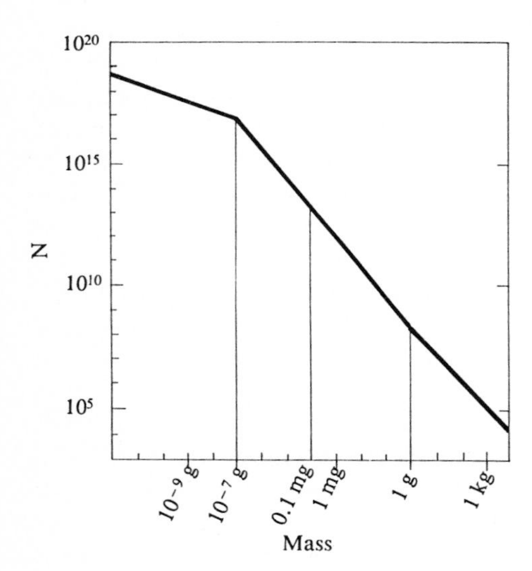

Fig. 11.1. A schematic representation of the mass flux of interplanetary grains

It can be seen that the contribution to the overall mass rapidly becomes negligible for particles where $m < 10^{-7}$g, and that most of the particles have masses close to 10^{-6} g, i.e. have a radius of about twenty micrometres. As regards the amount of material intercepted by the Earth, we obtain an overall mass of a few thousand tonnes per year. The number of bodies with masses larger than one kilogramme, which are the source of meteorites, is only of the order of 10 000 per year over the whole surface of the Earth. As for the largest impacts, it has been estimated that an object with a mass of about 65 000 tonnes, comparable to the one that formed Meteor Crater in Arizona, falls on dry land about once in every 25 000 years. For a body 1 km in diameter, there is an impact only once every one hundred million years. Although rare, such impacts may nevertheless play an important part, in particular by lifting into the atmosphere hundreds of cubic kilometres of material, or about one hundred times the most powerful volcanic eruption. Such a mass of particles, enveloping the Earth for several years, could have a profound influence on the climate by intercepting sunlight. An event of this type might have been responsible for the extinction of a large number of animal species at the end of the Cretaceous.

Has the flux of interplanetary particles and meteorites that we measure today remained the same since the birth of the Solar System? Observation of the surfaces of the inner planets, the Moon, and of some of the satellites of the giant planets, appears to indicate that the meteorite flux was much greater in the first few hundred million years after the formation of the Solar System. According to some authors, this flux may have been several thousand times greater than its current value. We are dealing with what is known as the *accretion tail*: once the formation of the planets (and the Moon) had finished, innumerable pieces of debris from the planetary accretion process remained. Those that intersected the orbits of the planets produced an intense meteoritic bombardment, which was responsible for the giant craters that are still visible on the Moon and Mercury. It is these craters which, when later filled by lava that welled up from below, became the dark lunar *maria*, which contrast with the surrounding light highlands. Once the Solar System had been swept clear of most of this debris – the meteorite flux decreasing exponentially with time – it would appear that over the last three thousand million years the rate has remained constant. This flux is considered to be of "secondary" origin, being derived from asteroids and comets. Meteorites, dust grains and large-size interplanetary objects are fragments of material produced by collisions between asteroidal bodies, whilst micrometeorites and small-size interplanetary particles result when solar radiation causes the icy nuclei of comets to sublime and frees dust grains trapped within them.

11.2 The Spatial Distribution of Interplanetary Material

Observation of planetary surfaces and the measurements made by satellites (the Pioneer, Voyager and Helios probes, for example), only tell us about the distribution of particles close to the plane of the ecliptic. Observation of sunlight scattered by interplanetary particles (the *zodiacal light*) in principle allows us to determine the distribution outside the ecliptic. The results obtained show that the spatial distribution depends strongly upon the size of the particles and upon the class of objects that is being considered.

For the smallest particles, which form the *zodiacal dust cloud*, some characteristics have been obtained by analysis of the zodiacal light, notably thanks to experiments on board the French D_2A and D_2B satellites and also on the Franco-Soviet space-flight in 1982 (the Piramig experiment). The most abundant particles have sizes of the order of a micron, or a few tens of microns at the very most. In total there are only about a few hundred particles per cubic kilometre close to the Earth's orbit, and the overall mass of the zodiacal dust cloud seems to lie between 10^{18} and 10^{23} g. The particles show a strong concentration towards the plane of the ecliptic, which increases the amount of light scattered in the zodiacal constellations. They are, of course, by no means absent outside that plane, but the distribution of particles is symmetrical about the ecliptic. The scattered light can also be seen in the antisolar direction, where it is known as the *Gegenschein*. There is a notable decrease in concentration of particles with heliocentric distance. This has also been established by space experiments, notably by the Pioneer probes, which showed no obvious increase in the flux of very small particles in crossing the asteroidal belt.

For particles with greater masses (10^{-2} g) and the objects responsible for the impact craters on Mercury, the Moon, and Mars, we find, on the other hand, no marked variation in concentration with distance from the Sun, except that there is an increase by a factor of about ten around the asteroidal belt.

This difference in the way in which the distributions of large and small particles vary with heliocentric distance agrees with there being two different sources (cometary and asteroidal), for these two populations.

The lifetimes of interplanetary particles depends on their mass and composition. Apart from the gravitational attraction of the Sun and the planets, the solar-radiation pressure tends to modify their orbits. It is obvious that the gravitational acceleration of a particle depends on its mass, whilst radiation pressure increases with surface area. The second factor will only affect the smallest particles, with radii lower than a few microns, becoming negligible for particles several tens of microns in diameter. On the other hand, for particles that have dimensions comparable, or less than, the wavelength (a few thousand angströms), the efficiency of collisions with photons decreases as a result of scattering, and theories such as that developed by Mie have to be applied to explain the overall balance of forces. Finally, the *Poynting-Robertson Effect* must be taken into account. This results from the fact that the – radial – force produced by the photons, which propagate with the velocity of light c, is exerted on a body moving at a velocity v, which is not co-linear with c. The transfer of momentum induces a decrease in the tangential component of the particle's velocity. In other words, as seen in the reference frame of the body in motion, the direction of the incident photons is always ahead of the direction of the Sun. It is as if the impact of the photons give rise to a braking effect, the strength of which relative to the radial force (the radiation pressure) is proportional to the ratio v/c between the velocity of the particle and that of the photons. This braking force continually reduces the eccentricity of the particle's orbit, reduces it to a circle and eventually causes the path to become a spiral. It has been calculated that a particle, one micrometre in size, at one astronomical unit, will fall into the Sun in a few thousand years. As its distance from the Sun decreases, a particle loses more and more mass under the combined effects of radiation and of the solar wind, which vaporize and partially pulverize it, and eventually lead to its disappearance.

For most of the particles with micron or sub-micron sizes that form the greater part of the zodiacal dust cloud, the lifetime, after their ejection from their parent body, is very short when compared with the age of the Solar System. It should be

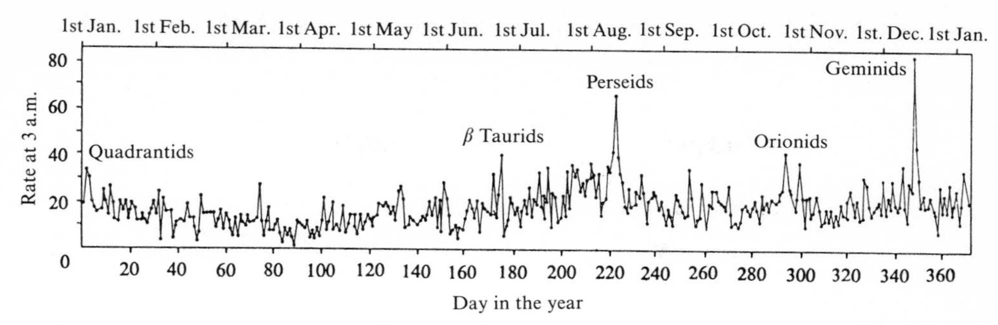

Fig. 11.2. Variation in the meteor hourly rate throughout the year. [After D.W. Hughes: *Cosmic Dust*, ed. by J.C. McDonnell (John Wiley and Sons 1978)]

noted that as they orbit between the planets the particles are in direct contact with particles from the Sun, the solar wind, and low-energy cosmic rays. It is through the effects of this irradiation that it is possible, in some cases, to identify particles collected on Earth as being of extraterrestrial origin.

If we consider the mass of the zodiacal dust cloud and the lifetimes of the particles of which it consists, it is obvious that it cannot be a cloud formed a long time ago that is slowly decaying, but that there must be a permanent source of replenishment. This must be adding tonnes, or even hundreds or thousands of tonnes of material to it every second. Several factors argue in favour of comets being the principal source of these small particles.

The first argument derives from the variation in the particle concentration as a factor of heliocentric distance, which reflects the spatial distribution of the perihelia of cometary orbits.

The second factor is the existence of meteor swarms, the cometary origin of which is now well-established. It is found that the number of *meteors* observed (i.e. those with magnitudes below 2), is not constant throughout the year (Fig. 11.2). Although the average rate is about ten meteors per hour, there are recurrent periods when the rate rises to hundreds, thousands or even more per hour during spectacular showers (such as that of the Leonids in 1966). It is also known that during these showers the trails appear to come from a single point on the sky, the *radiant*, which is given the name of the constellation in which it is located. Table 11.1 gives details of the most intense meteor showers, together with the names and periods of the comets with which they are associated.

These showers originate as follows. When a comet approaches perihelion, sublimation of the nucleus liberates particles trapped in the ice of which it is formed. Through the effects of radiation pressure, the smallest particles acquire hyperbolic trajectories and go to replenish the zodiacal dust cloud, losing their original orbital identity. Those with sizes (of at least a few micrometres) that are sufficiently large for their orbits not to be affected, except by their low, initial ejection velocities, follow paths close to that of the original comet. They orbit in a *swarm*, a region of space surrounding the cometary orbit (Fig. 11.3). When the Earth encounters such a swarm it intercepts the particles, the more the closer it passes to the orbit of the

Table 11.1. Dates the Earth passes through the principal meteor swarms and their associated periodic comets

Swarm	Date	Associated comet	Period
Lyrids	10–20 April	Thatcher	
Aquarids	1–8 June	Halley	76 yrs
Orionids	18–26 October	Halley	76 yrs
α Taurids	24 June – 6 July	Encke	3.3 yrs
Perseids	25 July – 17 August	Swift-Tuttle	120 yrs
Draconids	9–10 October	Giacobini-Zinner	6.4 yrs
Andromedids	2–22 November	Biela	
Leonids	14–21 November	Temple-Tuttle	33.2 yrs
Geminids	7–15 December	?	
Ursids	17–24 December	Tuttle	13.6 yrs

comet itself. All these particles have velocity vectors that are essentially the same as regards magnitude and direction. As the particles have velocities that may reach several kilometres per second, if they are larger than about one hundred micrometres in size, they are destroyed when they encounter the atmosphere. The majority of visible meteors arise from particles that are a few tenths of a millimetre to a few centimetres in radius.

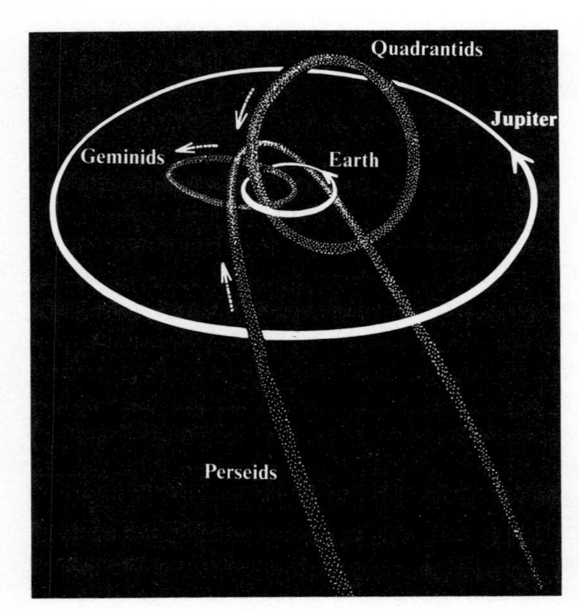

Fig. 11.3. A representation of the position in space of a few of the principal cometary swarms. [After D.W. Hughes: *Cosmic Dust*, ed. by J.C. McDonnell (John Wiley and Sons 1978)]

11.3 Meteorites

11.3.1 Introduction

As it sweeps through interplanetary space, the Earth encounters solid bodies of all sizes that are found within it. But because of the shield formed by the atmosphere, the size-distribution of objects recovered on the ground is very different to that prevailing in the interplanetary medium. All particles with sizes between 0.1 mm and a few centimetres burn up completely in the atmosphere, these are the meteors that we have just described. The smallest particles have a high surface/volume ratio, which allows them to radiate away the energy that they acquire from friction sufficiently rapidly for them to survive their passage through the atmosphere, and also to decelerate from more than 10 km/s to a few cm/s at the base of the stratosphere. They fall to the surface and are found in surroundings as different as polar ice-caps and oceanic sediments. As we have mentioned, comets are the principal source of these *micrometeorites*.

Bodies larger than about ten centimetres are properly known as *meteorites*. During their passage through the atmosphere their surfaces are also very strongly heated; despite *ablation* some of their mass remains and reaches the ground, with a velocity of about one hundred metres per second. It is only the very largest objects, several tens of metres in diameter, which create large impact craters, that reach the ground with little change in their velocity.

It was Chladni (1756–1827) and Biot (1774–1862) who showed that meteorites were of extraterrestrial origin. This began a period of systematic research about these objects, which are classed according to whether their fall was observed or not. For more than two centuries, meteorites were the sole source of extraterrestrial material available to the scientific community. Since 1969 lunar samples returned by the Apollo and Luna missions have been added, and one could say that the study of meteorites was rejuvenated by the exceptionally precise methods of analysis that were developed for use with lunar samples.

Nowadays more than 3000 meteorites have been catalogued, generally named after the site of their fall. This number is in the process of increasing rapidly: recently rich "fields" of meteorites have been discovered on the Antarctic ice-cap. These meteorites are all the more interesting in that they have been preserved essentially intact for thousands of years. The number of objects preserved in museums around the world greatly exceeds the 3000 meteorites catalogued, because each meteorite is usually fragmented during its entry into the atmosphere, and the pieces may be scattered over hundreds of square kilometres.

11.3.2 Classification

Mineralogical and chemical study of meteorites rapidly showed their great diversity. Some are almost entirely metallic (iron and nickel), other resemble terrestrial rocks, and yet others are rich in carbon and are very friable. In addition, many of them contain spherical inclusions, known as *chondrules*, which may be preserved to a greater or lesser degree. Finally, the size of the component crystals, which in terrestrial rocks reflects the faster or slower rate of cooling, may vary very greatly in meteorites of the same chemical and mineralogical composition. These three criteria gradually became the bases for the classification of meteorites that is normally used. Meteorites are therefore divided into three major classes according to their metal content: the *iron, stony-iron* (*lithosiderite*), and *stony* meteorites. About half the meteorites are *breccias*, conglomerates of materials having either the same composition (*monomict breccias*) or different compositions (*polymict breccias*). Among the breccias, some, known as *gas-rich* breccias, contain traces of rare gases implanted by the solar wind, and traces of irradiation by solar cosmic rays.

Iron meteorites consist almost entirely of iron and nickel in two principal forms: *kamacite*, with a nickel content less than 8 %, which crystallizes in the cubic system, and *taenite*, where the nickel content is greater than 20 %, and which crystallizes in the face-centred cubic system. The *hexahedrites*, which are low in nickel (< 6 %), consist of large cubic crystals of kamacite. When etched by acid, a fine network of lines appears: these *Neumann lines* were probably caused by the shock undergone by the meteorite when it was separated from its parent body. The *octahedrites*, richer in nickel (12 %) are more numerous and consist of plates of kamacite bordered by taenite, surrounding regions of plessite (interlocking crystals of kamacite and taenite). The plates form a geometrical pattern: the *Widmanstatten pattern*, which is orientated parallel to the faces of an octahedron, whence the name of the class. The higher the nickel content, the narrow the plates of kamacite. This very characteristic structure may be explained on the basis of extremely slow cooling, or, on the other hand, by sudden solidification from a phase relatively rich in nickel.

The *lithosiderites* consist of intimately mixed metallic and silicate phases. Two classes are distinguished: the *mesosiderites*, where the silicate phase primarily consists of feldspars and pyroxenes; and *pallasites*, where olivine is dominant. In the latter, a metallic matrix, similar to that in the octahedrites, encloses olivine crystals, whilst it is usually the converse in mesosiderites (silicate matrix, metallic inclusions). Pallasites have been formed through the invasion by a liquid metallic phase at the boundary between the olivine-rich mantle and the core of the parent body, whilst the metallic inclusions in the mesosiderites were probably incorporated in a solid form.

The *stony meteorites* are by far the most numerous group of meteorites (more than 90 % of observed falls). They are divided into *chondrites*, which contain chondrules, and *achondrites*, where they are absent. Chondrules are spherical inclusions about one millimetre in size, which are perfectly recognizable in some chondrites, such as the Saint-Mesmin meteorite, and are almost completely incorporated into the matrix in others, such as the Mezo-Adaras stone. Chondrules have been observed only in meteorites, and their origin is still poorly understood. A distinction is drawn between *carbonaceous chondrites*, *ordinary chondrites*, which mainly consist of olivines and pyroxenes, and *enstatite chondrites*, where almost all the iron is present as a metal or sulphide. The achondrites are the meteorites than most resemble certain terrestrial or lunar rocks.

The *carbonaceous chondrites* consist of a very fine-grain matrix, rich in carbon, and may contain inclusions of olivine, pyroxene, glasses or metal. These are the only meteorites in which silicates in sheets (like mica) are found. Generally these are in hydrated form and their water content may exceed 20 %. These meteorites are classified into Types I, II and III, in decreasing order of carbon, water, and iron-sulphide (troilite) content. The very friable Orgueil meteorite, for example, consists of very small grains. The carbon content exceeds 3 %, and that of FeS 15 %. It is a type I carbonaceous chondrite. Allende, which is the most massive carbonaceous chondrite (two tonnes), is type III. It contains very large numbers of light inclusions, which are refractory and in which the first *isotopic anomalies* (see Sect. 11.3.6) were found. These meteorites, which have no equivalent among terrestrial rocks, are usually considered to have undergone least alteration since their formation.

Ordinary chondrites are divided into H, L and LL chondrites, in according to their decreasing iron content (*High, Low* and *Low-Low*). In each of these three groups they are classified by their degree of recrystallization on a scale of III to VI. This may reflect an increasing depth of burial within the parent body (or bodies). The major minerals are olivines, pyroxenes and feldspars, like terrestrial or lunar rocks, and there is often a metallic phase.

The *achondrites* consist of minerals (pyroxenes and plagioclases) similar to those observed in lunar rocks and terrestrial basalts. They are divided into *aubrites*, which consist almost entirely of enstatite, a magnesium-rich pyroxene; and achondrites that consist of both pyroxenes and plagioclases: the *eucrites*, the *howardites* and the *diogenites*, classified according to their decreasing content of iron and calcium. The eucrites are the meteorites that most resemble terrestrial basalt. Howardites are polymict breccias, produced by the fragmentation and subsequent compaction of different types of rock, whilst the diogenites are monomict breccias.

11.3.3 The Origin of Meteorites

The great diversity of meteorites is directly linked to that of their origin and their history within their parent bodies. In particular, their mineralogical composition reflects differing degrees of thermal evolution: most micrometeorites, as well as certain carbonaceous chondrites are *primitive*, in the sense that they have not been subject to sufficiently high temperatures for alteration of their component minerals to have taken place. Ordinary chondrites may have be subjected to more or less pronounced metamorphism, producing solid-phase changes. Finally, the achondrites, the stony-iron and the iron meteorites have obviously been formed by complete melting, followed by recrystallization (or *differentiation*). Of the possible parent bodies, comets have remained at very low temperatures (below 300 K) since their formation. Although they could be the source of primitive meteorites, particularly those of very small sizes, they are obviously not the source of differentiated meteorites, the achondrites, iron or stony-iron meteorites. For this reason it is generally accepted nowadays that most meteorites derive from asteroids, whose range of types covers the different classes of meteorites in quite a satisfactory manner. The achondrites may be derived from the crust or mantle of a differentiated asteroid, iron meteorites from the core, and stony-irons from the core-mantle boundary. This plausible scheme requires a significant source of heat to melt the asteroid. Neither the energy liberated during its accretion, nor that released by radioactive elements with long half-lives (uranium, thorium, and potassium) are sufficient, because the small size of these bodies (the largest being a few hundred kilometres in diameter) causes them to cool rapidly. On the other hand, the energy provided by "short-lived" elements (a few million years at the most), such as aluminium-26, would suffice to melt bodies more than about ten kilometres in diameter. Cometary nuclei and the smallest asteroids would escape melting.

The brecciated structure in many meteorites is generally explained as resulting from violent impacts on the surface of the parent body, which would have both fractured and compacted the substrate through the compression waves that they produced. Among the breccias, those formed of grains exposed at the surface of a regolith would correspond to the so-called gas-rich meteorites, because the grains would be enriched in volatile elements implanted by the solar wind.

In distinction to a cometary origin, an asteroidal origin poses the dynamical problem of how meteorites were transferred from the main belt to the Earth. One possibility considered is the following: collisions within the main belt and subsequent gravitational perturbations create a specific family of asteroids, the *Apollo-Amor* group, the orbits of which come close to that of the Earth. This family may be one of the main reservoirs for meteorites.

The transit time between this reservoir and the Earth may be deduced from the degree of irradiation of these meteorites by cosmic rays, which induce nuclear reactions within them. From this it appears that iron meteorites have often passed several hundred million years in space. Stony meteorites have generally undergone a much shorter exposure time, less than one hundred million years. This difference is still poorly understood. In addition, exposure ages often appear very similar for meteorites of a particular class. This suggests that the reservoirs of different families of meteorites are repopulated by a small number of events, such as the fragmentation of a nearby asteroid.

11.3.4 Chemical Composition of Meteorites

The systematic study of meteorites shows one important result: all the chondrites have very similar chemical compositions. Table 11.2 indicates the elementary abundances (by number of atoms) in the Orgueil meteorite, normalized to that of silicon (line a).

In addition, the abundances of nearly all the elements are close to those observed in the atmosphere of the Sun (line b). The only exceptions relate to volatile elements (such as hydrogen, carbon, oxygen and the rare gases), which are very underabundant in chondrites, and lithium, which is burnt in the Sun by nuclear reactions. This

Table 11.2. Elemental abundances (by number of atoms) in the Orgueil meteorite (a) and in the solar atmosphere (b), normalized to that of silicon

	H	C	O	Na	Mg	Al	Si	S	Ca	Cr	Fe
a	6.2	0.7	8.2	0.07	1.1	0.1	1	0.5	0.06	0.01	0.9
b	30 000	12	21	0.06	1.1	0.09	1	0.5	0.07	0.01	0.8

result reinforces the theory that the chondrites are material that condensed from the primitive solar nebula and has undergone little subsequent chemical modification. The achondrites on the other hand have very varied chemical compositions because of the differentiation to which they have been subjected.

11.3.5 The Isotopic Composition of Meteorites

Overall, the isotopic composition of meteorites is similar to that of terrestrial rocks and lunar samples. Important departures are, however, observed for some elements, and these are attributed to three physical chemistry processes: mass fractionation; nuclear reactions induced when the grains are irradiated; and the decay of radioactive elements, which enriches the grains in daughter isotopes. This last process has a very important application: the concentration of daughter isotopes allows the principal stages in the evolution of the meteorites to be dated.

Mass Fractionation

This process, which modifies the isotopic ratios of an element, arises with any change of phase, diffusion or chemical reaction. For example, the three isotopes of oxygen are found in slightly different ratios in ice, liquid water, and water vapour. The changes in the isotopic ratios of an element depend, nevertheless, on a simple law:

$$N_{A+1}/N_A = (1+f)(N_{A+1}/N_A)_{\text{ref}} \tag{11.1}$$

where f is the fractionation coefficient per unit mass, N_{A+1}/N_A is the ratio between the numbers of isotopes of mass $A+1$ and A respectively, in a given sample, $(N_{A+1}/N_A)_{\text{ref}}$ being the corresponding ratio in a reference sample. In the same way one may derive:

$$N_{A+2}/N_A = (1+f)^2(N_{A+2}/N_A)_{\text{ref}} \tag{11.2}$$

and so on, with the result that the abundance ratios of isotopes of successive masses $A+i$, referred to the corresponding ratios in a reference sample, fall on a straight line, of slope f, and where for small values of f:

$$\frac{(N_{A+i}/N_A)}{(N_{A+i}/N_A)_{\text{ref}}} = (1+f)^i \sim (1+if) \quad . \tag{11.3}$$

Frequently the δ *notation* is used, where δ_i is defined by:

$$\delta_i = 1000\frac{(N_{A+i}/N_A) - (N_{A+i}/N_A)_{\text{ref}}}{(N_{A+i}/N_A)_{\text{ref}}} \quad \text{whence} \tag{11.4}$$

$$\delta_i = 10^3 if \quad . \tag{11.5}$$

If, using the δ notation, the isotopic ratios are represented by the pairs (δ_i/δ_j), it will be seen that the ratio δ_i/δ_j is equal to i/j, independent of f, which means that any sample, whatever the degree of isotopic mass-fractionation that it has undergone, will be represented on a (δ_i/δ_j) diagram by a point lying on a single straight line of slope i/j. Thus, for oxygen, all the samples that came from the same initial "reservoir" should lie on a line of slope 1/2, known as the *fractionation curve* (see Fig. 11.4).

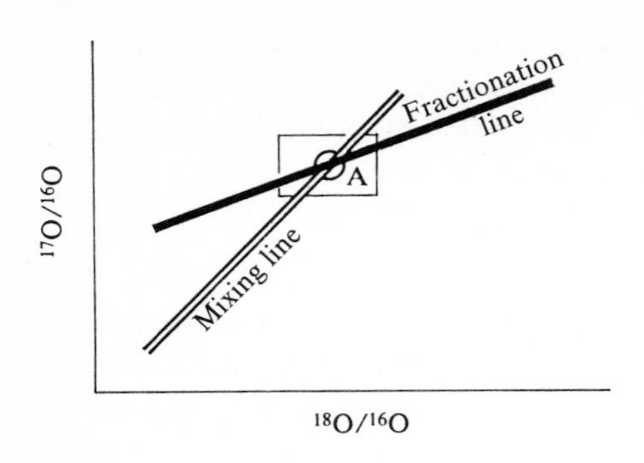

Fig. 11.4. Isotopic diagram for oxygen. [After R.N. Clayton et al.: Science **182**, 485–88 (© AAAS 1973)]

Cosmic-ray Induced Radioactivity

The bombardment of grains by solar and galactic cosmic rays produces isotopes that may be detected in some samples. Given that this radiation does not penetrate very deeply into the material, the concentration of daughter isotopes allows the length of time that the grains were exposed at the surface of their parent bodies, or were in orbit, to be measured. The main reactions are produced by cosmic-ray protons (H^+) and α-particles (He^{++}) in the most abundant atoms in the grains (C, N, O, Na, Mg, Al, Si, Fe). Among the stable daughter isotopes we may mention: ^{10}Be, ^{21}Ne, and ^{53}Mn. For example, the ^{21}Ne is produced by the chain of reactions:

$$^{24}\mathrm{Mg}\,(p,\alpha)^{21}\mathrm{Na}\,(\beta^+)^{21}\mathrm{Ne} \quad .$$

The decay of ^{21}Na is extremely rapid (23 s), so irradiation by cosmic-ray protons of the magnesium atoms in a grain produces ^{21}Ne that remains trapped within the grain.

Radioisotope Dating

Some of the isotopes trapped within grains as they crystallized are unstable, and have since decayed into daughter nuclei. The excess of the latter can, in certain cases, be detected and measured, allowing the date of formation of the grains to be established, once the radioactive decay rate λ[1] is known. Consider, for example, a system (P, D) of parent isotopes P that decay into daughter isotopes D, where $P(t)$ and $D(t)$ are the abundances of P and D isotopes respectively at time t. Between the times t and $t + dt$, $dD(t)$ isotopes D appear, and $dP(t)$ isotopes P disappear:

$$dD(t) = -dP(t) = \lambda P(t)dt \quad . \tag{11.6}$$

By integration, we obtain:

$$P(t) = P(0)\mathrm{e}^{-\lambda t} \quad \text{and} \tag{11.7}$$

[1] The radioactive decay constant is the inverse of the time T_e after which the number of parent isotopes is reduced by the factor e (e = 2.171828). In a similar way the half-life ($T_{1/2}$) is defined as the time after which the number of parent isotopes is reduced by half: $T_{1/2} = (\ln 2)T_e = 0.693\,T_e$.

$$D(t) = D(0) + P(0)(1 - e^{-\lambda t}) \quad \text{or else} \tag{11.8}$$

$$D(t) = D(0) + P(t)(e^{\lambda t} - 1) \tag{11.9}$$

equations in which $P(0)$ and $D(0)$ are the "initial" concentrations of the isotopes P and D respectively, i.e when $t = 0$.

Determining the "age" of a sample amounts to dating this initial time, that is, to determining t, which corresponds to the time separating it from the present day. It will be seen from (11.9) that measuring the P and D abundances in a sample does not suffice to determine t, given that we do not know $D(0)$. $D(0)$ depends on two factors: the abundance of the isotope in the reservoir from which the sample condensed, and the condensation process itself, which may trap certain isotopes in preference to others as a function of the physical conditions prevailing at the time (temperature, pressure, and kinetic energy). The indeterminacy of $D(0)$ may be avoided as follows: one assumes that the condensation process trapped different isotopes of the same element in identical ways, and studies pairs (P, D) where there is a stable isotope S of the same element as D, i.e. where $S(t) = S(0)$. Assuming that the two isotopes D and S were perfectly mixed in the initial reservoir, the ratio $D(0)/S(0)$ is identical from one sample to another. Equation (11.9) may therefore be written:

$$\frac{D(t)}{S(t)} = \frac{D(0)}{S(0)} + (e^{\lambda t} - 1)\frac{P(t)}{S(t)} \quad . \tag{11.10}$$

Measurements of the $D(t)$, $S(t)$ and $P(t)$ abundances present now at time t, from different samples, allow the $D(t)/S(t)$ ratios to be plotted as a function of the $P(t)/S(t)$ ratios. Points representing samples with the same age and with the same initial value of the ratio D/S fall on a straight line with slope $(\exp(\lambda t) - 1)$, and the ordinate at the left-hand side allows both the age t, and $D(0)/S(0)$ to be determined.

When rocks from different parent bodies are examined (in terrestrial, lunar and meteoritic samples), the age that is determined is that of the formation of the parent bodies themselves. The fact that on a diagram of the type that we have just described the representative points are aligned indicates that all the planetary bodies formed more or less simultaneously. The ratio $D(0)/S(0)$ that is obtained is the abundance ratio for D and S isotopes in the primordial solar nebula. When different minerals from the same rock are examined, the representative points lie on a different line, the (lesser) slope of which allow us to determine the date when the rock last crystallized.

The principal isotope pairs used are: rubidium-87 and its daughter isotope strontium-87, strontium-86 being the stable reference isotope (the S isotope in the preceding equations); potassium-40 and its daughter product argon-40, with argon-39 as a reference; uranium-238 and lead-206. All these pairs have characteristic decay periods that are measured in thousands of millions of years, allowing the formation of the planetary bodies to be dated. In the case of (Rb, Sr), for example, $T_{1/2}$ is 4.7×10^{10} years. On a diagram plotting the $^{87}Sr/^{86}Sr$ ratio as a function of the $^{87}Rb/^{86}Sr$ ratio, different meteoritic samples fall on a straight line, the slope of which corresponds to a common age, that at which their parent body differentiated, shortly after the formation of the Solar System (Fig. 11.5). The most important result

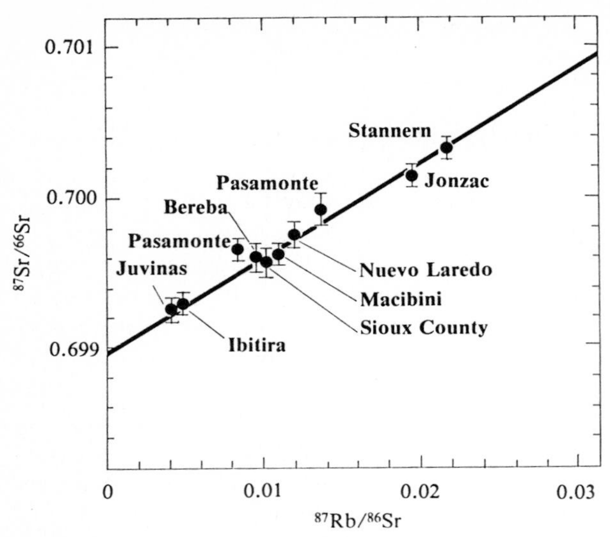

Fig. 11.5. Dating with the (Rb/Sr) system. [After C. Allègre et al.: Science **187**, 436 (© AAAS 1975)]

to come from these measurements of dates is probably that the Earth, the Moon, and the meteorite parent bodies condensed simultaneously, to within ten million years, some 4.55 thousand million years ago.

Other pairs of radionuclides, with shorter periods, provide information about the evolution of the primitive solar nebula. This is the case, for example, with ^{129}I, which decays into ^{129}Xe with a period of 2.5×10^7 years, as well as ^{244}Pu, which decays by fission to produce other isotopes of xenon ("fission xenon") with a much longer period of about 10^8 years. It is found that the $^{129}Xe/^{127}I$ ratio (^{127}I being a stable isotope, not produced by decay) is more or less constant in meteoritic samples in which there is, by contrast, an excess of fission xenon. This means that the period of time between the introduction of ^{129}I into the nebula – i.e. the end of this element's nucleosynthesis – and the accretion of the meteorites' parent bodies – when their temperature became sufficiently low for all the ^{129}Xe produced to remain trapped within them – was sufficiently long for all the ^{129}I nuclei to have decayed into ^{129}Xe. On the other hand this time could not have been too long, as otherwise all the ^{244}Pu would have disappeared through decay. Overall these two constraints indicate that about 10^8 years passed between the time the freshly synthesized nuclei were incorporated into the primordial interstellar cloud and the formation of the planetary bodies. This lapse of time may correspond to the period required for the interstellar cloud to pass from one spiral arm (where it was formed) to the next (where the protosolar nebula would have collapsed to give rise to the protoplanets). The parent bodies of meteorites condensed at about the same time, to within 10^7 years.

11.3.6 Isotopic Anomalies and the Formation of the Solar System

Until 1970, most of the measurements of isotopic ratios in meteorites could be interpreted in terms of a model where the Solar System condensed from a protosolar nebula with a homogeneous composition. This should not be taken to mean that the isotopic ratios for a single element, measured in different samples, were identical. There were discrepancies, but they were explained as having arisen through the three types of process that have just been described. The ability to carry out measurements with a much higher degree of precision on very small samples (smaller than a milligramme) completely changed this idea. Evidence has been found in some meteoritic samples of *isotopic anomalies*, i.e. isotopic ratios that cannot be explained by any modification via known physical-chemistry processes of the reference isotopic ratios that are determined in the analysis of any terrestrial or lunar samples. The anomalies are thought to indicate that nucleosynthesis occurred elsewhere than in the solar nebula. This is particularly the case with the Allende carbonaceous chondrite, which fell in 1969. It contains numerous light inclusions, consisting of very refractory minerals that are rich in calcium and aluminium. Analysis of the isotopic composition of these inclusions, specifically for oxygen, reveals that they are heterogeneous, reflecting the atmosphere from which they formed. They contain presolar material, i.e. material that condensed around another star, which has been preserved to this day. By way of illustration we will discuss the isotopic ratios for oxygen, measurements of which started systematic investigation of isotopic anomalies.

Terrestrial oxygen has three stable isotopes, with atomic masses of 16, 17, and 18, and with the following approximate abundances: 99.76 % oxygen-16; 0.04 % oxygen-17 and 0.2 % oxygen-18. These isotopic ratios are not strictly identical in all terrestrial material, however, whether we are dealing with the atmosphere, water in the oceans, or different rocks. As we have seen previously (Sect. 11.3.5), isotopic differentiation arises through chemical reactions or phase-changes such as vaporization or condensation. However, this differentiation follows a law that is proportional to the mass: for any variation in the isotopic ratio between oxygen-17 and oxygen-16 (separated by one atomic mass unit) there is twice the variation in the ratio between oxygen-18 and oxygen-16 (separated by two atomic mass units). This is shown by the fact that all points representing terrestrial samples fall on a single line of slope 1/2 (the *isotopic fractionation curve*), in a diagram plotting, in δ notation – see (11.4) – the $^{17}O/^{16}O$ ratios as a function of the $^{18}O/^{16}O$ ratios (Fig. 11.4). It should be noted that points from lunar samples also fall on this line.

Points from mineral phases taken from the refractory inclusions in the Allende meteorite line on a line of slope 1 (the *mixing line*). This means that in these mineral phases the ratio of oxygen-17 to oxygen-18 is constant, and that only the absolute quantity of oxygen-16 varies. No known radioactive decay, irradiation, or mass-fractionation process can account for this exclusive enrichment in oxygen-16.

This *isotopic anomaly*, discovered in 1973, may be explained as follows: these refractory inclusions have a variable proportion (up to 5 %) of grains with "normal" isotopic ratios (point A in Fig. 11.4) and grains that only contain the oxygen-16 isotope. These latter grains condensed from a gas that lacked ^{17}O and ^{18}O isotopes, outside the primordial solar nebula, and probably in a supernova envelope. They are

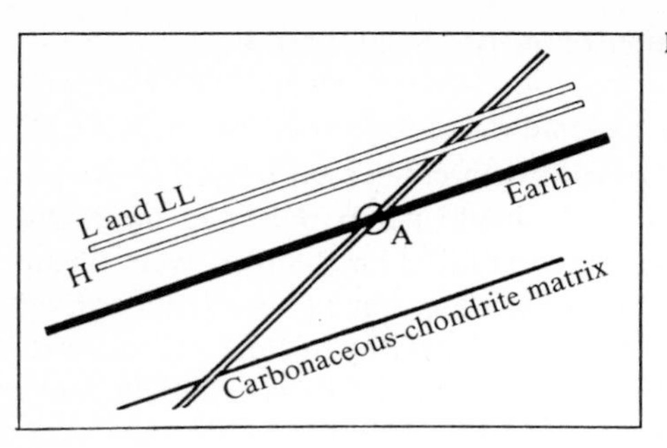

Fig. 11.6. Enlargement of region A in Fig. 11.4

therefore *presolar* grains, that have survived being incorporated into the protosolar nebula (which had a "normal" isotopic composition), and then their accretion into planetary bodies. The same finding refutes the theory that the primitive solar nebula was homogeneous, which would have implied the vaporization of all the presolar grains.

Enlarging the region at A in Fig. 11.4 shows the way in which the primitive solar nebula was spatially heterogeneous (Fig. 11.6). Terrestrial and lunar samples taken together, ordinary H chondrites, chondrites of types L and LL, and also the carbonaceous-chondrite matrix all fall on different straight lines. Each of these lines, of slope 1/2, corresponds to a different fractionation curve. On the other hand, the mass-fractionation process does not allow material to pass from one curve to another, implying that the parent bodies of the various families of samples condensed from different reservoirs, i.e. from reservoirs that contained different proportions of *presolar* grains.

Isotopic anomalies involving other elements have also been detected in meteorites. We may mention in particular that observed for magnesium isotopes. Some minerals contain an overabundance of magnesium-26, which is proportional to the sample's aluminium abundance. The excess of magnesium-26 probably arises from the radioactive decay of aluminium-26. Disintegration of sufficient amounts of aluminium-26 to explain the enrichment in magnesium-26 could have provided sufficient heat to cause the differentiation of bodies more than about ten kilometres across.

The decay period of aluminium-26 is 700 000 years, so the time separating the formation of that isotope, by nucleosynthesis in a supernova, from the condensation of the meteoritic grain could not have exceeded a few million years. So it would seem that a supernova explosion shortly preceded the formation of the primordial solar nebula. This may not have been a coincidence, but a general occurrence: the most massive stars, which evolve very rapidly, may induce the formation of new stellar systems through the shockwave produced by their final explosion.

Over the decade that has followed the discovery of an isotopic anomaly, indicating a presolar origin for part of the Allende meteorite, anomalies have been discovered for more than ten elements, in many different meteorites. Can all of these be interpreted in terms of a single model of the origin and evolution of the Solar System? The answer would appear to be that, in general, all these anomalies had different nucleosynthesis origins, and for the time being all attempts to isolate unaltered presolar grains have failed. It would seem that the mineral phases in which these anomalies are observed have been altered, at least from the mineralogical point of view, since the presolar grains condensed in a stellar atmosphere. Perhaps we should conclude that the growth of the parent bodies of the carbonaceous chondrites did not entirely preserve the presolar identity of even the most refractory of these grains. If this is the case, then cometary nuclei must be the most primitive samples of extraterrestrial material. Meteorites now turn out to be extremely valuable witnesses of the origin and early stages in the evolution of the Solar System; of the primitive solar nebula's stellar environment; of the probable survival of presolar grains; of the dynamical processes accompanying the condensation of the planetary bodies; and of the thermal evolution of the nebula.

11.4 The Collection of Cometary Material

As we have just mentioned, everything leads us to believe that the grains trapped in cometary nuclei, which have remained at a low temperature since the presolar nebula condensed, are the best-preserved of the early components of the nebula. Collection of this material is therefore of great interest, comparable with the difficulties that it presents. The most effective way would, of course, be to send a space probe to land on a cometary nucleus, to drill a deep core and to bring the sample back to Earth. Just such a project does exist, but it is scheduled for the first decade of the next century. For the moment other solutions have to be found; these may be divided into two categories.

The first relies on the fact that the smallest particles from comets pass through the atmosphere without being completely destroyed, being slowed down in the stratosphere and slowly descending into the lower atmosphere. So we can try to collect them, either in certain favoured sites on the ground, in particular in the ice and snow in the polar regions, or in the lower stratosphere, where terrestrial contamination, although predominant, does not prevent us from detecting extraterrestrial particles.

Since the middle of the 1960s, NASA, at the instigation of D. Brownlee, has collected particles in the stratosphere. Plates covered in a very viscous oil are deployed under the wings of NASA's U2 aircraft, which flies at a height of about twenty kilometres, sampling large volumes of rarified air. The particles collected are subsequently cleaned and sorted to isolate extraterrestrial particles from terrestrial contamination. More than half of the extraterrestrial particles are aggregates of grains that are smaller than 1000 Å in size, which have a composition approximately the same as that of the carbonaceous chondrites (Fig. 11.7). Their extraterrestrial origin is also indicated by the presence of ^{10}Be, which is produced by cosmic-ray bombardment. The texture of the particles does not resemble any type of meteoritic sample. It is possible that these particles are of cometary origin. Overall, several hundreds of grains have been catalogued, with diameters of 5 to 100 microns.

The obvious point about such collection is that it is non-destructive: we make use of the progressive deceleration of the particles by the atoms in the upper strato-

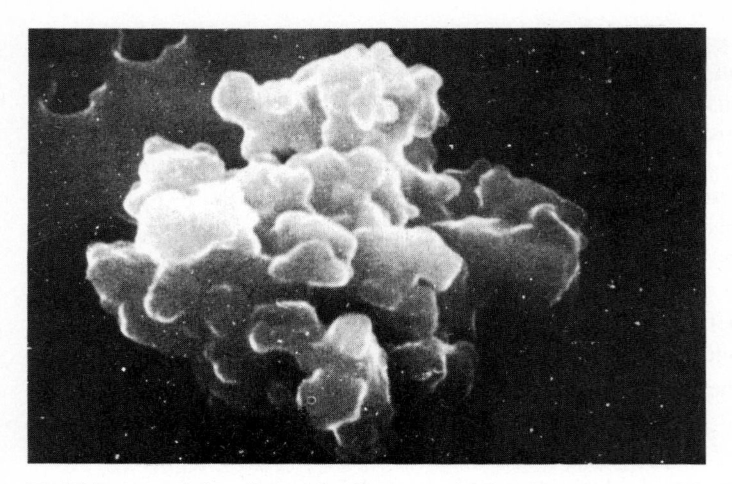

Fig. 11.7. A particle collected in the stratosphere. (By courtesy of D. Brownlee)

sphere. On the other hand, apart for the problems caused by contamination by terrestrial particles, such collections have the additional disadvantage in that they mix all extraterrestrial contributions together, preventing us from identifying the origins of the particles collected. What we need to do is to intercept cometary particles, outside the Earth's atmosphere, when the Earth encounters the tail of a comet, or crosses a meteor stream. It is very difficult to devise a non-destructive method of collection, given the very high relative velocity of the particles, amounting to several kilometres per second. On the other hand, it is possible to retain all the atoms forming the particles in order to be able subsequently to analyze, particle by particle, and comet by comet, the isotopic composition of this cometary material. Such an experiment, financed by the CNES, has been devised by a Franco-Soviet team. The first sampling instrument, built in France, was installed on the Salyut-7 space station. A second is planned to flow on the Mir space station. This will be the first time that identified samples of cometary material will be returned to Earth.

12. Conclusions

In this work we decided to discuss the Solar System as a series of successive reviews dealing with the different objects that it contains. What emerges is the astonishing diversity found in the Solar System. Examples of this diversity abound. For example, bodies with very similar initial sizes and heliocentric distances, such as the Galilean satellites of Jupiter, have undergone extremely different evolution. The same applies to the surface evidence for the internal activity within planets (volcanism and tectonics). On the other hand, traces of internal activity similar to that of the inner planets has been found on some of the icy satellites of the giant planets (Ganymede, Thetis, and Miranda). The evolution of planetary atmospheres, and their interactions with the interplanetary medium also differ from one object to another.

Overall, it is obvious that planetary evolution occurs through a large number of processes, governed by factors only a few of which are understood. The initial values for the mass and distance from the Sun are not sufficient to reconstruct the later history of bodies within the Solar System. Other features intervene, in the form of additional initial conditions: the content of "short-lived" radioactive isotopes (with typical lifetimes of about 10^6 years), the energy released during accretion, etc. This is why, in modern planetology, the systematic description of the various objects in the Solar System advances on two fronts:

1. *research into initial conditions*, which concerns the dynamical processes that applied (the fragmentation of the nebula, the accretion of protoplanets, and collisions) and the composition (both chemical and isotopic) of the original material;
2. *comparative planetology*, which aims to understand how a single physical process can take on the different forms that are observed on various objects within the Solar System, thereby reconstructing their evolution and gaining understanding of its fundamental causes. This would not be possible if study were restricted to a single object.

Such research, which has been made possible through recent progress in the exploration of the planets, is beginning to outline the factors governing the individual evolution of each of the bodies. One of these factors, and by no means the least important, is the existence of living organisms on Earth, which raises the wider problem of the existence of life elsewhere in the universe, which we shall discuss at the end of this chapter.

12.1 Establishing the Initial Conditions

Two types of information about the conditions under which the Solar System was formed are at present available for observation. One involves the dynamical and physical processes that occur in the accretion of protoplanetary bodies and their growth through collisions. The other involves the determination of the chemical and isotopic composition of the material (gas and dust grains) that formed the original solar nebula.

12.1.1 Dynamical and Physical Processes

The discovery of systems of rings, not just around Saturn but also around Jupiter, Uranus and Neptune, primarily thanks to the Voyager observations, enables us to examine systems of particles and small bodies that are evolving under dynamical conditions that are similar to those found in the Solar System when it was forming. In addition, the observational data obtained from the photographs of the rings can be entered directly into numerical simulations as well as be used in analytical discussions that describe the evolution of the Solar System, beginning with the overall composition of the original solar nebula.

A second focus of interest relative to the early history of the Solar System is the system of asteroids, which are probably the bodies remaining from a process that failed to form a planet (see Chap. 6). Within the family there are bodies with differing degrees of evolution, particularly as regards internal differentiation: they represent successive stages in the growth of a planet, which took place, over a time-scale that did not exceed ten million years, some 4.55 thousand million years ago. Although some of their properties have been altered through the effects of multiple collisions, they form a system that is unique in that it allows us to analyze the processes that took part in forming the planetary system. Given that the asteroids are the source of meteorites (Chap. 11), some of these objects are directly available for laboratory analysis. Such studies are, however, severely limited by our inability to link the properties of meteorites to those of a parent body (with specific mass, diameter, density, heliocentric distance, degree of differentiation, etc.). They need to be supplemented by in-situ observations of sufficient asteroids to be representative of the whole range existing within the family. Several space missions, planned for the 1990s, have the aim of flying past several asteroids.

12.1.2 The Primordial Material

When the original solar nebula's parent molecular cloud formed, about five thousand million years ago, it isolated a mass of gas and dust, whose overall properties were doubtless those of the interstellar medium at the time, but which has subsequently been divorced from the chemical evolution of the rest of the Galaxy. Naturally, the series of processes that led to the formation of the Solar System have profoundly modified most of those properties. Nevertheless, we now believe that favoured sites do exist, where intact, or little-altered vestiges of this primitive material may be found. Indeed there does exist a whole family of samples, both gaseous and solid, that acquired their properties at various stages in the Solar System's formation and early evolution. We are thus able to retrace these stages and study their dynamical evolution. Searching for, and analyzing primitive material has therefore become a specific planetary discipline that is now developing very rapidly.

As far as the gaseous components of the primordial nebula are concerned, we can expect to find them in two types of site:

1. in icy bodies, primarily comets, which may have trapped molecules from the pre-solar nebula as they condensed. The organic molecules that they contain may therefore be, at least in part, representative of the molecular components of the protosolar nebula;
2. in the giant planets, particularly Jupiter. All may have accreted gas directly from the nebula, and this would not have been affected by subsequent evolution of the Sun. In dealing with the volatile elements, therefore, the relative abundances should, to a first approximation, be representative of the interstellar medium five thousand million years ago. We have discussed, in Chap. 7, the possibly of using the values measured for various isotopic ratios, such as D/H, to determine the primordial amounts.

With regard to the solid phases, the situation has changed considerably over the last ten years, since it was established that some meteorites contain grains that are probably pre-solar, and which have not lost their volatiles during the processes of accretion into the meteorite parent bodies (see Chap. 11). Nowadays it is possible to isolate such phases and to analyze their isotopic composition. Such analysis reveals details about the sites of nucleosynthesis and condensation of these grains, and thus about the types of stars that surrounded the region where the Sun was to form. In addition, the existence of the radioactive-decay products of unstable elements, created in specific stars, sets severe limits on the dynamical processes governing the collapse of the nebula and the growth of the planets.

12.2 Comparative Study of the Atmospheres of the Inner Planets

Unlike the atmospheres of the giant planets, those of the inner planets have undergone radical changes since their formation. Their present-day properties reflect the differences in their initial conditions (mass, heliocentric distance, kinetics of their accretion, and their composition), as well as the physical and chemical processes that have influenced the later evolution of these objects. We shall first summarize the principal data concerning the atmospheres and then suggest the possible origins of the observed characteristics.

12.2.1 The Overall Chemical Composition of the Atmospheres of the Inner Planets

In the case of the inner planets, the temperature differences and the differences in evolution are such that a given element may occur in very different forms on one or other of the bodies. We should therefore take into consideration all the volatile elements that are present, whether they are in the atmosphere, the crust, the mantle, or the oceans.

Table 12.1. Atmospheric characteristics of the inner planets. (After D. Gautier, W. Hubbard, H. Reeves, *Planets – Their Origin, Interior and Atmosphere,* P. Bartholdi et al., eds., 14th Advanced Course, Observatoire de Genève, 1984)

Planet	Gravity ($cm\ s^{-2}$)	Surface pressure	Surface temperature	Principal atmospheric components
Venus	888	90	730	CO_2 (0.965) N_2 (0.035)
Earth	978	1	288	N_2 (0.77) O_2 (0.21) H_2O (0.017) Ar (0.0096)
Mars	373	0.007	218	CO_2 (0.95) N_2 (0.027) Ar (0.016)

Table 12.1 gives the overall physical and chemical properties of the inner planets. Mercury is not considered here: its mass and radius give rise to a very low escape velocity ($\sim$4.25 km/s), whilst the close proximity of the Sun gives the atmospheric components very high kinetic energies. Mercury has therefore lost all the volatile elements that were degassed during the period of the planet's internal activity, a thousand million years after its accretion. Since that time its very tenuous atmosphere, rich in helium, has consisted of solar-wind components only, continuously implanted into the regolith, and released by thermal effects.

Atmospheres are not the only reservoirs of planetary volatile elements. On the Earth, for example, H_2O is primarily held in liquid form. Carbon dioxide CO_2 is trapped in carbonate rocks: the existence of liquid water has enabled the CO_2 dissolved in the water to react with calcium oxide, forming calcium carbonate.

As far as Venus is concerned, the situation is different; its very high temperature prevents us from postulating the existence of ice or liquid water. The very low water-vapour content indicates that the water that has been degassed has been lost from the planet. A runaway greenhouse effect would have caused H_2O and CO_2 to be completely vaporized, and subsequently photodissociated by the solar ultraviolet flux. With H_2O, the dissociation products are unable to recombine because H, being light, escapes to space. The photodissociation of CO_2 into CO + O, on the other hand, may be followed by the inverse reaction; CO_2 is therefore generally stable under the conditions on Venus.

Table 12.2. Abundances of volatile elements in the inner planets (mass of the component per unit planetary mass). (After T. Donahue and J. Pollack, *Venus*, Univ. of Arizona Press, 1983)

	Venus	Earth	Mars
C	2.6 (−5)	3 (−5) (1.5 → 4.5)	1.2 (−8)
N	2.5 (−6)	2 (−6)	8 (−10)
^{4}He	1.1 (−10)	1.6 (−9)	
^{20}Ne	2.3 (−10)	1.1 (−11)	4.9 (−14)
^{36}Ar	2.5 (−9)	3.5 (−11)	2.1 (−13)
^{40}Ar	2.9 (−9)	1.14 (−8)	6.8 (−10)
Kr	9.2 (−12)	2.9 (−12)	2.7 (−14)
^{84}Kr	4.9 (−12)	1.6 (−12)	1.4 (−14)
Xe	<1.2 (−11)	3.5 (−13)	1.1 (−14)
^{129}Xe	<2.9 (−12)	8.9 (−14)	6.7 (−15)
^{132}Xe	<3.1 (−12)	9.5 (−14)	2.7 (−15)

Mars is a very interesting case: although it is the major atmospheric component, carbon dioxide is less abundant by a factor of about 10000 than in the atmosphere of Venus. Similarly it is only a small fraction of the amount of carbon present on Earth in the atmosphere and in carbonates. We may therefore conclude that a large portion of carbon on Mars is present not in the form of atmospheric CO_2, but as solid compounds (associated with calcium or iron). Forthcoming space missions should detect such compounds, at least in certain geological provinces. As to H_2O, the primary reservoir must be in the form of thick subterranean layers of ice (permafrost).

Overall, if we compare the abundances of volatile elements with respect to the masses of the planets (Tables 12.2 and 12.3), it will be seen that Earth and Venus have carbon and nitrogen concentrations that are very similar, and notably higher that those that have been measured in the atmosphere of Mars. In order to explain this, different pressure and temperature conditions in the primitive solar nebula at

Table 12.3. Abundance ratios for the inner planets. (After T. Donahue and J. Pollack, *Venus*, Univ. of Arizona Press, 1983)

	Venus/Earth	Earth/Mars
C	0.6–1.8	133–6.7
N	1.02–1.7	133–6.7
^{4}He	0.07–1.4	–
^{20}Ne	21 ± 5	200^{+280}_{-130}
^{36}Ar	72 ± 10	165 ± 45
^{40}Ar	0.25 ± 0.04	16 ± 3
^{84}Kr	$2.9^{+3}_{-1.5}$	115^{+185}_{-75}
^{129}Xe	<32	13 ± 4
^{132}Xe	<32	35 ± 9

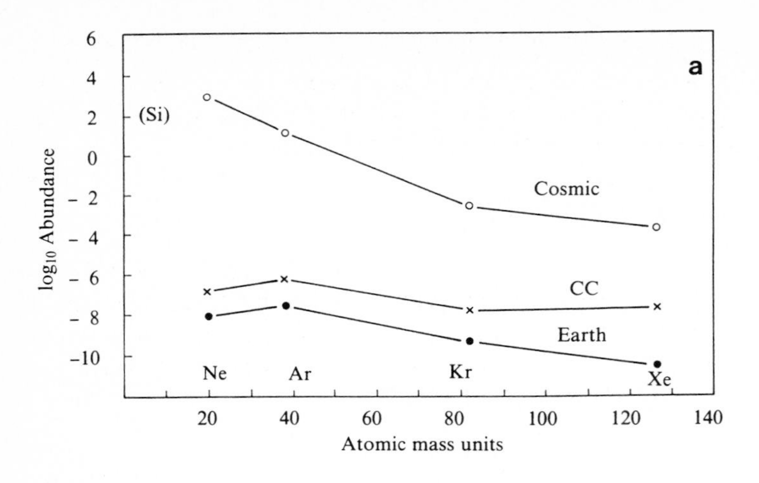

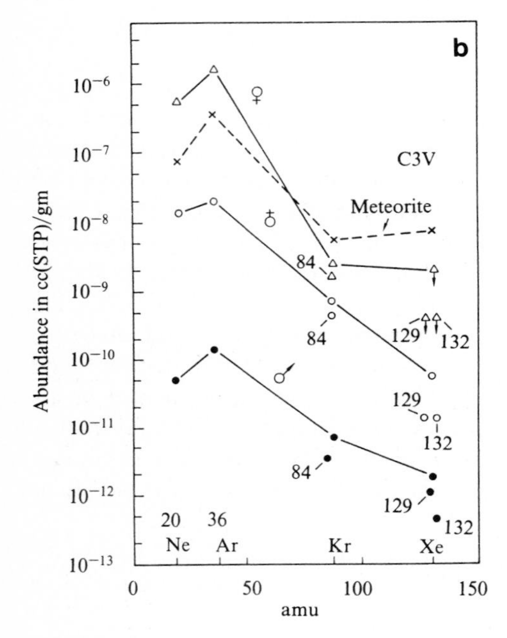

Fig. 12.1a,b. Rare-gas abundances for the inner planets, compared with cosmic abundances and those of carbonaceous chondrites. [After Donahue and Pollack (1983)]

the time when gas accreted into the planetesimals have been invoked. Pollack and Black have thus been able to account for the relative abundances of N_2, CO_2 and rare gases, by postulating a large variation in pressure and temperature in the primordial nebula between the orbits of Venus and Mars. Nevertheless, it does seem that for the latter planet, the main reservoirs of volatile elements are the surface and sub-surface layers.

The study of the variation in abundance of the rare gases as a function of atomic mass gives important complementary information. Figure 12.1 compares the cosmic abundances with those of carbonaceous chondrites (primitive meteorites) and of the inner planets. It is notable that the curve for the inner planets is much closer to that of the carbonaceous chondrites than to that for cosmic abundances. The simple explanation is that the atmospheres of the inner planets were not accreted from the primordial solar nebula, but were outgassed from the interior of the planets, where the gas was trapped when the planetesimals were formed.

12.2.2 Isotopic Ratios in the Inner Planets

Precise measurements of isotopic ratios have been obtained thanks to the Pioneer Venus and Venera probes for Venus, and the Viking probes for Mars. It is very striking (Table 12.4) that the isotopic ratios are very similar, with the exception of deuterium and nitrogen. It is probable that the deuterium-enrichment on Venus is connected with the absence of water on that planet.

Table 12.4. Isotopic ratios for the inner planets. (After T. Donahue and J. Pollack, *Venus,* Univ. of Arizona Press, 1983)

	Venus	Earth	Mars
$^{22}Ne/^{20}Ne$	0.07 ± 0.02	0.097	0.1 ± 0.03
$^{38}Ar/^{36}Ar$	0.18 ± 0.02	0.187	0.19 ± 0.02
$^{13}C/^{12}C$	$\leqslant 1.19\ (-2)$	1.11 (−2)	$(1.1 \pm 0.1)\ (-2)$
$^{18}O/^{16}O$	$(2.0 +\pm 0.1)\ (-3)$	2.04 (−3)	$(2 \pm 0.2)\ (-3)$
$^{15}N/^{14}N$		3.7 (−4)	5.7 (−4)
D/H	$(1.6 \pm 0.2)\ (-2)$	1.6 (−4)	–

For nitrogen, the enrichment in ^{15}N by a factor of 1.7 on Mars may perhaps be explained by the differential escape of nitrogen from the martian atmosphere during the planet's lifetime. For the rare gases, the two species measured are "primordial". It will be noted in particular that the $^{22}Ne/^{20}Ne$ ratio increases progressively as the heliocentric distance increases, the value for Venus being the same as that of the solar wind. The abundance of ^{36}Ar decreases sharply with increasing distance from the Sun, from Venus to Mars. Yet ^{40}Ar is most abundant on the Earth. We should recall that argon-36 (and argon-38) is produced in stars by thermonuclear reactions; the material found in the Solar System is therefore of primordial origin. The argon-40 observed in the terrestrial-type atmospheres, on the other hand, is produced by the radioactive decay of potassium-40 within rocks.

It has been suggested that the solar wind played an important part in the formation of the planetesimals (Donahue, Pollack). According to this theory, solar-wind implantation in the surface grains would have had the effect of enriching the rare-gas abundance on Venus (which is closer to the Sun, and therefore affected to a greater degree). This process explains the relative Venus/Earth and Earth/Mars abundances shown in Table 12.3 for the rare gases, as well as the variation in the $^{22}Ne/^{20}Ne$ ratio with distance from the Sun. It does not, however, explain the differences between the relative abundances of rare gases shown in Fig. 12.1. This is why it may be said that the process by which the inner planets were formed is still not completely understood.

12.2.3 The Effects of Internal Activity

It is generally agreed that the atmospheres of the inner planets have been outgassed from the interior of the planets. They are therefore a direct result of planetary activity, the extent of which is linked to the planetary masses. The latter therefore have two effects on present-day planetary properties: on the one hand, they determine the escape velocities, which govern the dynamical equilibrium of the various species as a function of temperature. On the other hand, they determine the degree of planetary activity, which contributes, mainly by volcanism, to the abundance of atmosphere components. So the more massive a planet, the higher the temperature at its centre, T_c: T_c is proportional to the square of the planetary radius (see Sect. 3.3.1). Given that the source of energy (radioactive decay of long-lived isotopes) decreases with time, the least massive planets become inactive soonest. In qualitative terms, the evolution of internal activity may be described as follows: after accretion, the energy released by radioactivity raises the internal temperature – this is the markedly non-stationary regime when the energy is released essentially in proportion to the increase in temperature with time, dT/dt. The surface radiative losses limit the rise in the internal temperature to a maximum central value T_c, which is reached several hundreds of millions of years after accretion. Magmatic activity is then at a maximum. This takes place on all the bodies, including the Moon and Mercury: it is, for example, responsible for the filling of the large impact basins that gave rise to the lunar maria. Radioactive decay causes this activity to decline, and eventually to cease when the central temperature becomes too low. Thus the Moon and Mercury became inactive at the end of the first thousand million years after their accretion. Mars, more massive, remained active for longer, giving rise to very extensive volcanism. The martian atmosphere was then dense; the partial pressure of water and the temperature were above the values at the triple point: intense flooding would have been possible, giving rise to a widespread network of channels. Over the last thousand million years, Mars has, in its turn, become extinct. What about Venus today? Its mass is only slightly less than that of the Earth. Has it yet become geologically extinct?

12.2.4 The Effects of Surface Temperature

With Venus, water was probably in the gaseous state when it emerged from the interior, and, as it makes a significant contribution to the infrared opacity, a greenhouse effect was created, which progressively increased the temperature of the surface. Several processes have been proposed to explain the later disappearance of water from Venus (such as dissociation by solar UV radiation, and reaction with the surface or with atmospheric CO). Whatever the process may have been, the high value of the D/H ratio on Venus certainly appears to indicate that originally Venus must have contained a considerable quantity of water; this tends to support the model just described.

For the Earth, the initial surface temperature was such that water degassed from the interior must have condensed immediately and formed the oceans. The present-day temperature of the Earth's surface has therefore remained close to the initial temperature. The evolution of the terrestrial atmosphere subsequently took place in four stages (Walker): (1) the formation of the atmosphere (by degassing); (2) the physical and chemical era (the pre-biological era); (3) the microbiological era (the development of photosynthesis and the accumulation of oxygen); (4) the geological era (the formation of the protective ozone layer).

Finally, for Mars, as soon as water was degassed, it ought to have been precipitated in the form of ice. Yet analysis of the martian surface appears to show traces of the effects of liquid water in the past; we must therefore conclude that Mars once had a higher temperature, caused by a denser atmosphere's greenhouse effect. This type of model has been studied by Pollack. Whatever the initial composition of the martian atmosphere, the various components that were degassed must have disappeared one after another (CO_2 in the form of carbonates in the surface rocks, N_2 in the form of nitrates and by escape). As a result, the greenhouse effect vanished, the temperature fell again, and water could have become frozen within the surface layers. Only a tenuous atmosphere of CO_2 and N_2 remains.

In conclusion then, it is possible that the three inner planets may have had similar primordial abundances of volatiles (CO_2, H_2O and N_2). The early atmosphere were probably acquired though degassing; the very different evolution of the three atmospheres was caused by differences in the surface temperatures.

12.3 Comparative Study of the Atmospheres of the Giant Planets

We have already seen in Chap. 7 that study of the atmospheres of the giant planets lends itself to comparative analysis. We will summarize the main results of such analysis in a more comprehensive manner.

The most significant mechanisms governing planetary formation and evolution are: gravitational accretion and escape; degassing of components that were trapped as solids when the grains condensed, or else implanted into grains by the solar wind; chemical or photochemical modifications to the atmosphere; and finally changes of state linked to temperature variations.

The processes that are capable of creating or modifying an atmosphere are numerous and complex. In nearly every case, the gaseous envelopes should be considered as systems that may evolve on very short time-scales (which may be as low as a few hours for comets), or on very long ones (comparable with the age of the Solar System for the planets themselves).

The complexity of the mechanisms involved also indicates that the distinction between primary and secondary atmospheres, which we described in the first chapter (Sect. 1.2.2), is only an approximate classification, designed to give some qualitative expression to the degree of evolution that is found.

There are striking differences between the four giant planets. The most important of these are:

1. for Uranus, the abundance of helium relative to H seems compatible with the solar value. It might be slightly less for Jupiter, and reduced by a factor of two for Saturn. The values for Neptune are not yet known;

2. the carbon enrichment relative to hydrogen, with reference to the solar value, reaches a factor of two for Jupiter; from two to six for Saturn; and two to thirty for Uranus and Neptune;
3. Jupiter, Saturn and Neptune have internal sources of energy. The ratios of this internal energy to the solar energy that is reradiated are 1.7, 1.8 and 2.6 for the three planets respectively. For Uranus, on the other hand, the internal energy detected is extremely weak (see Table 7.6). This peculiarity of Uranus is probably related to the fact that its thermal profile shows only a very weak rise above the temperature-minimum, unlike the other three planets. It is also possible that this peculiarity of Uranus is related to its geometrical configuration: Uranus is, in fact, the only Solar-System planet with an axis of rotation lying in the plane of its orbit.

We have seen (Sect. 7.1.4) that there were two broad classes of models that were capable of describing the formation of the giant planets: the homogeneous-accretion model, where the composition of the planets ought to reflect that of the primordial nebula; and the nucleation model, which assumes that a core of silicates, metals and ices first formed, and then ices accreted around this core. We have seen that models of the second type are preferable, according to modern work, in particular because they account for the enrichment in carbon found in the giant planets by the degassing of hydrated clathrates. Another argument favours the nucleation model: the gravitational moment J_2 has been measured for each of the four giant planets, and from this the value for the mass of the central core has been derived. In every case this has been found to be around ten to thirty Earth masses. This similarity of the cores may be an indication of a method of formation that was unique to the giant planets. The model also has the advantage of explaining the increasing enrichment in carbon from Jupiter to Neptune, which would be a consequence of the increase in the ratio of the core's mass to the total mass with increasing distance from the Sun.

If we are able to describe the formation of the giant planets by a single process, the differences that we see today ought to correspond to different stages of evolution. We have shown earlier (Sect. 7.1.4) that Saturn's energy-excess and the depletion in helium in its outer layers may be explained by the condensation of helium and its migration towards the centre. It is therefore easy to understand why this process has occurred preferentially on Saturn. Starting at a lower temperature than Jupiter, Saturn has cooled sufficiently for the adiabatic curve $P(T)$ to cut the curve for helium saturation in metallic hydrogen (see Fig. 7.23). With Jupiter, the time required for this to occur is longer, because the initial temperature was higher. It is nevertheless possible that helium differentiation has started in Jupiter, and this would explain the slight enrichment in helium that is observed for the planet. Jupiter's energy-excess would have a small contribution from this process, but would mainly result from the thermal energy released in cooling. As for Uranus and Neptune, according to this model, their interiors would not include a metallic phase (see Fig. 7.23), condensation of helium would not take place, and the H_2/He ratio should be closer to the primordial value. This has been confirmed by Voyager in the case of Uranus; and the H_2/He ratio for Neptune will be measured as a result of the fly-by in

1989. According to the model described earlier, Neptune's measured energy-excess is the thermal energy produced by the planet's cooling.

12.4 Organic Material in the Solar System and the Search for Extraterrestrial Life

The first ideas about the chemical evolution of life date from the twenties, when Oparin advanced the first theory that microorganisms may have appeared at the end of a long series of steps that involved complex organic molecules, all taking place in a reducing medium. The first laboratory experiments were carried out by Miller and Urey. They succeeded in synthesizing amino acids from a reducing mixture of gases (H_2, CH_4, NH_3) subjected to electrical discharges. This experiment opened the way to important and fruitful laboratory research, all of which confirmed the possibility of synthesizing complex organic molecules from mixtures of simple gases, with the energy required being supplied by ultraviolet radiation or electrical discharges. At the beginning of the seventies, radio-astronomy and space exploration of the planets discovered a very large number of complex molecules in various astrophysical sites with very different temperatures and pressures (the interstellar medium, circumstellar envelopes, and planetary surfaces and atmospheres). This confirmed ideas that the synthesis of complex molecules is not an exceptional process in the universe. One of the latest discoveries, in 1981, has been that of half-a-dozen, complex, "prebiotic" molecules on Titan (Chap. 9). Yet the step from this organic chemistry to life has not been detected, even in fossil form, anywhere except the Earth. The problems regarding the conditions that favour the appearance of life remain unanswered.

12.4.1 Laboratory Experiments

The most important work has been on the synthesis of amino acids. It has been carried out with a medium simulating the conditions on the early Earth, have synthesized fourteen amino acids directly. The first synthesis of adenine, one of the bases of DNA, was achieved by Oro in 1963, starting with a concentrated solution of ammonium cyanide. A little later, Ponnamperuma and his collaborators obtained it by irradiating a mixture of CH_4, NH_3 and H_2O, with HCN being produced as an intermediate step. Guanine and xanthine were also synthesized by Oro. The synthesis of monosaccharides has been more difficult. Gabel and Ponnamperuma have shown that formaldehyde might be the basis for the synthesis of sugars.

12.4.2 The Search for Complex Molecules in the Solar System

The first traces of life on Earth have been found in the form of microfossils that may be as old as the most ancient rocks on Earth, i.e. about 3.6×10^9 years. Attempts have been made to look for traces of prebiotic processes in sites that are free from Earth contamination: the Moon, meteorites, comets and Mars.

Analysis of Lunar Samples

Lunar samples returned by the Apollo missions have been analyzed in the laboratory with the aim of detecting traces of complex organic molecules. An early finding was that the carbon concentration in grains from the regolith was astonishingly high, and that it arose from implantation of solar-wind ions (which strike the surface of the Moon at their full velocity). But this carbon is mainly in the form of small compounds (CO_2 being predominant), synthesized as a result of irradiation of the grains. Concentrations of the order of 10^{-9} have been found for all the more complex components examined (fatty acids, amino acids, sugars, and nucleic-acid bases). These experiments show that the synthesis of simple organic molecules is a continuing process, which continues even under the most hostile conditions – with an intense solar ultraviolet flux – but that it is not necessarily followed by the growth of large molecules.

Carbonaceous Chondrites and Comets

Study of organic molecules in meteorites dates back more than a century. It is now generally accepted that the carbonaceous chondrites contain complex organic molecules. In the Orgueil meteorite, and then in the Murchison meteorite, which fell in 1969, a large number of amino acids have been found, and these may have had an extraterrestrial origin. Many results should be forthcoming in future from collections made in Antarctica, which provide meteorites that have been better protected from terrestrial biological contamination. When an Antarctic carbonaceous chondrite can be examined, we shall have a sample of primordial extraterrestrial material that condensed at a low temperature and which has been conserved ever since.

On the other hand, comets, which originated from the freezing of the presolar nebula constituents, might represent the end products of the chemical evolution of a typical molecular cloud. The analysis of the grains and gas released from the nucleus of Halley during the Vega and Giotto encounters in 1986, gave strong indication that a high proportion of the carbon was trapped in the form of large refractory molecules, detected as grains containing C, H, O and N atoms.

Mars

In Sect. 5.5 we discussed the experiments carried out by the Viking missions, beginning in 1976. It turns out that analysis of the soil at the two sites, which were a long way from one another, showed only very low concentrations of carbon, and revealed no traces of either current or past biological activity. Naturally one cannot draw definite conclusions from just these two experiments. One of the major problems that has yet to be resolved is the existence of water in the past. It seems to be generally accepted from laboratory experiments that reactions are greatly accelerated when the molecules are free to diffuse throughout liquid water. In addition, the presence of "canyons" seems to bear witness to flowing water in the past – although it does not form conclusive proof. It would seem that the partial pressure of water was once considerably higher than it would be today, even if the polar caps were to evaporate completely. We would expect to find this water trapped in subterranean ice, in the form of thick layers of *permafrost*, confirmation of which must be one of the primary objectives of any future exploration of Mars.

The Giant Planets

The reducing atmosphere of the giant planets, consisting of hydrogen, helium, methane and ammonia, seems particularly suited to the synthesis of complex organic molecules. Indeed, numerous carbon compounds have been detected on Jupiter, in particular HCN, as well as a large number of hydrocarbons (Chap. 7). The synthesis of more complex molecules ought to be possible, and these may be the source of the reddish-orange colour of some of Jupiter's cloud structures, which remains unexplained. There is still the question of the stability of any such large molecules: if gravity causes them to fall into the deeper layers they run the risk of being destroyed by the high temperature and pressure.

Titan

The discovery from Earth of an atmosphere containing methane around Titan meant that it was the latest site to be considered as possibly being suitable for the formation of complex organic compounds. In particular, it has sometimes been suggested that there may be a greenhouse effect permitting the existence of liquid water. This was sufficient justification for Voyager 1 being placed in a path that took it past the satellite at a very low altitude (4000 km), in order for the surface to be examined at a very high spatial resolution.

As it happened, Titan's atmosphere proved to be opaque and to consist mainly of molecular nitrogen, with a large concentration of aerosols, preventing direct observation of the surface. On the other hand, spectroscopic observations revealed a very complex organic chemistry (Chap. 9), and some of the compounds detected form part of a "prebiotic" chain. Although the presence of a surface – and perhaps even seas – may favour the development of higher molecules, the low temperature is a handicap that would slow down any chemical reactions considerably.

To conclude, we now know that living matter, even in the most elementary form – i.e. being able to reproduce itself at the very least – has practically no chance of existing anywhere in the Solar System except on Earth itself. But on the other hand, we do know of numerous sites, such as in the interstellar medium, where complex organic molecules may be synthesized.

12.5 Possibilities for Life Elsewhere in the Universe

We have just seen that there is very unlikely to be any form of life in the Solar System, except on Earth. But what about the universe? As yet, astronomers are not able to give any form of definitive answer to this very fundamental question, because, at present, we have no idea of the probability of life arising from prebiotic molecules. A link is missing that neither laboratory experiments nor astronomical observation have been able to supply. All astronomers can say is that there is no known reason for the appearance of life to be confined to the Earth. This statement rests on the simple fact that the Sun is a perfectly ordinary star, and that there are about one hundred thousand million stars in our Galaxy alone.

12.5.1 What Is the Probability of Life Existing in the Universe?

What probability is there that some planet outside our Solar System could shelter life? Even if we cannot solve this problem, we can at least try to devise some formula to expresses it. If N is the number of planets in the universe with life, we may write:

$$N = N' \times f \times n \times f' \quad . \tag{12.1}$$

In this equation, which was first drawn up by Drake, N' is the number of stars of similar spectral class to the Sun; f is the fraction of stars that have planetary systems; n is the number of planets that have an environment suitable for life; and f' is the fraction of those planets where life has actually developed. At present we have no idea of the values that n and f' may take. Stellar astronomy, however, tells us that N' ought to be of the order of 10^9 to 10^{10} in the Galaxy. Further than that, we can only hope that in a few years or decades, developments in research aimed at discovering planetary systems around other stars will enable us to estimate the value of f.

12.5.2 The Search for Extraterrestrial Life

Taking his idea even further, Drake asked the question of the chance inhabitants of the Earth have for detecting the presence of a possible extraterrestrial civilization. For this, he introduced into the equation that is named after him (12.1), two further factors f_1 and f_2: f_1 is the fraction of planets that have evolved an "intelligent" civilization, and f_2 is the fraction of the planet's lifetime during which the degree of technology attained would allow some form of communication. Once again, we have no way of estimating f_1 and f_2. However, the problem has appeared sufficiently important to a number of scientists for them to devote a systematic programme of research to it. The idea is as follows. Given the distance of any extraterrestrial civilizations, communication can only take place by electromagnetic radiation. (Remember that light takes several years to reach even the closest stars.) Searches have therefore been made for non-random radiation emitted by any planets in the Galaxy. The spectral region chosen is the radio region, because it has the largest effective radius for any given power source; it is also a spectral region where the Earth's atmosphere is transparent. It only remains to choose a specific wavelength. The hyperfine transition of neutral hydrogen (present in all molecular clouds), at 21 cm, has been selected. For more than twenty years, American astronomers, particularly with the SETI (Search for Extra Terrestrial Intelligence) programme, and Soviet astronomers have monitored radio emission from galactic sources, in the hope of detecting an "unusual" signal. So far, no really decisive result has been obtained .....

Appendix 1. Mobility and Conductivity of Ionospheric Plasma

A1.1 Transport Equations for Ionospheric Ions and Electrons

In the ionosphere, where plasma is a minor component within the neutral atmosphere, the equations of motion for the ion gas are:

$$m_i\left(\frac{\partial \boldsymbol{V}_i}{\partial t} + (\boldsymbol{V}_i \cdot \nabla)\boldsymbol{V}_i - g\right) = -\frac{1}{N_i}\nabla P_i + e(\boldsymbol{E} + \boldsymbol{V}_i \times \boldsymbol{B}) + m_i \nu_{in}(\boldsymbol{V}_n - \boldsymbol{V}_i) \quad . \tag{A1.1}$$

The left-hand term represents the total (inertial and gravitational) acceleration acting on the ions. The three terms on the right-hand side represent – from left to right – the pressure term, the electromagnetic force and friction with the neutral gas. A similar equation (with an index e) applies to the electrons. ν_{in} and ν_{en} are the *collision frequencies in sense of the momentum transfer* of the ions and electrons with neutral atoms. They specify the efficiency of their frictional coupling with the neutral atmosphere.

Let us arrange the terms to simplify this equation.

1. The acceleration term in (A1.1) is negligible for a subsonic-flow regime because

$$\frac{\partial}{\partial t} = 0$$

and $(\boldsymbol{V}_i \cdot \nabla)\boldsymbol{V}_i$ is small relative to $-(1/N_i)\nabla P_i$. In fact, if L is the characteristic scale of variations in the macroscopic parameters, we have

$$|(\boldsymbol{V}_i \cdot \nabla)\boldsymbol{V}_i| \propto \frac{V_i^2}{L} \quad \text{and} \quad \left|\frac{1}{N}\nabla P_i\right| \propto \frac{kT_i}{L}$$

so their ratio is of the order of V_i^2/kT_i, i.e. $(V_i/C_{si})^2 = M_i^2$, C_{si} being the thermal velocity of the species being considered, and $M_i = V_i/C_{si}$ its Mach number. For values of the Mach number well below one (subsonic flow), the acceleration term is therefore negligible relative to the pressure term.

For electrons, the gravitational term can also be ignored, because of the low mass m_e. The stationary subsonic equations are further simplified in the rest-frame of the neutral gas, which is moving at a velocity $\boldsymbol{V}_n$ with respect to that of the observer. If $\boldsymbol{E}'$ is the electrical field within this frame,

$$\boldsymbol{E}' = \boldsymbol{E} + \boldsymbol{V}_n \times \boldsymbol{B} \tag{A1.2}$$

is the Lorentz transformation equation for the electrical field in $V_n \ll C$. The equations of motion reduce to:

$$m_i \nu_{in} \boldsymbol{V}_i - e\boldsymbol{V}_i \times \boldsymbol{B} = -\frac{1}{N}\nabla P_i + m_i \boldsymbol{g} + e\boldsymbol{E}' \qquad \text{(A1.3)}$$

in other words:

a linear function of $\boldsymbol{V}_i = \boldsymbol{F}_i$.

Dividing by eB, we have

$$\frac{\nu_{in}}{\Omega_i}\boldsymbol{V}_i - \boldsymbol{V}_i \times \boldsymbol{b} = -\frac{\nabla P_i}{NeB} + \frac{\boldsymbol{g}}{\Omega_i} + \frac{\boldsymbol{E}'}{B} \quad \text{or} \qquad \text{(A1.4)}$$

$$\frac{\nu_{in}}{\Omega_i}\boldsymbol{V}_i - \boldsymbol{V}_i \times \boldsymbol{b} = \frac{\boldsymbol{F}_i}{eB} \qquad \text{(A1.5)}$$

where $\boldsymbol{b} = \boldsymbol{B}/|\boldsymbol{B}|$, the unit vector co-linear with $\boldsymbol{B}$, and Ω_i is the gyrofrequency of the species i. If we set $r_i = \nu_{in}/\Omega_i$ and multiply (A1.5) vectorially to the left by $\boldsymbol{b}$, we find:

$$r_i \boldsymbol{V}_i \times \boldsymbol{b} - (\boldsymbol{V}_i \times \boldsymbol{b}) \times \boldsymbol{b} = \frac{\boldsymbol{F}_i \times \boldsymbol{b}}{eB} \quad \text{or again} \qquad \text{(A1.6)}$$

$$r_i \boldsymbol{V}_i \times \boldsymbol{b} + \boldsymbol{V}_i - \boldsymbol{b}(\boldsymbol{V}_i \cdot \boldsymbol{b}) = \frac{\boldsymbol{F}_i \times \boldsymbol{b}}{eB} \qquad \text{(A1.7)}$$

We see the appearance of $\boldsymbol{V}_i \cdot \boldsymbol{b} = \boldsymbol{V}_{i\|}$, the velocity component along B, and $\boldsymbol{V}_{i\perp} = \boldsymbol{V}_i - \boldsymbol{b}V_{i\|}$, the velocity component orthogonal to $\boldsymbol{B}$. This allows the motion to be decomposed into components parallel and perpendicular to $\boldsymbol{B}$.

A1.2 Mobility and Conductivity Perpendicular to the Magnetic Field

Perpendicular to $\boldsymbol{B}$, the pair of equations (A1.5) and (A1.7) reduce to (A1.8a) and (A1.8b)

$$r_i \boldsymbol{V}_{i\perp} - \boldsymbol{V}_i \times b = \frac{\boldsymbol{F}_{i\perp}}{eB} \qquad \text{(A1.8a)}$$

$$\boldsymbol{V}_{i\perp} + r_i \boldsymbol{V}_i \times \boldsymbol{b} = \frac{\boldsymbol{F}_i \times \boldsymbol{b}}{eB} \quad . \qquad \text{(A1.8b)}$$

We only have to solve this system directly by eliminating $\boldsymbol{V}_i \times \boldsymbol{b}$, which allows us to write the velocity of the species i in the form

$$\boldsymbol{V}_{i\perp} = \frac{1}{eB}\left(\frac{r_i}{1+r_i^2}\boldsymbol{F}_{i\perp} + \frac{1}{1+r_i^2}\boldsymbol{F}_{i\perp} \times b\right) \quad \text{or} \qquad \text{(A1.9)}$$

$$\boldsymbol{V}_{i\perp} = \mu_{iP}\boldsymbol{F}_{i\perp} + \mu_{iH}\boldsymbol{F}_{i\perp} \times b \quad \text{with} \qquad \text{(A1.10)}$$

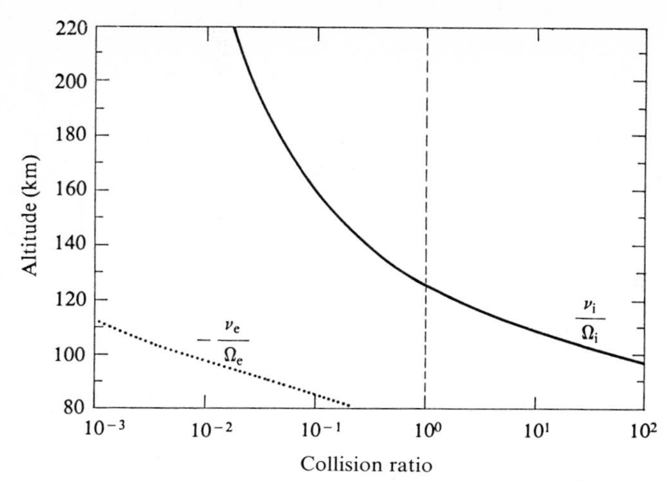

Fig. A1.1. The variation in velocity of the diffusion of ions or electrons through the neutral gas in the plane perpendicular to the magnetic field, as a function of $r_i = \nu_{in}/\Omega_i$, the ratio of the frequency of collisions with neutral atoms to the gyrofrequency of the species considered. It will be seen that diffusion is parallel to the force applied $F_{i\perp}$ for high values of r_i (the dynamics then being governed by collisions with neutral atoms), and becomes perpendicular to it for low values of this ratio (the dynamics are then governed by the gyration in the magnetic field)

$$\mu_{iP} = \frac{1}{eB}\,\frac{r_i}{1 + r_i^2} \tag{A1.11}$$

the *Pedersen mobility*, parallel to the force applied, and

$$\mu_{iH} = \frac{1}{eB}\,\frac{1}{1 + r_i^2} \tag{A1.12}$$

the *Hall mobility*, perpendicular to the force applied and to the static magnetic field.

We have

$$|\boldsymbol{V}_{i\perp}| = \frac{|\boldsymbol{F}_{i\perp}|}{eB}\,\frac{1}{\sqrt{1 + r_i^2}} = \frac{|F_{i\perp}|}{eB}\cos\alpha \tag{A1.13}$$

if we introduce the angle α such that $r_i = \tan\alpha$.

Therefore when r_i varies from 0 to infinity, the vector $\boldsymbol{V}_i$ describes a semicircle, the diameter of which is given by the vector $\boldsymbol{F}_{i\perp} \times \boldsymbol{b}/eB$ (Fig. A1.1).

The collision ratio r_i for each species generally varies exponentially with altitude, like the concentration of neutral atoms to which it is proportional. (The terrestrial ionospheric layers are shown by way of example in Fig. A1.2.) We should therefore envisage the circle shown in Fig. A1.1 as being traced out with a change in altitude through the ionospheric layers. To a first approximation, it is the position of each ionospheric layer on this circle, for ions and electrons, that defines its mobility and electrical conductivity.

In the lowest layers, for the highest values of ν_{in}, r_i has reached the limiting value of infinity: the plasma is dominated by collisions with neutral atoms. $\boldsymbol{V}_i$ is then co-linear with the force applied, and at that limit we may write:

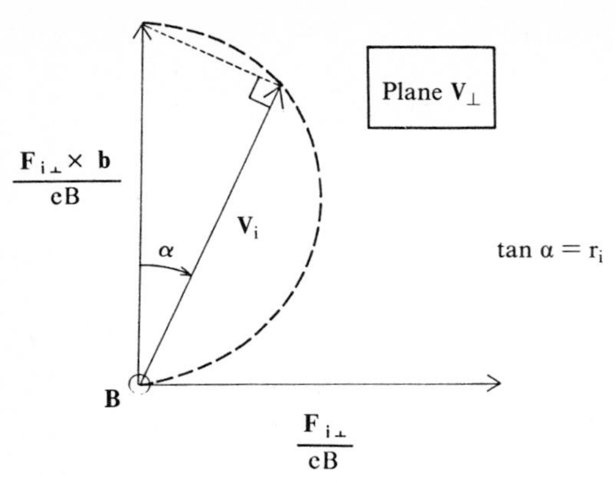

Fig. A1.2. Variation with altitude of the ratios $r_e = \nu_{en}/\Omega_e$ and $r_i = \nu_{in}/\Omega_i$ in the terrestrial ionosphere. It will be seen that at the height of the ionospheric layers ($h > 90$ km) we always find $r_e \ll 1$. The conducting ionospheric layer occurs at the height where r_i is of the order of unity. [After A.D. Richmond: "Thermospheric Dynamics and Electrodynamics", *Solar-Terrestrial Physics,* ed. by R.L. Carovillano, J.M. Forbes (D. Reidel Publishing Company 1983)]

$$V_{i\perp} = \frac{F_{i\perp}}{m_i \nu_{in}} \tag{A1.14}$$

It will be seen that the diffusion rate, inversely proportional to ν_{in}, decreases with the increase in atmospheric concentration, generally becoming negligible at the base of the ionosphere.

With increasing altitude, on the other hand, collisions rapidly decrease in importance. When the limit $r_i = 0$ is reached (no collisions), $V_{i\perp}$ becomes orthogonal to the total force applied (which therefore does no work). Taking the principal forces, we have:

$$V_{i\perp} \simeq \frac{E \times B}{B^2} + \frac{g \times b}{\Omega_i} - \frac{\nabla P_i \times b}{NeB} \quad . \tag{A1.15}$$

In addition to the dominant drift velocity in $E \times B$ induced by the electrical field (which is that of ideal MHD fluids), there are two correction terms, generally several order of magnitude weaker, caused by gravity and pressure gradients in the plasma.

Summing the equations over all the charged species, we can find the constant of proportionality between the applied electrical field and the resulting electrical current $j_\perp = Ne(V_{i\perp} - V_{e\perp})$. The expression reduces to:

$$j_\perp = \sigma_P E'_\perp + \sigma_H E'_\perp \times b \quad . \tag{A1.16}$$

The Pedersen and Hall conductivities, σ_P and σ_H are defined by:

$$\sigma_{P,H} = \frac{e}{B}\left(\sum_j n_j \mu_{jP,H} - N_e \mu_{eP,H}\right) \tag{A1.17}$$

where the summation over j applies to all the ionic species present.

In many situations, because of the very high value of Ω_e relative to Ω_i, resulting from the low mass of the electrons, r_e is negligible compared to unity and to r_i (as Fig. A1.2 shows for all altitudes greater than one hundred kilometres). The conductivity equations can then be simplified by setting $\mu_{eP} = 0$, and $\mu_{eH} = 1$ (the electrons are constrained to follow the drift $\boldsymbol{E} \times \boldsymbol{B}$ exactly), and we have

$$\sigma_P \simeq \frac{e}{B} \sum_j \frac{n_j r_j}{1 + r_j^2} \tag{A1.18a}$$

$$\sigma_H \simeq \frac{e}{B} \sum_j \frac{n_j r_j^2}{1 + r_j^2} \tag{A1.18b}$$

and it will be noted that these values have been increased by $n_e e/B$, which represents the maximum conductivity that may be attained at right angles to the magnetic field.

A1.3 Mobility and Conductivity Parallel to the Magnetic Field

Solving (A1.7) for $V_{i\|}$, the drift velocity component parallel to the magnetic field, immediately gives

$$V_{i\|} = \frac{1}{m_i \nu_{in}} \left(eE_\| - \frac{1}{n_i} \nabla P_i - m_i g \sin I \right) \tag{A1.19}$$

where I is the inclination of the local, planetary magnetic field to the horizontal.

Again summing over all the charged species, we find the expression for *parallel conductivity*, which is defined by

$$j_\| = \sigma_\| E_\| \tag{A1.20}$$

in the form

$$\sigma_\| = \frac{e}{B} \left(\sum_j \frac{n_j}{r_j} - \frac{n_e}{r_e} \right) \quad . \tag{A1.21}$$

As $r_i \ll r_e$, this time it is the electronic term that dominates the parallel conductivity, because the electrons, being less massive, move far more rapidly along the lines of force than the ions. We can retain

$$\sigma_\| \simeq \frac{-n_e e}{B r_e} = \frac{n_e e^2}{m_e \nu_{en}} \tag{A1.22}$$

(where, by convention, r_e is negative).

Unlike the perpendicular conductivity, parallel conductivity may easily become greater than n_e/B, as soon as $\nu_{en} < \Omega_e$, which occurs in the atmospheres of all the planets known to possess magnetospheres.

A1.4 The Ionospheric Ohm's Law

The conductivities of the ionosphere as expressed in (A1.16) and (A1.20) may be incorporated into a single equation, which is known as the ionospheric Ohm's law:

$$\boldsymbol{j} = \sigma_\| \boldsymbol{b}(\boldsymbol{E} \cdot \boldsymbol{b}) + \sigma_P \boldsymbol{E}'_\perp + \sigma_H \boldsymbol{E}' \times \boldsymbol{b} \quad . \tag{A1.23}$$

Ionospheric currents are produced not only by the electrical field $\boldsymbol{E}$, but equally by the electromotive force $\boldsymbol{V}_\mathrm{n} \times \boldsymbol{B}$ induced by the motion of the atmosphere's conducting fluid across the planetary magnetic field's lines of force.

A1.5 Ambipolar Diffusion of the Plasma Across the Neutral Atmosphere

In the absence of electrical currents crossing the ionospheric layers, the motions of the ions and electrons are linked by the quasi-neutral state of the plasma. Along the magnetic lines of force, they are therefore constrained to diffuse at the same velocity $\boldsymbol{V}_\mathrm{i} = \boldsymbol{V}_\mathrm{e}$ (the transverse diffusion velocities may be neglected). Vertical transport of the ionospheric layers arises from this diffusion along the magnetic field lines, except, of course, in the immediate vicinity of the planets' magnetic equators. The vertical diffusion velocity V_iz of the plasma (this velocity being the same for ions and electrons, which for this reason is called the *ambipolar diffusion* velocity, by eliminating the electrical field $E_{||}$ from the two equations (A1.19) describing the parallel diffusion of ions and electrons. To simplify the calculation, let us assume that we have just one ionic species. The condition $V_{\mathrm{i}||} = V_{\mathrm{e}||}$ gives, with the help of (A1.19):

$$-eE_{||} - \frac{1}{n_\mathrm{e}}\nabla_{||}P_\mathrm{e} - m_\mathrm{e}g \sin I = \frac{r_\mathrm{e}}{r_\mathrm{i}}\left(eE_{||}\frac{1}{n_\mathrm{i}}\nabla_{||}P_\mathrm{i} - m_\mathrm{i}g \sin I\right) \quad . \tag{A1.24}$$

If the very low values of the ratio $r_\mathrm{e}/r_\mathrm{i}$ and of the electron's mass are taken into account, this condition is reduced, in practice, to:

$$E_{||} \simeq -\frac{1}{n_\mathrm{e}}\nabla_{||}P_\mathrm{e} \quad . \tag{A1.25}$$

The electronic pressure gradient is balanced by an electrostatic field called the *ambipolar diffusion field*, which is the force coupling the ions and electrons and maintaining the quasi-neutral state in a medium that is vertically inhomogeneous. By applying (A1.25) to each ion's transport equation, we can find the diffusion velocity along $\boldsymbol{B}$, i.e. vertically, assuming that ∇P is vertical, as may be expected for ionospheric layers that are horizontally stratified. We find, for $P_\mathrm{j} = n_\mathrm{j}kT_\mathrm{j}$:

$$V_\mathrm{iz} = -D_\mathrm{a} \sin^2 I\left[\frac{1}{n_\mathrm{i}k(T_\mathrm{e}+T_\mathrm{i})}\frac{\partial}{\partial z}\{n_\mathrm{e}k(T_\mathrm{i}+T_\mathrm{e})\} - \frac{m_\mathrm{i}g}{k(T_\mathrm{i}+T_\mathrm{e})}\right] \tag{A1.26}$$

or again

$$V_\mathrm{iz} = -D_\mathrm{a} \sin^2 I\left[\frac{\partial}{\partial z}\mathrm{Log}\, n_\mathrm{e}(T_\mathrm{i}+T_\mathrm{e}) + \frac{1}{H_\mathrm{p}}\right] \tag{A1.27}$$

the *coefficient of ambipolar diffusion* D_a being defined by

$$D_\mathrm{a} = \frac{k(T_\mathrm{i}+T_\mathrm{e})}{m_\mathrm{i}\nu_\mathrm{in}} \quad \text{and the } \textit{plasma scale height } H_\mathrm{p} \text{ by} \tag{A1.28}$$

$$H_\mathrm{p} = \frac{k(T_\mathrm{i}+T_\mathrm{e})}{m_\mathrm{i}g} \tag{A1.29}$$

an expression similar to the scale height for a neutral component, but where allowance is made for the *plasma temperature* $T_\mathrm{p} = T_\mathrm{e} + T_\mathrm{i}$.

Appendix 2. Local Equations for the Magnetic Equilibrium of a Magnetopause

The aim of this appendix is to give the equations that allow one to determine the position and spatial extent of a planetary magnetosphere (i.e for finding the position of the *magnetopause*. For any planet with a magnetic field, the magnetopause forms the boundary between the field and material in interplanetary space, and those arising from the planet itself (see Chap. 4).

To begin with, we give the conservation equations for the system comprising the electromagnetic field + the gravitational field. Then applying this conservative form to the magnetopause enables us to establish the fundamental equation governing its equilibrium.

A2.1 The MHD Field and Mass-conservation Equations

The mass-continuity equation is simply:

$$\varrho\frac{\partial \boldsymbol{V}}{\partial t}+\nabla\cdot(\varrho \boldsymbol{V})=0 \qquad \text{(A2.1)}$$

for the conducting fluid of interest. The conservation equation for its momentum is:

$$\varrho\left(\frac{\partial \boldsymbol{V}}{\partial t}+[\boldsymbol{V}\cdot\nabla]\boldsymbol{V}\right)=-\nabla\cdot\overline{\overline{P}}+\boldsymbol{j}\times\boldsymbol{B}+\varrho_{\mathrm{c}}\boldsymbol{E}+\varrho\boldsymbol{g} \qquad \text{(A2.2)}$$

where ϱ is the mass per unit volume, ϱ_{c} and $\boldsymbol{j}$ the charge per unit volume and the current density, $\overline{\overline{P}}$ the pressure tensor, $\boldsymbol{g}$ the gravity, $\boldsymbol{V}$ the overall fluid velocity, and $\boldsymbol{E}$ and $\boldsymbol{B}$ are the two components of the electromagnetic field.

This equation may be written in conservative form. Regrouping the terms applying to matter on the left-hand side, we have:

$$\frac{\partial}{\partial t}(\varrho\boldsymbol{V})+\nabla\cdot(\overline{\overline{P}}+\varrho\boldsymbol{V}\otimes\boldsymbol{V})=\varrho_{\mathrm{c}}\boldsymbol{E}+\boldsymbol{j}\times\boldsymbol{B}+\varrho\boldsymbol{g}\quad . \qquad \text{(A2.3)}$$

Similarly, using the field equations, we can put the terms on the right-hand in conservative form. The electromagnetic force can be written:

$$\varrho_{\mathrm{c}}\boldsymbol{E}+\boldsymbol{j}\times\boldsymbol{B}=-\frac{\partial}{\partial t}(\varepsilon_0\boldsymbol{E}\times\boldsymbol{B})-\nabla\cdot\overline{\overline{T}} \qquad \text{(A2.4)}$$

an expression where, as for the corresponding matter terms, $\varepsilon_0\boldsymbol{E}\times\boldsymbol{B}$ represents the field's momentum density, and $\overline{\overline{T}}$ the Maxwell tensor, defined by

$$\overline{\overline{T}}=\frac{1}{2}\left(\varepsilon_0\boldsymbol{E}^2+\frac{\boldsymbol{B}^2}{\mu_0}\right)I-\varepsilon_0\boldsymbol{E}\otimes\boldsymbol{E}-\frac{1}{\mu_0}\boldsymbol{B}\otimes\boldsymbol{B} \qquad \text{(A2.5)}$$

represents the field's momentum flux. The identity matrix is designated I and the tensor operator product of two vectors is $\otimes$.

In the same way, the gravitational force may be rewritten as:

$$\varrho \boldsymbol{g} = -\nabla \cdot \overline{\overline{\Gamma}} \tag{A2.6}$$

where $\overline{\overline{\Gamma}}$ is the tensor constraining the gravitational field:

$$\overline{\overline{\Gamma}} = \frac{1}{4\pi G}\left(\boldsymbol{g} \otimes \boldsymbol{g} - \frac{1}{2} g^2 I\right) \quad . \tag{A2.7}$$

So the whole of the equation for the conservation of momentum may be put in conservative form, by summing equations (A2.3), (A2.4) and (A2.6), obtaining:

$$\frac{\partial}{\partial t}(\varrho \boldsymbol{V} + \varepsilon_0 \boldsymbol{E} \times \boldsymbol{B}) + \nabla \cdot (\varrho \boldsymbol{V} \otimes \boldsymbol{V} + \overline{\overline{P}} + \overline{\overline{T}} + \overline{\overline{\Gamma}}) = 0 \quad . \tag{A2.8}$$

A2.2 The Equations Describing the Jump in Momentum at a Discontinuity Surface

Let us assume that the macroscopic quantities specifying the fields and mass undergo a jump across a surface (S), with local unit normal $\boldsymbol{n}$ at point $\boldsymbol{x}$. What conditions must be satisfied by the macroscopic quantities on both side of (S) so that, for a stationary regime, mass and momentum are conserved?

The equations are put in the conservative form

$$\frac{\partial \overline{\overline{D}}}{\partial t} + \nabla \cdot \overline{\overline{F}} = 0 \tag{A2.9}$$

where $\overline{\overline{D}}$, the density of a specific macroscopic magnitude, is a tensor of order n, and $\overline{\overline{F}}$, the flux of the same magnitude, is of the order $n+1$. For a stationary regime $(\partial/\partial t = 0)$ the equations describing the jump are written as

$$\overline{\overline{F}} \cdot \boldsymbol{n}\,(\text{side } 1) = \overline{\overline{F}} \cdot \boldsymbol{n}\,(\text{side } 2) \tag{A2.10}$$

in other words, the flux is continuous across the surface (S) between side (1) and side (2). We may apply this to the various quantities for which we want to calculate the jump across (S).

1. *Conservation of Mass*. We then have:

$$\overline{\overline{F}} = \varrho \boldsymbol{V} \quad \text{and} \tag{A2.11}$$

$$\overline{\overline{D}} = \varrho \quad . \tag{A2.12}$$

So $\varrho \boldsymbol{V} \cdot \boldsymbol{n} = \varrho V_{\mathrm{n}}$ is continuous: the mass flux is conserved across the discontinuity surface (S).

2. *Conservation of Momentum*. We have:

$$\boldsymbol{F} = \varrho \boldsymbol{V} \otimes \boldsymbol{V} + \overline{\overline{P}} + \overline{\overline{T}} \quad \text{whence} \tag{A2.13}$$

$$\varrho \boldsymbol{V}(\boldsymbol{V} \cdot \boldsymbol{n}) + (\overline{\overline{P}} + \overline{\overline{T}}) \cdot \boldsymbol{n}$$

is continuous.

3. *The Maxwell Equations*. Still in a stationary regime, we have

$$\nabla \cdot \boldsymbol{E} = 0 \tag{A2.14}$$

so E_{T} is continuous (T being the tangential component at the discontinuity), and

$$\nabla \cdot \boldsymbol{B} = 0 \tag{A2.15}$$

so B_{N} is continuous.

A2.3 An Example Magnetopause: Confinement of Earth's Magnetic Field by the Dynamic Pressure of the Solar Wind

Let us write the momentum flux on each side of the magnetopause. On side (1), the solar wind is essentially a supersonic flow of charged particles (neutral overall) that are not magnetized, with velocity $\boldsymbol{V}_{\mathrm{S}}$, which undergo specular reflection at the surface (S), which results in their having a velocity $\boldsymbol{V}_{\mathrm{R}}$. We therefore have

$$\overline{\overline{F}}_1 = \varrho(\boldsymbol{V}_{\mathrm{S}} \otimes \boldsymbol{V}_{\mathrm{S}} + \boldsymbol{V}_{\mathrm{R}} \otimes \boldsymbol{V}_{\mathrm{R}}) \tag{A2.16}$$

because the magnetic field is assumed to be zero, and the flow is supersonic, so the dynamical pressure $\varrho \boldsymbol{V}_{\mathrm{S}} \otimes \boldsymbol{V}$ is large compared with the pressure P.

On side (2), there is no matter, but there is a magnetic field $\boldsymbol{B}$ produced by the planet. As Maxwell's equations force $B_{\mathrm{N}} = 0$, we have $\boldsymbol{B} = \boldsymbol{B}_{\mathrm{T}}$. The momentum flux reduces to

$$F_2 = \frac{B_{\mathrm{T}}^2}{2\mu_0} I - \frac{1}{\mu_0} \boldsymbol{B}_{\mathrm{T}} \otimes \boldsymbol{B}_{\mathrm{T}} \quad . \tag{A2.17}$$

Equilibrium is therefore written as:

$$\varrho \boldsymbol{V}_{\mathrm{S}}(\boldsymbol{V}_{\mathrm{S}} \cdot \boldsymbol{n}) + \varrho \boldsymbol{V}_{\mathrm{R}}(\boldsymbol{V}_{\mathrm{R}} \cdot \boldsymbol{n}) = \frac{B_{\mathrm{T}}^2}{2\mu_0} \boldsymbol{n} - \frac{1}{\mu_0} \boldsymbol{B}_{\mathrm{T}}(\boldsymbol{B}_{\mathrm{T}} \cdot \boldsymbol{n}) \quad . \tag{A2.18}$$

As, by definition

$$\boldsymbol{B}_{\mathrm{T}} \cdot \boldsymbol{n} = 0 \tag{A2.19}$$

and, for specular reflection

$$\boldsymbol{V}_{\mathrm{R}} \cdot \boldsymbol{n} = -\boldsymbol{V}_{\mathrm{S}} \cdot \boldsymbol{n} \tag{A2.20}$$

we have

$$\varrho(\boldsymbol{V}_{\rm S} - \boldsymbol{V}_{\rm R})\boldsymbol{V}_{\rm S} \cdot \boldsymbol{n} = \frac{B_{\rm T}^2}{2\mu_0}\boldsymbol{n} \tag{A2.21}$$

i.e. (as $\boldsymbol{V}_{\rm S} \cdot \boldsymbol{V}_{\rm S} \cos\chi$):

$$2\varrho V_{\rm S}^2 \cos^2\chi = \frac{B_{\rm T}^2}{2\mu_0} \quad . \tag{A2.22}$$

This is the equation used, with success, to calculate the position of the magnetopause of a planet as a function of the mass per unit volume and of the velocity of the solar wind (see Chap. 4).

Bibliography

Chapter 1

Audouze, J., Israël, G. (eds.): *Cambridge Atlas of Astronomy,* 2nd ed. (Cambridge University Press, Cambridge 1988)

Bureau des longitudes: *Encyclopédie scientifique de l'Univers: les étoiles, le Système Solaire* (Gauthier-Villars, Paris 1986)

Jones, B.W.: *The Solar System* (Pergamon Press, Oxford 1984)

O'Leary, B., Beatty, J.K., Chaikin, A. (eds.): *The New Solar System* (Cambridge University Press, Cambridge and Sky Publishing Corp., Cambridge, Mass. 1982)

Chapter 2

Chandrasekhar, S.: *Radiative Transfer* (Dover Publications, New York 1960)

Goody, R.M.: *Atmospheric Radiation, Vol.I: Theoretical Basis* (Clarendon Press, Oxford 1962)

Léna, P.: *Observational Astrophysics* (Springer-Verlag, Berlin, Heidelberg 1988)

Chapter 3

Brahic, A. (ed.): *Formation of Planetary Systems/Formation des systèmes planétaires* (CNES/Cepadues, Paris 1982)

Reeves, H. (ed.): *Origin of the Solar System* (Symposium CNRS, Paris 1972)

Williams, I.P.: *The Origin of the Planets* (Adam Hilger, Bristol 1978)

Chapter 4

Carovillano, R.L., Forbes, J.M. (eds.): *Solar-Terrestrial Physics* (D. Reidel Publishing Company, Dordrecht 1983)

Giraud, A., Petit, M.: *Ionospheric Techniques and Phenomena,* Geophysics and Astrophysics Monographs (D. Reidel Publishing Company, Dordrecht 1978)

Kennel, C.F., Lanzerotti, L.J., Parker, E.N. (eds.): *Solar System Plasma Physics* (North-Holland, Amsterdam 1979)

Nomes, E.W.Jr. (ed.): *Magnetic Reconnection in Space and Laboratory Plasmas,* Geophysical Monograph 30 (American Geophysical Union Publications 1984)

Chapter 5

Hunten, D.M., Colin, L., Donahue, T.M., Moroz, V.I. (eds.): *Venus* (University of Arizona Press, Tucson 1983)
Kondratyev, K.S., Hunt, G.E.: *Weather and Climate on Planets* (Pergamon Press, Oxford 1982)
Krasnopolsky, V.A.: *Photochemistry of the Atmospheres of Mars and Venus* (Springer-Verlag, Berlin, Heidelberg 1986)
Murray, B., Malin, M.C., Greeley, R.: *Earthlike Planets: Surfaces of Mercury, Earth, Moon, Mars* (W.H. Freeman and Co., San Francisco 1981)

Chapter 6

Gehrels, T. (ed.): *Asteroids* (University of Arizona Press, Tucson 1979)

Chapter 7

Atreya, S.K.: *Atmospheres and Ionospheres of the Outer Planets and their Satellites* (Springer-Verlag, Berlin, Heidelberg 1986)
Gautier, D., Hubbard, W., Reeves, H.: Planets: Their Origin, Interior and Atmosphere, in *An Introduction to the Evolution of Planetary Atmospheres*, ed. by Bartholdi, P., Bochsler, P., Chmielewski, Y., 14th Advanced Course, (Observatoire de Genève, Geneva 1984)
Gehrels, T. (ed.): *Jupiter* (University of Arizona Press, Tucson 1976)
Gehrels, T. (ed.): *Saturn* (University of Arizona Press, Tucson 1984)
Lewis, J., Prinn, R.G.: *Planets and their Atmospheres: Origin and Evolution* (Academic Press, New York 1984)

Chapter 8

Brahic, A. (ed.): *Planetary Rings/Anneaux des planètes* (CNES/Cepadues, Paris 1985)
Burns, J.A. (ed.): *Planetary Satellites* (University of Arizona Press, Tucson 1977)
Burns, J.A., Mathews, M.S. (eds.): *Satellites* (University of Arizona Press, Tucson 1986)
Morrison, D. (ed.): *Satellites of Jupiter* (University of Arizona Press, Tucson 1981)

Chapter 9

Gehrels, T. (ed.): *Saturn* (University of Arizona Press, Tucson 1984)

Chapter 10

Grewing, M., Praderie, F., Reinhard, R. (eds.): *Exploration of Halley's Comet* (Springer-Verlag, Berlin, Heidelberg 1988)
Wilkening, L. (ed.): *Comets* (University of Arizona Press, Tucson 1982)

Chapter 11

Brahic, A. (ed.): *Formation of Planetary Systems/Formation des systèmes planetaires* (CNES/Cepadues, Paris 1982)
Gautier, D. (ed.): *Rapports isotopiques dans le Système Solaire/Isotopic Ratios in the Solar System* (CNES/Cepadues, Paris 1985)

Subject and Name Index